TOWARD

A UNIFIED

ECOLOGY

*Complexity
in Ecological Systems
Series*

Complexity in Ecological Systems Series

T. F. H. Allen and David W. Roberts, EDITORS
Robert V. O'Neill, ADVISER

Robert Rosen
*Life Itself: A Comprehensive Inquiry Into the Nature,
Origin, and Fabrication of Life*

T. F. H. Allen and Thomas W. Hoekstra
Toward a Unified Ecology

TOWARD
A UNIFIED
ECOLOGY

T. F. H. Allen and

Thomas W. Hoekstra

COLUMBIA UNIVERSITY PRESS

New York

Columbia University Press
New York   Chichester, West Sussex
Copyright (©) 1992 Columbia University Press
All rights reserved

Library of Congress Cataloging-in-Publication Data
Allen, T. F. H.
Toward a unified ecology / T. F. H. Allen and Thomas W. Hoekstra.
p.   cm. — (Complexity in ecological systems series)
Includes bibliographical references and index.
ISBN 0-231-06918-9
ISBN 0-231-06919-7 (pbk.)

1. Ecology—Philosophy.   I. Hoekstra, T. W.   II. Title.
III. Series.
QH540.5.A55   1992   91-47209
574.5'01—dc20   CIP

Printed in the United States of America

c 10 9 8 7 6 5 4 3 2 1
p 10 9 8 7 6 5 4 3 2 1

The research reported here was funded by the USDA Forest Service Rocky
Mountain Forest and Range Experiment Station at Fort Collins, Colorado
under contract with T. F. H. Allen through Management Assistance
Corporation of America, 8600 Boeing, El Paso, Texas 79925.

To Clyde Fasick,
*for his quiet courage
and trust*

———◆·◆———

CONTENTS

ACKNOWLEDGMENTS

The dedication of this book is to the late Clyde Fasick, but we must explain his crucial role beyond the statement of the dedication itself. He earned it. While the spines of books and headings of papers bear the names of those who wrote the words, there may stand behind those names critical individuals without whom nothing, or something very much less important, would have been written. Some of these critical people are very special administrators. Often the services of administrators are forgotten or taken for granted as "only administration." In this project we cannot say enough about the role of Clyde in making this happen. We feel that this book has important contributions to make to the framing of policy and action for the U.S. Forest Service and other agencies dealing with natural resources in a changing society. However, it is to a degree iconoclastic, not the sort of thing one might necessarily expect to be supported by a federal agency which, as it should, has both its feet on the ground. Clyde, as Assistant Director for Research at the Rocky Mountain Forest and Range Experiment Station, had faith in us and extended to us a trust that was the hallmark of his special kind of administrative facility. He had to pry some cracks open, where lesser men might have taken easier passages. He gave us space to move, to follow our commitment to a new, more integrative approach to the management of natural resources. He was also a very kind friend to us, and we miss him.

The various tasks of manuscript preparation were reviewed for the U.S. Forest Service by Dr. Linda Joyce. Her comments, detailed reviews, and encouragement kept us moving forward as the size of the task mounted, and we are most grateful. Some of the ideas herein, good ones, are certainly Dr. Joyce's. The authors take full responsibility for any poor ideas between these covers. We wish to thank expressly Mr. A. Johnson of Management Assistance Corporation of America for his facilitating the progress of the project. Both Dr. Joyce and Mr. Johnson had to be very patient with us as the project grew larger than anticipated.

Some of Allen's effort surrounding the project, taking it from full second draft to a book in hand, was supported by National Science Foundation awards BSR 85-14330, BSR 90-11660, and a supplemental award for intersite studies to the University of Wisconsin. These awards support the Long-Term Ecological Research, Northern Lakes project. We are grateful to all those involved in that project for their support and encouragement, particularly John Magnuson, Tim Kratz, Barbara Benson, and David Egger.

The groundwork theory at the project's inception was supported by the Wisconsin Alumni Research Foundation through the Research Committee of the Graduate School of the University of Wisconsin while allowing Allen to get off campus long enough to think his way past blocks on the road. This is an example of the boldness characteristic of the Research Committee in making awards. We are saddened by the passing of Dean Robert M. Bock who fashioned that style of research support at the University of Wisconsin through the 1970s and 1980s. Allen is thankful for the privilege of knowing him.

Some of the ideas here, particularly on landscape ecology, were first raised in R. V. O'Neill's war room, at Oak Ridge National Laboratory, when Allen was supported by subcontract 11X-57599V from Martin Marietta. It is impossible to tease apart whose ideas were whose in the rough and tumble, but some of what is new here comes from participants in the summer of 1986 and after. Critical contestants in roughly the order of their regularity of attendance at ringside or in the ring were: R. V. O'Neill, Dean Urban, Tony King, Alan Johnson, Steve Bartell, Bob Gardner, and Mac Post. Many other residents of Building 1505 threw in their two cents' worth, and we thank them all. R. V. O'Neill has had an enormous influence on our thinking over the years. We are grateful to Steve Bartell for allowing us to present his results on changes in control in his aquatic ecosystem model before he has had a chance to publish them himself.

In November 1989 the U.S. Forest Service supported a workshop on land management planning in Madison, Wisconsin. Full-time attendees at what has become affectionately called "The Intergalactic Workshop" were: Steve Bartell, Linda Joyce, Tony King, Tim Kratz, Bruce Milne, Ron Neilson, Steve Pacala, Dave Roberts, Gene Robkin, and Dean Urban, along with the present authors. Ideas on managing from the context come from that group discussion. Everyone there has some ownership of those notions. We are grateful for the comments of attendees on the management and research chapters which were background reading for the workshop.

The manuscript has been reviewed in its entirety by Ken MacKay, Bruce Milne, Jim Gosz, and Dave Roberts. Ken MacKay co-taught with Allen a seminar at San Jose State University on the 1989 version of the manuscript. The students there, as well as those in a parallel seminar at the University of Wisconsin, Madison, deserve much gratitude for their criticism. All reviewers helped us a lot.

The logistics of getting pages typed, files copied, and manuscripts mailed were performed with dispatch and willingness by the secretarial staff of the University of Wisconsin Botany Department. Willow Ealy, Barbara Schaack, and Sarah Rau played the largest role there, but our thanks go to all our secretarial assistance. Allen also thanks his Botany Department colleagues for their patience as this work diverted him from some routine matters (particularly administrative), whereupon they took up the slack.

Beyond the photographs for which she is credited in the figure legends, Claudia Lipke graciously reproduced many slides as plates for inclusion here. We thank her. J. Magnuson, A. D. Bradshaw, Gene E. Likens, R. F. Evert, Brad Musick, B. McCure, L. Tyrell, D. Roberts, B. Milne, S. Levin, S. M. Bartell, and E. W. Beals went out of their way to supply figures from their files and sent them to us for inclusion here.

Hoekstra owes a debt of gratitude to Hank Montrey, Director of U.S. Forestry Service, Rocky Mountain Forest and Range Experiment Station, and Ronald Landmark, Director of U.S. Forest Service North Central Forest Experiment Station for allowing him the time required to complete this work. Thomas Hamilton, Thomas Mills, and H. Fred Kaiser, U.S. Forest Service National Office Staff directors in Washington, D.C., receive thanks for their patience, encouragement, and support. Curtis Flather was our companion at the Rocky Mountain Station as we started to move down the path that led to this book. He kept us honest.

Deserving special mention is Kandis Elliot, who did the artwork and graphs. Her great facility with computer graphics is evidenced by the figures she created. Allen would come down to the graphics workshop with chicken scratches on paper or sometimes only a verbal description, and Kandis would produce images beyond his imagination. She allowed this project to eat into her time beyond the call of duty. We thank her profoundly.

We are particularly grateful to our wives, Valerie Ahl and Sharon Hoekstra, for their support and patience. They have made real sacrifices in this process; we have noticed them all, if often in silence.

FOREWORD

T. F. H. Allen and David W. Roberts

Complexity in ecology is not so much a matter of what occurs in nature as it is a consequence of how we choose to describe ecological situations. That description is often only implied by the questions we ask. Therefore, complexity is a function of the terms in which we wish to understand nature. Systems become complex when, in seeking understanding or prediction, we invoke levels of organization that are distant in temporospatial scale, or are characterized by entities of disparate types. Large systems are not necessarily more complex. For example, a model for the entire planet that needs only one number for the atmosphere, say carbon dioxide concentration, clearly casts the world as a simple system. Models for forest succession can be simple or complex depending on the level and type of explanation that is required.

If it is decisions of the observer that make a system complex, then attention must focus on the subjective end of doing science. The focus in ecology heretofore has been upon the observed side of the observer-observed duality. The study of complex systems requires a more even treatment that dissects observer decisions as much as it addresses the world beyond the observer. We appear to have to recast our system descriptions to deal with complexity. Trying harder to get more detail and rigor in system mathematization appears not to help in the face of significantly complex systems.

This is the second book in the series *Complexity in Ecological Systems*. The first volume in the series is the contribution of Robert Rosen, *Life Itself*. The two books, in their different ways, set the purview of the series as a whole. Both address a certain epistemology and style of model building that is set apart from linear causality and mechanism. *Life Itself* captures the world view of complex systems by driving to whatever level of abstraction is necessary. The present volume is still abstract but is anchored in the practice of ecology, emphasizing examples of what ecologists do and why they do it.

It is anticipated that the common spirit of both these books will be maintained as the series proceeds. In particular, the present book suggests areas that deserve amplification in the same mode of operation.

The various chapters herein are invitations to expand each one to a book-length contribution in itself. Several authors are preparing further volumes on landscape ecology, community analysis, ecosystem processes, and economic and management issues. All these facets of ecology need development in light of new insights into the nature of complexity. Complex systems analysis is a new way of maintaining coherence during challenging ecological investigations that follow from pressing contemporary issues. The series will proceed with an emphatically holistic view that we hope will, nevertheless, be hard-nosed and practical.

INTRODUCTION

Ecology is one of a handful of disciplines whose material study is part of everyday encounters: birds, bees, trees, and rivers. It is, however, a mistake to imagine that this familiarity makes ecology an easy pursuit. It is wonderful to be outdoors simply enjoying nature, but a formal study is not a simple matter. From the outset we state that the very familiarity of ecological objects presents the difficulties. Perhaps we can make that point with more force if we reflect on another discipline that works on even more commonplace experience.

Sociology studies the familiar. In our own circle or in our families we collect sociological data with ease. Even so, it is hard to solve our personal social problems. We may have a detailed knowledge of the quirks of our own family members but somehow it does not always help. This might indicate the need to distinguish between the acquisition of data, on the one hand, and a firm grasp of the problem, on the other. In a formal analysis of a sociological disorder, the social scientist collects data, develops understanding, and uses it to build a predictive model. With a predictive model at hand, the consequences of doing this as opposed to that become apparent and the choice of remedial action is then clear. The data collection, or at least the identification of the form of the problem, appears easy. Finding the solution is a different matter. We can see on the evening news, or even firsthand, the homeless or the consequences of drug abuse, but we have learned that it is unreasonable to expect a solution to such problems in the short term.

It is as difficult to pursue an ecological question as to get our private lives in order or to eliminate the social ills of our time. We all try hard, but most of us are pleased with achieving an adequate private life. As in sociology, the root of the problem in ecology is the very facility with which we see ecological things and identify pressing ecological questions. The ecologist, the sociologist, and ourselves in our personal lives are all awash with data. The hard part is seeing through the walls of data to achieve a powerful summary. In ecology there is a need for a framework that the scientist can use to organize experience; that is the challenge of ecology. This book hopes to erect that framework.

Central to that framework is the notion of scale. The concept is rich, requiring this whole book for a complete accounting of scale in ecology. However, at this early stage we need to introduce briefly what we mean by scale, so that the word can pass from jargon to working vocabulary. Scale pertains to size in both time and space; size is a matter of measurement, so scale does not exist independent of the scientists' measuring scheme. Something is large-scale if perceiving it requires observations over relatively long periods of time or across large parcels of space, or both. With all else equal, the more heterogeneous is something the larger its scale; for example, comparing vegetated tracts of the same size, the vegetation that has more types of plants in more varied microhabitats more evenly represented is larger scale. Not only do things that behave slowly generally occupy larger spaces, but usually they are also more inclusive and so heterogeneous.

In several topic areas of the ecological literature there is confusion because of opposite meanings between vernacular and technical terms. For example, as we shall explain later, low-frequency behavior refers to behavior at high levels of organization. Also, organisms spread evenly across the ground are technically called "underdispersed," because the dispersion refers not to how the organisms are dispersed across the landscape, but rather to the statistical dispersion. Evenly spread means low variation between locales, hence underdispersion between samples. The difference is that vernacular meanings refer directly to the thing or behavior, whereas the technical meaning refers to how one views the situation from some standard device or unit (figure 1).

So it is with "scale." Something which is big we call large-scale because it is large in and of itself. That is the way we couple the words "large" and "scale" throughout this book. Accordingly, we say that small things are small-scale. However, cartographers reading this book, and anyone else using their terminology, will be tearing their hair out. Our choice is between ignoring the sensitivities of a group of specialists or using "scale" in a counterintuitive way. We choose the former, but for

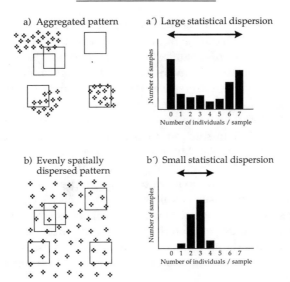

Figure 1. Two patterns of spatial distribution and their corresponding frequency distributions. Sampling clumped individuals with a quadrat gives mostly either empty or very full samples, but with few in between. The variability of such a collection of samples is large, giving an overdispersed frequency distribution. With a population that is evenly distributed, all samples contain about the same number of individuals with little variation; the frequency distribution is accordingly underdispersed. Quadrats a) and b) are the patterns on the ground, while a′) and b′) are the respective statistical distributions.

clarity need to present the geographer's point of view. A small-scale map in geography indicates that a unit measure like a mile will be very small on the map. Obversely, a large-scale map shows small things on the ground with clarity, because a large-scale map makes them large. Therefore, a large-scale map must be of a relatively small area. A map of the entire globe would be of the order of 1:50,000,000, a geographer's small-scale map of what we in this book will call a large-scale structure. The technical meaning to a cartographer refers to the smallness of the one in relation to the fifty million. The vernacular meaning, the one we will use throughout this book, indicates that the fifty million is a big number and the whole world is a big place; it is large-scale.

Our reference in this book is first to the material world, and only then to the devices needed to observe it. Material systems are scale-dependent because, with all else equal, bigger material systems manifest different relationships between themselves and the outside world than do smaller material systems; ants can ride on ants, but we squash them flat. Also, bigger things manifest different relationships between their parts than do equivalent smaller things. If something is scale-dependent it necessarily manifests quantitative changes with a shift in scale used to observe it. Conversely, something is scale-dependent if a certain scale is required to give it a certain appearance. For example, the form of organisms is scale-dependent because large organisms are required to be stocky, or somehow accommodate their large mass. Also, large and small organisms require a different scale of observation to make them appear similar. Mites require a microscope to make them appear as a body on legs; elephants require us to stand back from them some distance if they are to leave the same impression. However, the concept of organisms itself is scale-independent because both mites and mammoths qualify. Thus both big and small objects can be organisms, and the relationships between the parts of the concept of organism remain the same although individual material organisms may require very different scales in time and particularly space to observe them (figure 2).

In summary, then, according to our usage: 1) big, slow things are large-scale, while small, ephemeral things are small-scale; 2) scale-dependent entities require a certain scale of perception to make them appear a certain way; 3) we must treat material systems as scale-dependent; 4) scale-independent entities do not change their qualities when perceived at different scales; and 5) conceptual devices, the name of classes (community, organism, etc.), are scale-independent. This will suffice as a first pass at the concept of scale.

Ecology includes material and processes ranging from the physiology and genetics of small organisms to carbon balance in the entire biosphere. Between the largest and smallest ecological systems are systems of intermediate scope. At all scales there are many ways to study the material systems of ecology. Let us emphasize that the physical size of the system in time and space does not prescribe the pertinent conceptual devices. Each set of devices or point of view embodies a different set of relationships. One ecologist might choose to emphasize physiological considerations while another might look at relationships that make an organism part of a population. It is mistaken to suppose that the physiologist is necessarily reductionist or concerned only with ephemeral systems of narrow scope. Elephant physiology includes more matter than does a whole population of nematode worms, and it is much bigger than the entire community or ecosystem in a small tide pool or pothole (figure 3a–d). The physiological differences in photosynthetic mechanisms between grasses define entire biomes in the dry western United States, so physiologists can think as big as almost any sort of

Figure 2. Both fleas and elephants are organisms, but their different sizes demand observation from very different distances. Note that the organismal form is very different with change of size, even though there is a head and body and legs in both cases. Those differences relate to scale.

ecologist. Brian Chabot and Harold Mooney have published an entire book on the physiological ecology of communities and biomes.

Underlying mechanisms involve subsystems more local and ephemeral than the entire entity showing the phenomenon. Although explanatory mechanisms are necessarily smaller scale than the whole and so

A

B

C

D

Figure 3. Pockets of water much smaller than large organisms represent fully functional self-contained ecosystems: A. the insect traps leaves of pitcher plants; B. the pools in the middle of epiphytes (photo C. Lipke); and C. and D. pothole ecosystems from inches to meters across made by boulders trapped in eddies in the glacial outwash of the St. Croix River, Minnesota (photo T. Allen).

operate at a lower level as defined by scale, there is no guarantee as to the type of ecological subsystem they must be. Types of ecological system are often ranked: biosphere, biome, landscape, ecosystem, community, population, organism, cell. That ranking we will call the conventional biological or ecological hierarchy; each level therein we call a conventional level of organization. When seeking mechanisms it is certainly a mistake to assume that explanatory subsystems must come

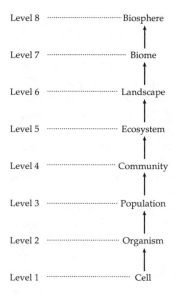

Level 8 ···························· Biosphere

Level 7 ···························· Biome

Level 6 ···························· Landscape

Level 5 ···························· Ecosystem

Level 4 ···························· Community

Level 3 ···························· Population

Level 2 ···························· Organism

Level 1 ···························· Cell

Figure 4. The conventional hierarchy of levels of organization from cell to biosphere.

from lower down the conventional ranking of levels of organization (figure 4).

In the literature of biological levels of organization, there are some branched variants of the conventional scheme, but the simple hierarchy captures the prevailing paradigm for grand, unifying designs for biology. The orthodox levels of organization define the various subdisciplines inside ecology at large. Although many ecologists view themselves as working at a level of organization in the grand hierarchy, the levels defining the different types of ecologists do not strictly depend on the scale used by the respective scientists. We would suggest some caution in the use of the word "level," for the conventional levels of organization do not correspond to levels defined by scale of the observed material system. The relationships between the conventional levels of biological organization in fact offer relatively few explanations for the configurations which we seem to find in nature.

If the ordering of conventional levels is often unhelpful, where then can ecologists find powerful explanations for what they observe? It seems that constraints and limitations are the stuff of which powerful explanations are made. What of the constraints that appear to work in the material world we observe in ecology? Most of those constraints give the order of scale-defined levels of organization. The ordered sequence from cell to biosphere receives lip service as a grand scheme, but it is not the driver of ecological research activity. The conceptual devices that ecologists actually use in practice invoke explicitly scaled structures, not the generalized entities from the conventional hierarchy. For example, tree form is understood by applying scale-based allometry to loading; the individual tree as a population or community member is more esoteric and less predictive. Scale-defined levels are far more directly connected to the functioning of the material system than are the relationships in the conventional hierarchy. We emphasize that conceptions invoked by conventional levels of organization are very important for ecological understanding, particularly when each is given autonomy from the grand conventional scheme. However, they play a very different role from that of the scale-ordered relationships in a strictly temporally or spatially defined hierarchy.

We will call the levels of the conventional biological hierarchy "criteria for observation," or just "criteria," to distinguish them from scale-defined levels. Criteria are the basis upon which one makes a decision as to what relationships are important in an ecological observation. The principal criteria in this book are: organism, population, community, landscape, ecosystem, biome, and biosphere. However, we will not use them as ordered levels per se. Our comprehensive organization puts them in a scale-ordered framework independent from the type of ecological entity under consideration. We prefer a scheme that explains what ecologists find in practice, even if it lacks the intellectual tidiness of the conventional hierarchy. Rather than a grand ordering scheme, we see the criteria as the prevailing means whereby ecologists categorize themselves, for example, population biologists as opposed to community ecologists. Each type of ecology appears to involve its own style of investigation that follows naturally from the critical characteristics of the preferred conception. The contrast between those stylistic differences gives the relationships between the parts of this book.

Many general texts on ecology either start with organisms or physiology and then proceed to focus on the way that aggregations of those considerations build up to larger systems, extending to large ecosystems or the biosphere. Other texts start with the biggest ecological systems and work down through a process of disaggregation, looking for mechanisms and explanatory principles going down the hierarchy. Each of these approaches have their respective advantages, but they both invite a misplaced emphasis. By choosing either order, up to biosphere or down to organisms and cells, the ecologist can easily be led away from considering interlevel relationships in the other respective direction. For any level of aggregation, it is necessary to look both to larger scales to understand the context and to smaller scales to understand mechanism; anything else would be incomplete.

For an adequate understanding leading to robust prediction, it is necessary to consider at least three levels at once: 1) the level in question; 2) the level below that gives mechanisms; and 3) the level above that gives context, role, or significance. A full account of mechanisms becomes irrelevant if the context changes. For example, some plants have attributes or mechanisms, like thick bark or release of seeds from scorched cones, both of which are clear adaptations to the context brought about by fire. A perfect understanding of these attributes is only really useful when the ecology is set in a particular context. Should there be a change in the fire regime through fire suppression or protection, those adaptations become at best irrelevant in the new habitat, and at worst a lethal cost. Conversely, an understanding of the role of an ecological entity is of limited utility if one has no idea how it works. The insight that prairies represent fire-adapted communities is empty if the mechanisms of fire-adaptedness are ignored. We see the orthodox way of organizing ecological texts as inviting one or the other of these types of errors of omission: either a lack of mechanistic explanations or an insufficient attention to context and broader significance.

Given our reservations for either a top-down or bottom-up approach separately, we feel that a different organization is appropriate, one which works with types of systems as alternative conceptual devices with equal status. We do not feel compelled to deal with a sequence of levels ordered by definitions that emphasize degrees of inclusiveness; for example, organisms are included in populations. The conventional ecological hierarchy is not wrong, but it is far too particular to serve as a framework for an undertaking as broad as we have in mind. There are many other facets to the conception of population beyond the fact that it consists of organisms; it is that only by definition. There is no requirement that population phenomena be necessarily explained by the behavior of the constituent organisms; there are other explanations, some of which pertain to the larger-scale context. For example, the mite populations on our bodies might be healthy because they have a role to play in the context; sometimes when we get a rash, it might be that our mites are too sick to clean the pores of our skin.

This book offers a cohesive intellectual framework for ecology. We will show how to link the various parts of ecology into a natural whole. The prevailing lack of unity that we address comes from ecologists resorting to telling stories about special cases instead of rigorously defining the general condition. There are too many tangibly different cases in ecological subject matter for us to retreat into description of all the differences that come to mind. We offer an emphasis on what is similar across ecology, so order can emerge from a wide-ranging pattern. Only then can the important differences stand out from a background of incidental distinctions.

The body of ideas we use has been gaining credence in ecology for over a decade under the rubrics of hierarchy theory, patch dynamics, scale questions, general systems theory, multiple stable points, surprise, chaos, catastrophe, complexity, and self-organizing systems. Although computers are not always used in the application of all these ideas, the mind set that they have in common came from computer-based modeling of complexity. These collective conceptions are sufficiently mature for us to pull them together, with some new material, into a cohesive theory for ecological systems in general.

The principles we use are those of hierarchy: a formal approach to the relationship between upper-level control over lower-level possibilities. Always the observer has a scale of perception and a level of analysis that deals with the system as a complex observable. Hierarchy theory is not a set of esoteric speculations about ontological reality. It provides a hard-nosed protocol for observing complexity without confusion; it is an epistemology.

There are two separate aspects to observation. In hierarchies, content and context together generate significant behavior at each scale-defined level. We find the level by using a certain scale of observation. However, at a given scale it is possible to recognize many different types of things. Which types are recognized and which are ignored comes from the observer's decisions as to what is to be considered important. "Criteria for observation" is the name we give to whatever it is that makes something important enough to be recognized in an observation or set of observations. Our hierarchical framework thus focuses on scale on the one hand and criteria for observation on the other.

In ecology the criteria for observation give rich perceptions, above and beyond the fact that many scales of perception are necessary to do justice to ecological material. For example, the organism can be conceived in many ways: the consequences of and housing of a genome; a collection of internal regulated processes; an input/output system showing irritability; a system in a loop of action on, and response to, the world; a structural mechanical system with scaling problems; and so on. This book will investigate the richness of character of the objects of study that define the principal subdisciplines in ecology. We distinguish between landscape, community, ecosystem, and physiological ecologists by their respective criteria for observation. We will unite ecologists by the common strategies for observation that each group has developed parallel to their colleagues in other ecological subdisciplines.

As we apply hierarchy theory to ecology, we pursue relationships of a functional sort. In and around the material that ecology studies there is a tangle of flows of material and energy. We organize our treatment of ecology around those fluxes and the connections they embody. The conceptual and perceptual devices of ecologists address a world that is bound by scaled physical possibilities but which becomes ecologically interesting when only certain special configurations appear to be allowed by the ecology of the situation. It is thus the unexpected limits on material flows that make ecology more than complicated physics and chemistry. This book will look at the material system, identifying fresh conceptions to account for the limits that appear to be ecological in nature.

A minimal set of premises underlie the organization of this book. They are:

1) All ecological processes and types of ecological structure are multiscaled. Each particular structure relates to a particular scale used to observe it such that, at that scale of perception, the entity appears most cohesive, explicable, and predictable. The scale of a process becomes fixed only once the associated scaled structures are prescribed and set in their scaled context. Scaling is done by the observer; it is not a matter of nature independent of observation.

2) The structures that match human scales of unaided perception are the most well known and are the most frequently discussed. Ecological processes are usually couched in terms of those familiar structures. The scale of those processes is prescribed by a) those tangible structures and b) a context that is also scaled so as to be readily observable. A common error is to leave the context undefined and so unscaled, at which point discussion becomes ambiguous. The principles derivable from observing tangibles deserve to be applied to the unfamiliar and the intangible.

3) At some scales of perception, phenomena become simpler than at others; the material system thus indicates powerful scales of perception. Predictability is improved if the scale so suggested by the material system anchors the investigation. The attributes of the particular type of ecological system distinguish foreground from background or the whole from its context. The criteria that distinguish foreground from background can be independent of scale. It is therefore sensible to determine the appropriate scale of perception separately from choosing the type of system.

Our approach will be to focus the theoretical basis for scaling and linking ecological criteria for observation. We will characterize as richly as we can the major types of ecological entities. When appropriate, we will expose the hidden agenda or anthropomorphism behind the conventional criteria for ecological observation. We will not always challenge the conventional conceptual structures, but we will unveil their hidden implications. Rather than some grand truth, this is a treatise on ecology as it is done by ecologists. There is an observer in the system; only by knowing the location and activities of the observer can we avoid self-deception and start to make ecology a predictive science.

1. THE PRINCIPLES
OF ECOLOGICAL
INTEGRATION

————◆◆————

Some scientific disciplines study objects distant from commonplace human experience. The stars are literally far away, and the quarks might as well be for all the direct experience we have of them. By contrast, ecology studies a bundle of rich and direct human encounters. Most ecologists have fond memories of some childhood place or activity that not only stimulated a first interest in field biology, but now determines what in particular they study. While limnologists might remember a pond behind the house and their first microscopes, oceanographers may sit in their offices smelling imaginary sea air and wandering along sea cliffs of summers long ago. Ecology is a very "hands on" study, where often the scientist goes to natural places and looks at other living things in their own habitats.

The things we study in ecology seem very real. Nevertheless, ecology is a science and is therefore about observation and measurement more than about nature independent of observation. It is easy for a physicist tracking subatomic particles in a bubble chamber to remember that science must work through observation and has no direct access to ontological reality. Science is not about truth and reality, it is about organizing experience and predictive power. Almost everything a physicist measures comes through the filter of some gadget. For the ecologist it is harder to remember that measurement is not reality, for out in the field where birds sing and flowers bloom, all the human senses are flooded with experiences that have an ecological basis (figure 1.1). Even so, ecology is a matter of organizing and challenging perception, with reality

Figure 1.1. Ecology is a matter of primary human experience.

always at least one step behind the screen. Ecology may be dealing with a fair reflection of what is behind the veil of our observations, but we have no way to know that. As scientists, we will deal only with observations, observables, and their implications. We will not rely on assertions that any ecological entity is real in an ultimate sense. We will try not to be biased in favor of observables in tune with unaided human perception. However, we do acknowledge that understanding in ecology, as in all science, involves an accommodation between measurements and models that are couched in distinctly human terms.

The Observer in the System

If a physicist studying quantum mechanics chooses to suppress the role of the observer in the system, the consequence is wrong predictions. This is called the observation problem, the dilemma of reliance on observation to gain insight into the world which is above and beyond the specifics of the observation. Biologists also have their observation problem, but they prefer, for the most part, to postpone dealing with it. In this volume we will tackle the observation head on. We will discuss ecology driven by observation. At this juncture ecology needs a generally acceptable body of theory, for which we propose a theory of observation. If ecology is to become more predictive it will have to be more careful in recognizing the implications of its observation protocols. Without the ecological observer there can be no study of ecology. Even at the grossest level of decision making, when the ecologist chooses what to study, that act influences the outcome of the investigation. When one chooses to study shrews, there is an implicit decision not to study everything else. In that implicit decision most other things ecological, such as trees, rivers, or ants, are excluded from the data.

Ecological Phenomena and Definitions

Phenomena in ecology, as in science in general, are manifestations of change; there can be no phenomenon if everything is constant. Recognizing those changes that constitute a phenomenon must be preceded by observer decisions about what constitutes structure. To accommodate change, there must be some defined structure, a thing to change state. Our observations involve arbitrary structural decisions, many of which revolve around making or choosing definitions. Definitions are not right or wrong, but some give us more leverage against nature's secrets than others. For example, scientific species names or their vernacular counterparts seem to be powerful ways to categorize living things. Nevertheless, the things included under those names are as arbitrary as any other named set. An investigator may come to a conclusion that implies a definition different from the one actually used during the study. In that case, the investigator used the wrong definition, but such a definition is not fundamentally wrong, only wrong for the purposes at hand. A misidentified plant involves using the wrong definition; the species mistakenly used may be a good species, but it was the wrong one on that occasion.

One might argue that species are abstractions and that their very abstractness is the source of their arbitrariness. However, even something tangible as a tree is, in fact, arbitrary. The entire army of Alexander the Great camped under a single banyan tree. The army was large, but the tree was old and had grown, as is the nature of the species, by sending roots from its limbs down to the ground. At first the roots are threadlike, but after a long time they thicken to become tree trunks in their own right. Thus one tree can become a forest. Contrast this with a clone of aspen. As the stem establishing the clone becomes large enough to spare reserves, it spreads out its roots. These long roots periodically send above the ground a branch which then grows into an apparently separate new plant. Nevertheless, the organic connections between the trees in a clone are quite as strong as the limbs connecting the trunks of an old banyan tree. The only difference between a clone of aspen and a banyan tree is our perception. A worm's-eye view might see the banyan as a grove of separate trees but the aspens as all one organism. Thus, even what constitutes a single tree is a matter of arbitrary human judgment (figure 1.2).

Definitions are generally based on discontinuities that have been experienced or at least conjectured. In the case of the banyan tree and the aspen clone, there are two critical discontinuities. One is the separation between tree trunks; the other is the separation of whole banyan groves or whole aspen clones. Note that the separation between the tree trunks in both banyans and aspens is only a matter of degree. That is why a formal definition is so important, because without it there is ambiguity as to whether it is the tree trunks or the whole interconnected collection of tree trunks that constitute the organism. Note that neither the separate trunks nor the collection of them all are truly the proper level of aggregation to assign to the class "organism." However, a given discourse about aspens or banyan trees has to be consistent in the meaning of the words it employs.

A definition is a formal description of a discontinuity that makes it easy to assign subsequent experience to the definition. The observer experiences the world and decides whether or not the experience fits the definition. With a new definition, some experiences that were once within the defined class are now excluded while others may be added. By some definitions a banana plant is a tree. By other definitions, even a

Aspen   Banyan

Figure 1.2. The army of Alexander the Great camped under a single banyan tree. Both banyans and aspen clones have stems connected: banyans above the ground, aspen below ground.

banana plant ten meters high would not be a tree. If "tree" is defined as a plant above a certain height, bananas can be trees. The "trunk" of a banana plant is mostly fleshy sheathing leaf bases, not woody stem (figure 1.3). Therefore, a definition of tree that insisted on woody stems would leave even the tallest banana out of the class "tree." In a sense, a banana is a tall herb, but the distinction has a human origin; nature does not care what we call it. Behind all acts of naming are implicit definitions. For all we know, nature itself is continuous, but to describe change, we must use definitions to slice the world into sectors. The world either fits into our definitions or not. Either way, all definitions are human devices, not parts of nature independent of human activity.

As stated above, there can be no phenomenon without change. Once the things to be measured and the measurement techniques are identified, the measured changes of state are relatively objective. The ecologist can choose to look at lions, and can choose to measure spatial position; these aspects of the observations are subjective. The relative objectivity comes from what the lions are observed to do after the subjective decisions have been made. The ecologist has nothing to say about where the beasts will go. However, even with a set of movements recorded, more subjectivity must enter the picture before the changes of state, that is, position, generate a full-blown phenomenon. That subjectivity enters when the scientist chooses which changes are significant;

Figure 1.3. A cross-section of a banana tree identifies that the "trunk" is almost entirely fleshy sheathing leaf bases, not woody stem (photo C. Lipke).

only these become phenomenal. A phenomenon is a significant change that stands out from a background of meaningless fluctuation.

Consider the example of the phenomenon of lake acidification due to acid precipitation. The pH may change in a lake, but lake acidification requires crossing a certain arbitrary threshold. Most lakes have an alkaline buffer which holds the degree of acidity of the lake water within a certain range. Thus a lake with an alkaline buffer does not become significantly more acid with the addition of moderate amounts of acid. However, add enough acid and the neutralizing power of the buffer is consumed. When the calcium buffer in the lake is gone, adding more acid rain causes an irrevocable lowering of the pH. That precipitous downward turn in the value of the pH makes the event phenomenal as opposed to happenstance. Smaller reversible changes are insignificant and amount only to daily fluctuation. Phenomena are usually associated with changes that exceed the normal range of background variation. Some lakes have larger buffering capacity than others. That is why those with small buffering capacity are at high risk of acidification. Intelligent environmental policy takes into account not just the acid in rain but also the level of risk for the lakes in each region.

Definitions, naming, and identifying critical change are not the only arbitrariness in scientific observation. The observer uses a filter to engage the world. The filter chosen by the observer is as much a matter of human decision as is the definition of structures. We cannot measure anything in infinite detail, and so differences too small for the instrument to detect are filtered out of the data. In this case the small differences fall below the level of resolution of the study. Furthermore, any difference that takes too long a time or is too large spatially to fit into the entire sweep of the data will also not survive the filter of the data-collecting protocol. Beyond this, any signal that the instrument cannot detect will be missed. For example, a light meter will not measure pH. Sometimes one hopes for a surrogate signal, as occurred when Carol Wessman used remotely sensed radiation as a measure of lignin concentration in forest canopy, which in turn is a correlate of soil nitrogen. All of these aspects of the scientist's input filters are arbitrarily chosen, and they all influence what the ecologist experiences.

ECOLOGICAL GRAIN AND EXTENT

There are two aspects of filters that are particularly important for the present treatment. One is the limit of the resolving power of individual measurements, and the other is spatiotemporal extent of the data. The resolving power of the data we associate with the term "grain." Grain determines the smallest and most ephemeral entities that can be found in the data. For example, the limnologist without a microscope cannot obtain species lists of unicellular algae. Furthermore, not only are individual algal cells small, but in a lake they also divide to make new individuals in a matter of hours and days, not months. The student of microscopic algae usually samples the lake every few days, a very fine-grained sampling regime for an ecologist. Even collecting lake water once a week can miss entire large populations. Low levels of organization are generally occupied by small transient entities like phytoplankton cells. The fineness of grain therefore limits the lowest level of organization that can be accessed in the data.

Ecologically relevant grain is not always fine relative to the human scale, but even coarse grains limit what can be detected. Certain very high-resolution images of remotely sensed satellite data are unavailable to the public. This is because at that scale it is possible, amongst other things, to see the wake of submarines. The coarser grained, publicly available channels have a grain coarser than ten meters. Such large units cannot resolve plain water from water with a submarine in it. The grain may be coarse but it still limits perception. Thus grain does not always involve minutiae.

Scientists must also set the larger limits to which their work applies. Limnologists would state clearly that the predictions they make only apply to lakes of such and such a type in a given region. Researchers often state that their results are only relevant to certain limited time frames, and so there are both temporal and spatial limits to the situations that are involved in their discussions. These limits that scientists must impose on themselves define the universe of discourse. The universe of discourse indicates what is and is not fair game for discussion in a given argument.

The scope of the data fixes the widest extent of the universe of discourse. We use the term "extent" for the aspects of a data set that are related to scope. Extent determines the highest level of organization that can be accessed. This is because higher levels of organization are generally occupied by larger persistent entities. If the size of an entity is larger than the spatial extent of the data collection, then phenomena associated with it cannot be observed because the change between the entity and its setting will not fit within the scope of the study. For example, the high level of resolution of a microscope comes at the price of a narrow extent. That is why one cannot conduct a full study of trees through a microscope. The universe of the investigation must have a scope wide enough to include all relevant subsystems relating to the upper-level phenomenon to be studied. For example, forests and prairies each have their own mechanisms of expanding: forests shade smaller plants on their margins while prairies employ fire to open the forest edge for colonization. The struggle between forest and prairie could not be investigated if the study area fails to include either one.

If one waits long enough almost all processes that at first appear to be a linear progression will emerge as cyclical. Fire might seem a one-way process of destruction, but it does open the vegetation so that plants demanding high light levels can rebuild fuel. Eventually fire returns as the recovering vegetation burns again. Individual fires are directional and have a before and after; nevertheless, fires return in a fire cycle. Should the temporal framework of a study be too short to include a full cycle, then phenomena associated with those temporal extents cannot be seen. For example, sampling in only one summer does not allow the limnologist to look at phenomena associated with the annual cycle, much as sampling for one year cannot accommodate comparison of variation from year to year. Extent thus refers to both the spatial and the temporal universe of discourse embodied in the data collection.

LEVELS OF ORGANIZATION

With some background behind us, let us now turn more specifically to what we mean by a level of organization. At the outset let us say that we do not view levels of organization as an attribute of nature alone. In the framework we erect, levels are a property that only emerges from observation. Levels emerge from the interaction between decisions of the observer and the part of the universe observed.

Levels of organization are occupied by entities, and these entities are responsible for the characteristics of the level in question. Whether a level is considered to be higher or lower depends on the scale of the resident entities. Scale of a level is determined by the grain and extent that are required to see the entities that characterize the level. Upper levels are occupied by large-scale entities and therefore have to be addressed with an observation protocol involving wide extent. Because a wide extent would contain too many very fine-grain observations to be tractable, a wide extent implies a corresponding coarse grain.

As a thought experiment, compare the relationship between the population and organism. Intuition says that populations occupy a higher level than organisms, and this certain can be so, but is not necessarily the case. The exceptions would be when an organism is so large relative to others that it has a larger spatiotemporal scale than the population of the other organisms. However, when the organisms are all about the same size and of the same species, then the population of those organisms does occupy a higher level by our standards. We will use the comparison of populations and organisms to illustrate how our more formal procedure for identifying relative level can correspond to intuition. Intuitively comfortable examples are easy to follow and so save the attention of the reader for the subtle distinctions of the general condition that might include counter- or nonintuitive cases. To the extent that populations do come from a higher level than organisms, they make a useful vehicle for conveying an understanding of the roles of grain and extent in fixing the level.

If data only include measurement on one organism, the set of observations will not cross a universe big enough to allow for populations of that organism. To see the population one must have at least a collection of individuals. Note also that it is possible to miss some of the properties of a population if one focuses on individuals as separate entities. Signals compete for the observer's input channels and memory, so accounting for the details of all individuals in a large population is likely to obscure critical population phenomena. The solution is to ignore some aspects of the individuals, even their separate identities, to save memory and attention span. Ignoring individuals amounts to coarsening the grain. It overcomes the problem of missing the forest for the trees.

Conversely, to get a clear picture of phenomena associated with individuals, it helps to narrow the extent of the observations so that only a few individuals compete for input channels and memory. To observe behavior of individuals the grain must be fine enough to discriminate between the different states of each individual. To collect data from individuals, the observer narrows the extent and uses finer-grain distinctions. For populations of those same organisms, the observer expands the extent and uses coarser-grain distinctions. All this is often done intuitively without consciously invoking formalities of grain and extent.

To see the full set of properties of a whole population, the extent of the observation set should stretch beyond the bounds of the population into a region that the population does not occupy. Only then will the set of observations contain empty space to contrast with the space occupied by the population; remember that change or contrast is required for phenomena. If the extent of the set of observations does not include space outside the population, then one has no way to know if it contains all the population. If the whole population is not included in the universe of discourse, then one must question what is the entity that is contained in the set of observations if it is not a population.

A parallel geographical example can answer the dilemma. Consider a region containing only some of the contiguous United States. Depending on which states were missing, the incomplete set would only have a few of the characteristics of the whole country. Even so, the collection would still meet the general condition of people living in a region. Smaller than the nation, it might still have interesting properties, say those of the agricultural heartland. The ecologies of corn and wheat constrain and characterize that neighborly hard-working subculture, much in the way that low rainfall in the West necessitates large spreads that engender cowboy self-sufficiency and independence. Thus the subset could be a natural entity, but it would be of smaller scale than the lower forty-eight states, and so belong to a lower level of organization between individual states and the nation. As the geographic area is reduced, the prevailing culture in the region enclosed is governed by more local social processes. The smaller the region the less it resembles the entire nation. The central point to the above is if one reduces the extent so that a large entity no longer quite fits inside the universe, then what emerges is the next largest entity that can still fit along with some background for contrast.

The same considerations apply to ecological populations. Except in situations where populations are exceptionally discrete, it makes little sense to insist that population corresponds to only particular scale-defined level. Conventional wisdom recognizes "population" as one particular level. However, the above discussion of human populations of various sizes occupying different levels raises a point of tension. There are many levels, as defined by scale, that still appear to have the general properties of a population. Entities are called populations not because they are of a size that defines the population level, but because they show properties of the population type. Populations can be variously inclusive, even within one species. Between species, populations can be even more differently scaled.

Note the human populations above blend into each other but it is still possible to identify the distinct character of successively smaller populations. Early in this century Wallace characterized the Dakotas by climate, ecology of plants and animals, the landscape, the Native American culture and that of the invading Europeans. In ecological populations, there will usually be stragglers who blur the edge of the population, and whether one includes them or not is a judgment call made by the scientist. Are the outsiders just insignificant distant members of the population, or do they belong to a larger entity by virtue of being outsiders, say immigrants, emigrants, and passers-by? Sometimes concentrations of individuals that are called single populations for many purposes will, nevertheless, smear into each other. As more parts are excluded, what remains in the universe of discourse continuously takes on properties of successively lower-level populations.

Thus, grain and extent used to identify scale and rank levels of organization can be applied to the relationship between the organism and its aggregation, the population. Our formal scaling operations can be used to capture the intuitive, commonsense ranking of the conventional hierarchy, in this case organism to population. One therefore loses nothing of the conventional hierarchy of ecological levels by insisting on scale as the organizing principle for levels, but one gains generality. However, generality is lost by submitting to any insistence that the population is singular. Our precision scaling accommodates populations of any degree of inclusion, and allows not only a ranking of them, but also a means of connecting the different sized populations.

Populations of various sizes also differ in other characteristics. The distinctive character of a given population comes from processes operating inside. At a larger scale more extensive processes can be included, and formerly ubiquitous processes become local and inconspicuous. At a smaller scale, local processes come to the fore and characterize the population at that scale. Scale is thus a continuous variable that can be used to expose the relationships between different qualitative types, like national character as opposed to local color, or endemic species characteristics as opposed to the character of a family group therein.

Patterns of Explanation

For the most part the ecologist looks for predictive models that will facilitate finding the answer to ecological questions. The trick is to model at the right level of organization; then predictions follow easily. When a scientist claims to understand a system there are two separate considerations: 1) scale and 2) qualitative type. The "right level of organization" can only be found if both are employed appropriately. Qualitative type is well known to be crucial; it amounts to identifying what is important by asserting the appropriate relationships. Although scientists know that scale makes a difference, it is not generally understood that scale is an equal partner. Operating at a wrong scale gives results as misleading as asserting and acting on incorrect relationships.

Humans reason intuitively and effectively in the realm of qualitative types, for that is how we follow and develop lines of argument. Scale is a very different matter; intuition only helps a little and logical necessity is hard-won in a scaling problem. That is why engineers cannot readily tell how good their small-scale models are, even though they take into account as many factors as they can afford. For example, a boat designer can make a scale model for a new hull design to be tested in a tank, but the physical properties of water stay the same and could cause spurious results. One can substitute a fluid with lower density, surface tension, and viscosity than water to account for the hull being smaller than the real thing, but no fluid exists that keeps the proportions of those factors the same as in water, except water itself. There is no way to correct for all scaling mismatches. In ecology, looking for the right thing is easier than looking for the right size.

Some problems in biology have yielded only to reduction to very fine-grain, low levels of organization invoking biochemical explanations. Nevertheless, in ecology errors in reduction more often employ levels of organization that are too low. It is important to ignore fine-grain data when trying to solve large-scale problems. At best, too fine-grain considerations are irrelevant to large-scale questions. For example, pairwise competition in homogeneous environments does not often pertain to community function. Worse, overly fine grain brings the research to a halt by burying the scientist in an unmanageable amount of bookkeeping. At the very worst, it confuses the scientist because it generates the wrong sort of data; remember, sometimes finer grain is not just smaller, it can be different.

When there is danger of overreduction, the concept of the minimal model is helpful. The minimal model gives predictions from the smallest number of explanatory principles. The system to be explained requires a universe of a given extent. Inside that universe various levels of organization can be found. Since low levels of organization are occupied by relatively small entities, the lower the level the larger is the number of entities from that level that can fit into a universe of a given size. Thus the small number of explanatory principles in a minimal model comes from using the highest level that can be contained inside the system to be explained. "Occam's Razor" invokes the principle of parsimony, where one should pursue the simplest explanation. The razor cuts away everything else as superfluous. The minimal model as we define it above may or may not conform to Occam's Razor, for the latter also insists on parsimony even after the scaling to the level of the minimal model has been achieved. It is not that the minimal model is singularly correct, it is just that it offers the most powerful explanation for the problem as defined by the upper level, given the current state of knowledge. Of course, the model must be consistent with the data, and scientific progress is made when data invalidate the model. However, until it is invalidated, the minimal model is the most efficiently predictive.

As a thought experiment, consider a forest as a collection of individual trees. If one knew what all the trees were going to do, then a simple summing of all those behaviors would yield a prediction about the forest. Of course the predictions on all the trees could not be made, so we are not talking about what any ecologist does or even could do. The point is that even in an ideal world where one had the predictions on the individuals, the prediction about the forest could not be applied to any other forest because it is focused on a particular collection of trees. Without general understanding as an underpinning, the prediction would lack generality. Thus, cumbersome models not only lack aesthetic appeal, they also lack generality. Employing minimal models is not an arbitrary matter of taste as to the appropriate modeling strategy. Minimal models give generality and that is the hallmark of good science.

The relationship between prediction and ultimate reality is not at all clear, so justifying cumbersome modeling by reference to ultimate reality should not be accepted as an excuse for overreduction. It would be a mistake to argue that the forest really consists of trees and so modeling with individual trees is just one of many ways of finding the truth. The very human feeling that tangibles like trees are part of an ultimate reality is seductive. Yielding to that seduction gives models based on faith about reality instead of one based on hard-nosed predictive power and general application. Science is about organizing experience in a manageable way, the more manageable the better, and it may or may not relate to ultimate truth. This argument about effective modeling is another manifestation of the problem we raised at the outset; ecology is plagued, not helped, by the familiarity of its subject matter.

The Theoretical Basis for Scaling and Integrating Ecology

STRUCTURE AND PROCESS

Up to this point we have emphasized structure and entities. However, another facet of levels and scale uses a process-oriented conception of entities and patterns. Biological systems are very much a matter of process. In fact, what appears to be distinctly structural in biology can often be seen as part of a process. For example, the human body consists of material that flushes through in about a seven-year cycle. There is very little left of the you of seven years ago. In biology often one process reinforces one or several others; in turn, the first process is reinforced by those it has influenced. That self-reinforcement leads to persistent configurations of processes. That persistence of process clusters explains how biological structure can appear concrete, although the substance of that structure is in constant flux.

Let us start with something tangible and identify its underlying processes before moving to more abstract biological structures. Solid, concrete things are surrounded by surfaces. The surface is all that we see of most things because it is the part through which the whole communicates with the rest of the universe. Although one might conceive of surfaces as passive and having nothing to do with dynamic processes, surfaces are places where the dynamic forces dominating the internal functions of an entity reach their functional limits. The skin of an organism corresponds with the furthest extent of the internal circulation system. The skin also coincides with the limits of many other fluxes.

Science looks for surfaces that define things with generality. We seek things relevant to systems according to many criteria, things which are detectable even when one looks at the system a different way. An entity which shows this persistence we call "robust to transformation." Ecology is full of such entities because it deals with tangibles. Tangibles, like trees, are robust to changes in observation protocol from sight to touch or even smell for an insect and sound for a bat. The reason for the large number of windows on a tree is that the tree is an integrated system with many processes. The integration of the tree produces a tree surface that coincides with the limits to a large number of processes. Each new way of looking at the same surface defining the same robust entity is a reflection of one or more of the processes limited at the surface.

Processes held inside a stable surface usually reinforce each other. In social systems, language and commerce reinforce the limits in each other at an international boundary. Trade uses language and a common language facilitates trade. In the tree, growth puts leaves in the light, then photosynthesis provides material for growth; growth and light each refer to the limits on the other. Processes are held within surfaces; in fact, the limits of a process define where the surface shall be. These mutually reinforcing processes give a set of surfaces that can also be seen as mutually reinforcing, one surface for each process. For example, in lakes in summer the surface between warm water above and cold water below is the surface at which oxygen and nutrient status coincide; the process of oxygen use depletes nutrients, so there is high oxygen concentration above the thermocline and high nutrient status below (figure 1.4). The multiplicity of devices that can be used to detect the surface of an entity that is robust to transformation reflects the multiplicity of the processes responsible for that surface.

Some plants grow and establish apparently new individuals by vegetative growth. A collection of individuals formed in that way have connections between genetically identical individuals. The aspen clones discussed earlier are a good example. The surface of a clone, as with many surfaces, is identifiable by sets of processes that press the surface outward. For example, the process of water transport within the clone is rapid. Internal processes only reach the surface but then attenuate rapidly. Fluxes associated with processes inside slow down at the sur-

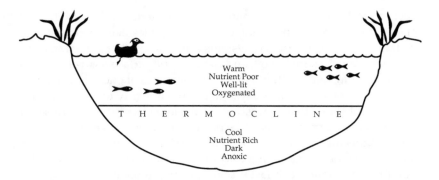

Figure 1.4. In north temperate lakes in the summertime, warm water mixes above, while cold water mixes below, separated at the thermocline where there is a very steep temperature gradient.

face; this can be used as a demonstration of the presence of a surface. For example, a radioactive tracer moves rapidly through a clone by virtue of the interconnections between individuals. The radioactivity moves rapidly along the surface of the clone, but only very slowly across it into the outside world. Surfaces are places where signal is either stopped or changed. We see the surface because it bounces back photons to give a visual image; the surface stops most light particles. Surfaces can be detected by observing changes in fluxes of the processes they influence. This may seem an elaborate way of looking at surfaces of tangibles, and it is. However, the surfaces of intangibles that we cannot experience directly also yield to this approach to surface detection. Tangibles illustrate the point which we will use later when we have to deal with intractable entities, like ecosystems, that we cannot see in a literal sense.

As we have seen above, surfaces disconnect the internal functioning of entities from the outside world. The disconnection is significant but not complete. Therefore, the observer has to judge whether or not the disconnection is sufficient to warrant designating a surface. That judgment is what makes all surfaces arbitrary, even natural surfaces that are robust to transformation. Surfaces are a matter of relative disconnection. Observations of a higher level of organization ignore the distinctions embodied in the surfaces of the entities at lower levels. For example, consideration of multicellular organisms generally ignores the membrane surfaces that separate the individual cells.

The inside of a natural entity is strongly interconnected, as in the case of the trees in the clone. The inside is relatively disconnected from the outside by the surface. The strong connections inside and the weak connections to the outside are a matter of relative rates of fluxes. This applies not only to tangibles, but also to intangibles like functional ecosystems. Large terrestrial ecosystems are composed of the interaction between plants, animals, soil, and climate. The surface of an ecosystem is not tangible, but is rather defined by the cycles of energy, water, nutrients, and carbon. If the ecosystem has the integrity to make it a worthy object of study, then it must have a stable surface in some terms that reflects that integrity. The surface of the ecosystem makes it an entity. Connections inside the cycles are strong relative to the connections to the outside world. Tracers put into an ecosystem will move around inside the ecosystem much faster than they move out of the ecosystem. The relative rates of movement define the ecosystem surface, even if the ecologist cannot see it literally.

Other intangible entities, like the regional cultures within the United States are also a matter of relative movement. There are more social contacts within a cultural region than between the region and the outside world. Because of the separation by water, the islands off the Carolinas each had their own dialect, until the establishment of just one school to serve all the islands. Even with the inroads of mass culture through television, the islands as a whole still maintain an accent that sounds more southwestern English than American. The reason is that the social contacts are strong locally but weak to the mainland. The ocean defines the cultural surface.

Many ecological entities are identifiable less by tangible surfaces and more by the strength of processes connecting the parts. Communities are less identified by physical placement of their boundaries on the ground, and are more recognized by the mutual facilitation of the species that constitute the community biota. For example, prairies may peter into savannas and not manifest a clear boundary on the ground. Meanwhile they can be readily identified as a community of pyromaniacs who together burn out invaders or weaken adjacent woody communities. Fires recycle nutrients to the mutual benefit of all community members. The integration of prairie into a fire-adapted community defines the entity by the process of repeated burning. Similarly, populations are held together and are recognizable less because one can see the whole population and more because of the mutually reinforcing processes that bind the individuals together. The process of breeding involves finding mates in a shared habitat. The process of survival in a habitat is often heritable. The shared responses of individuals to habitat define and are defined by what is inherited.

Although the strong bonds inside an entity hold it together, the weak connections across the outer surface characterize the entity. One observes things from their outsides. Signals passing to the outside are caused by what happens inside, so internal functioning is related to the properties manifested to the outside world. However, the final characterization of the whole can only relate directly to the output of the whole entity, and output has to pass the surface. For example, the personality of a person may be influenced by being well or ill, an internal consideration, but in the final analysis the personality is a matter of how the individual relates to the outside environment through words and action.

HIGHER AND LOWER LEVELS

A relationship between levels can be considered as the relationship between the internal functioning of an entity and the behavior of the whole. Lower levels are characterized by internal functioning, while the upper level relates to the whole entity. Many problems can be translated into the relationships between levels. Evolution by natural selection is an example of a concept that links levels. The lower level is occupied by the individuals that reproduce with varying amounts of success in between birth and death. A Darwinian view sees the aggregate of those successes and failures as determining the character of the upper level, the population.

We are now in a position to discuss the principles that govern the relationship between higher and lower levels. We recognize five interrelated criteria that order higher levels over lower levels.

1. Bond Strength. In the discussion of surfaces, we emphasized that the connections inside an entity are stronger than connections across its surface. We also implied that the surfaces which separate the entities of a lower level can be part of the bonding that unites those entities as parts of the upper level (figure 1.5). Relative to the strong connections within the parts, only weak signals pass out through their surfaces. The parts can only communicate with the relatively weak signals that pass out through their own surfaces. The weak connections between the lower-level entities, between the parts, become the strong connections that give integrity to the entity at the upper level, the whole. This explains the principle of bond strength. The higher the level the weaker is the strength of the bonds that hold entities at that level together. Sometimes it is possible to see the bond strength at a given level by breaking the bonds and measuring the energy released. Note that breaking chemical bonds in a fire or an explosion releases much less energy than breaking the bonds inside atoms in an atom bomb or a uranium fuel rod.

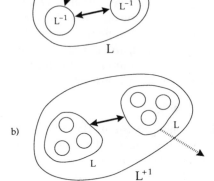

Figure 1.5. L is the level in question; L−1 is the next level down; L+1 is the level above. The weak connections of L−1 to the outside world beyond L become the strong connections within level L+1.

2. *Relative Frequency.* We established earlier that behaviors which might appear to be directional can be seen as cyclical if the set of observations has a wider extent. For example, eating can be described with much more generality as an activity that is repeated at a certain frequency, rather than as a sequence of items consumed. In fact, describing all sorts of human activities as recursions allows a completely new set of insights. Herbert Simon once noted that if a resource is to be found everywhere we keep only a small amount of it in our bodies and replenish it often. Simon, an economist, sees this as a matter of inventory; only keep a short inventory of readily available resources, but keep a long inventory of scarce material whose supply is erratic. There is no point in planning too far ahead with something as universal as air, so we breathe at a relatively high frequency and keep a three-minute inventory of oxygen. Water is common but not ubiquitous, so we drink many times a day and keep an inventory of a few days. Our biological requirement for food has been set by the hunt, and so we eat somewhere between once and five times a day. We have an inventory of food in our bodies that could last a few weeks, at least. Time to exhaust supply should be a good measure of past selective pressures.

Levels of organization are ordered by the frequency of the return time for the critical behavior of the entity in question. Higher levels have a longer return time, that is, they behave at a lower frequency. This is not so much a matter of nature, for any system that does not involve such behavior would be hard to know. We can only deal effectively with systems that are so defined. Since our experience is only coherent when couched in those terms, the temptation is to believe that nature must work that way. However, it does not follow that nature is hierarchical in itself. Nature only appears regular if what behaves at lower frequency is *defined* as occupying an upper level. Thus, in a coherent account of eco-systems many nutrients cycle once a year, but the upper-level whole ecosystem accumulates nutrients over centuries and millennia. Any other contradictory conception of system scaling would be incoherent.

3. *Context*. Low-frequency behavior allows the upper level to be the context of the lower level. The critical aspect of a context is that it either be spatially larger or more constant over time than the lower level for which it is context. It is not always transparent what are the critical cons-tancies which offer the context. Sometimes the context involves change but that change always happens; the constancy is the always of "always happens." Survival of the fittest is in fact survival of the ones that fit the context. Life is characterized by its capacity to operate predictively. Seeds germinate based on a favorable growing season that has not yet happened but which is predictable from warming moist conditions in the spring. Summer always follows spring, and that change is itself a constant on which temperate plants rely. If the environment is always changing, then what becomes dependable is the constancy of the change. Weedy plants thrive in disturbed habitats, and it is the constant upheaval that they find reliable and persistent (figure 1.6a).

4. *Containment*. In hierarchical systems, upper levels behave more slowly and are the context of lower levels. Consider pecking orders where upper-level animals are the context of animals lower in the hier-archy. Also, lower-level animals behave faster and make more local movement to keep out of the way of dominant individuals. The impor-tant point here is that despite being unequivocally upper-level entities in the social hierarchy, dominant animals do not contain lower-level in-dividuals. However, there is a special class of important systems where such containment is a requirement for existing at a high level. These sys-tems are nested systems where the upper level is composed of the lower levels. Organisms are nested in that they consist of cells, tissues, and organs. Of necessity, they also contain the parts of which they are made (figure 1.6b).

In nested systems the whole turns over more slowly than its parts, and the whole is clearly the context of its parts, so the nesting criterion

A

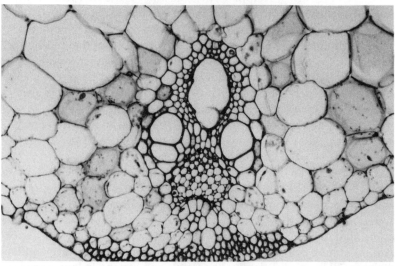

B

Figure 1.6. A. Upper levels are a context that is constant relative to the lower level. Thus the context of a seed germinating is not so much the changing season as it is that spring is *always* followed by a favorable summer season. It is the regularity of seasons which is contextual (photo C. Lipke). B. Cells aggregated to form a tissue are an example of a lower level being contained by a higher level (photo R. Evert). C. Food chain hierarchies represent non-nested systems where the higher levels do not contain the lower levels.

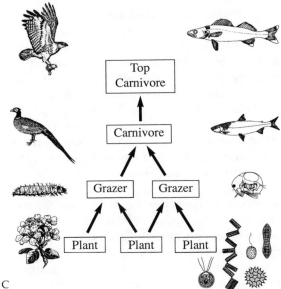

C

Figure 1.6. (*Continued*)

aligns with the frequency and contextual criteria. However, the reverse is not true, for the criteria of frequency and context do not depend on the system being nested. For example, a reliable food supply behaves more slowly than the animals which depend upon it, but it does not contain or consist of the animals it feeds. Unlike frequency and context, the bond strength criterion probably does apply reliably only to nested systems.

Nested systems are very robustly hierarchical in that the containment criterion corresponds to many other considerations. The nesting keeps the order of the levels constant even when the observer changes the rules for relationship between levels. Consider a hierarchy that goes from cells to communities. Plants may be seen as the synthesis of cellular interaction by physiological processes: cells to plant on physiological rules. However, the whole plant can be seen as a part of the ecological community on rules of assembly that bear no simple relationship to the physiological processes that built the plant as an individual. The switch from physiology to species associations causes no confusion because the nesting keeps things straight. By contrast, in non-nested systems the containment criterion does not apply, and so it cannot be used to keep the system ordered; accordingly, non-nested systems have to rely on some other criterion, and they must use that criterion from top to bottom of the hierarchy. A change in criterion in a non-nested system creates a new hierarchy that only incidentally shares with the old hierarchy the entity under consideration when the criterion was changed. Pecking orders or food chains are non-nested in that the top of the hierarchy does not contain the lower levels (figure 1.6c). The relationship between the plants and the grazer is the same as the relationship between the herbivore and carnivore; once again higher levels do not contain lower levels. If one takes an individual in a food chain, and then considers that individual's place in the pecking order, this amounts to a switch to a new non-nested hierarchy that only incidentally articulates with the first. From top to bottom of the food chain the ordering criterion is always eat and be eaten; the social hierarchy of dogs at a kill is a different matter that needs separate consideration.

5. Constraint. For our purposes, frequency and constraint are the most important criteria for ordering levels. Upper levels constrain lower levels by behaving at a lower frequency. Constraint should not be seen as an active condition. Upper levels constrain lower levels by doing nothing or even refusing to act. For example, one should never underestimate the power of impregnable stupidity. High-frequency manipulations of elegant ideas can be held in the vice of persistent misapprehension. Contexts are generally unresponsive to the insistent efforts and communications of things held in the context. Constraint is always scaled to the time frames used by that which is to be constrained. Any limitation that is temporary relative to that time frame is an inconsequential consideration.

In this regard consider an advancing dune system with plenty of sand still available for further advance and a suitable prevailing wind. Dune systems do not start easily from nothing; they mostly expand from established dunes which interfere with airflow. Dunes lose sand from their windward sides that is deposited by the slack air on their leeward side. A flat terrain will allow the sand to blow away without forming self-perpetuating irregularities in airflow. If an intermittent river cuts the front of the dune system back, then that is a temporary limitation that is not a serious constraint. In time the slow process of dune building will recoup the loss. However, if the river always returns before the dune system has had time to build past the riverbed, then the dune system is contained and constrained, and can never cross and establish a system on the far side. The individual floods are not the constraint; it is the "always" in the return of the river that makes the river a constraint. The river is always in the final analysis unresponsive to the dune system; it keeps coming back and so acts as a constraint. If the dune system could affect the course of the river, then there would not be constraint. Note in this example that constraint can only be described at the right level of analysis.

Constraint is so important in the ordering of levels in ecology because it allows systems to be predictable. Prediction appears to depend less on the details of the material system and more on the mode of description. Some system descriptions slice the natural world along patterns of constraint that are reliable over the period of the forecast. These are the effective descriptions that allow prediction. A system viewed in a fashion that ignores system constraints involves too many forces bearing on behavior for the system to be reliable or understandable. That is to say, a material system has too many potentialities for us to grasp the situation out of context. However, experience tells us that there is always a context and there are always constraints. Therefore, most of the potentialities are moot.

Predictability comes from the level in question being constrained by an envelope of permissible behavior. Predictions are made in the vicinity of those constraining limits. When a system is unpredictable, it has been posited in a form that does not involve reliable constraints. Any situation can be made to appear unpredictable, so predictability or otherwise is not a property of nature, it is a property of description. The name of the game in science is finding those helpful constraints that allow important predictions. Science would appear to be less about nature and more about finding adept descriptions.

In ecology, constraints are sometimes called limiting factors. An impossibly large number of factors could influence the growth of algae in a lake. Predicting population size is only possible when some critical known factor becomes constraining. Diatoms are microscopic plants that require silicon for their glass cell covering (figure 1.7). If a lake has low silicon concentrations, then diatoms are reliably a minor component of the plankton. When the constraint is lifted and silicon is abundant, diatoms may become abundant only if other factors are not limiting. If they are to grow abundantly when silicon is available, showing that silicon was the critical constraint, then understanding the silicon budget will allow us to predict population densities. If silicon was low but not constraining, nothing happens when more is added to the system; no release occurs because there never was a silicon constraint. Behavior of the system depends on some other factors that were and are still constraints. Predictions can come only from finding those constraints.

If we look at a system at too low a level of organization, then the scope of the observations will probably not extend to the principal constraints. Far away from the main constraints, there is ambiguity in the

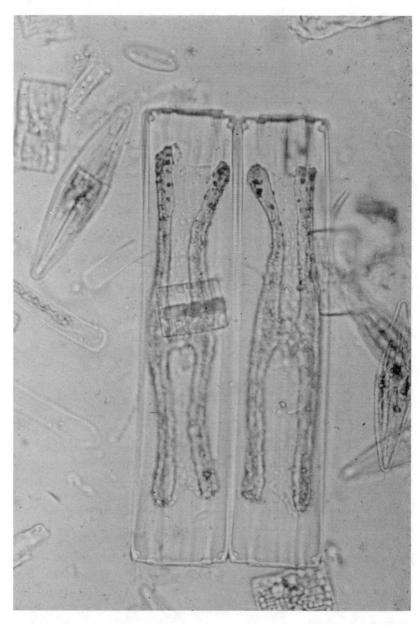

Figure 1.7. Diatoms develop a cell covering of glass. Accordingly, they are sometimes constrained by silicon supply in the water (photo T. Allen).

patterns of control. Any of a large number of factors could gain fleeting control, only to yield to some other factor before a clear and predictable pattern is established. When control of a system changes, the situation becomes unpredictable. With a change in constraint new factors take over the upper level.

A good example is the switch in control that occurs in an epidemic of spruce budworm which can be seen as a switch in constraints. For decades the budworms are held under the constant control of birds who eat them. Any increase in budworms feeds larger populations of birds that crop the budworm population back down again (figure 1.8). Budworms eat the young parts of shoots of conifer trees. The avian constraint on the budworms allows the trees to grow. As the trees grow, the budworms have more food and can handle the losses to bird predation more easily. The budworms are still constrained by the birds, but their population creeps higher. Eventually the trees are sufficiently large that they support a population of budworms which is still relatively small, but is almost large enough to saturate the constraining power of the birds. Any bird can only eat so much and bird density has upper limits controlled by factors like nesting sites or territorial behavior which have nothing to do with bird food. Some time during a critical period when bird populations are at their maximum, a chance event like a pulse in worm immigration can increase the worm population. Being limited by some other factor, the birds cannot respond as before by increasing their numbers. The effect is to increase the budworm population a small but critical amount. At that slightly higher level the worms can suddenly grow faster than the birds can eat them. The constraint on the budworm population is broken as the birds lose control of the increasing population of worms (figure 1.9).

Since there has been breakdown of constraint, there are problems of predictability. When exactly the constraint will be broken is unpredictable, although it can be generally expected during certain time windows. The whole system is no longer under the reliable constraint of the birds, and it takes on the explosive dynamics of the epidemic. The well-behaved system under the bird constraint is of no help in predicting what happens when that constraint no longer applies. Constraints allow prediction by being a constant in the system; thus they offer no predictive power when they are no longer constants. During an epidemic a new constraint takes over, the growth rate of the uncontrolled budworm population. Soon yet another constraint ousts the budworm growth curve as the limiting factor. All the food disappears as all the mature trees are killed. Worm starvation constrains the system and it returns to low budworm population densities.

A

B

C

Figure 1.8. The spruce budworm feeds
on foliage and can produce epidemic
die-off of several conifer species. The
light areas in the figure are a consequence of a budworm outbreak (photo
T. Allen).

38

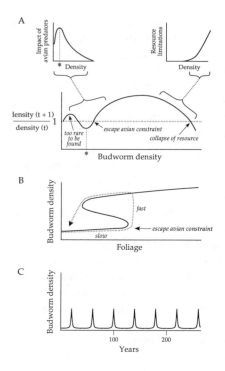

Figure 1.9. A. When the density in the next time unit is greater than the density in the present time unit the tendency will be for the population to increase in numbers. That is the tendency of the system when it is above the unitary line. It will proceed to higher densities until there is a constraint to hold the density in the next time unit the same as the present. That unitary balance creeps slowly toward higher densities as host foliage increases in quantity. At a critical stage the population can reach a high enough density so as to escape predation as it slips above the unitary line. The population increases uncontrolled until the pest eats itself out of its resources. B. The increased levels in pest population with increased foliage is gradual until there is a short-term rapid outbreak of the pest that finally destroys the resource base. C. Plotted on a time axis the population accordingly shows intermittent outbreaks approximately every thirty-five years (after Holling 1986).

The constraint of the birds gives the trees time to grow; the system behaves slowly. When the constraint changes, the budworms become the upper constraining level for the system dynamics. The system suddenly starts behaving very differently and according to the fast dynamics of the budworms. From the vantage point of the old avian constraint, the new constraint comes as a complete surprise. A reordering of levels to give control to a new upper level gives unpredictable behavior.

THE CONVENTIONAL FRAMEWORK: A POINT OF DEPARTURE

All scientific vocabularies reflect a conceptual framework. It is characteristic of paradigm replacement that the old questions are not answered differently, but are viewed as not worth asking any more. The old paradigm is embodied in its special vocabulary. When new conceptual frameworks emerge the protagonists ignore the old vocabulary, to the dismay of the old school. The new school needs a new terminology or a sharpening of the old vocabulary to reflect its new concern.

Related logical types refer to the same general sector of the universe, but are differently inclusive; for example, a particular book, a collection of books, a library, or the abstract concept of "book" are all related logical types. New logical types demand new shades of meaning, and sometimes new words. For example, children have to learn that there is not "more" water when water from a wide beaker poured into a thin beaker reaches higher up the sides; "higher" refers to a logical type different from "more." That realization allows the child to understand the principle of conservation of matter. Similarly, scientific advances amount to recognizing a new logical type that allows a set of distinctions not theretofore possible. New paradigms subtly redefine terms to recognize new logical types. That is why new scientific movements are so often resented and accused of being nothing but empty jargon, of introducing new terms for what was already known. The new terms seem superfluous to someone committed to the old paradigm, for they make distinctions that are not recognized in the old paradigm.

In ecology there are many distinctions to be made and so there is a great weight of vocabulary. Because of this, attempts to struggle free from prevailing paradigms are not such unequivocal successes or failures as occur in other disciplines. Also prediction, the acid test of a paradigm, is not the usual mode of ecological investigation, so new paradigms in ecology often go off half-cocked. The Clementsian notion of the plant association and formation (read here "community") was a shift of paradigmatic dimensions, but only a few other advances in ecology have been more than incremental. Most of the important ideas in modern ecology were considered in passing or clumsily by those early ecologists at the turn of the century. For example, enigmatic as it might be, Frank Eggler's effort at hierarchical integration in ecology was a burgeoning hierarchy theory. Accordingly, Robert MacIntosh and other commentators on modern ecology are usually justified when they assert that most that is new in ecology has been considered before. Ecology, more than many disciplines, cycles through ideas, sometimes refining concepts, sometimes only restating them in equally fuzzy terms. This pattern justifies cries of "Jargon!" but it also distracts ecologists' attention from genuinely new distinctions; it can obscure new logical types. Of course, we, like so many others, are convinced that our new terminology is justified and that it raises new logical types.

The term "constraint" is sufficiently similar to "limiting factor" for us to have used the latter term earlier in this chapter as a vehicle for introducing the idea of constraint. Let us now draw the distinction between the two terms as an example of the sort of change in view embodied in hierarchy theory. Some might resent the new term because they see no need for the contrast. Logical typing is always subtle, otherwise the distinction would have been seen before.

"Limiting factor" has a more static usage than "constraint," for the scientist is seeking to isolate the limiting factor as if somehow there were only one. Of course, those seeking limiting factors are aware that there are other limiting factors which they are not addressing just now. However, when such scientists turn to one of those other limiting factors, the first is often put aside and is no longer considered. By contrast, the notion of constraint encourages a focus on more than one constraint at a time. "Constraint" emphasizes that limiting factors supersede each other, and that they function in the context of each other. At the cost of introducing *nomena confusa*, we can say that limiting factors are reductionist tools, while constraints are their holistic counterpart. Let us press the point further.

As we said earlier, sometimes a constraint is an identifiable limiting factor. However, constraints can still be very useful even if they cannot be attached to an identifiable limiting factor. Along with the term "limiting factor" comes a need to know what it is. The emphasis is on the mechanism whereby a limiting factor limits. Constraint is more a flexible notion, one where we are interested in constraints breaking as much as constraints being identifiable limiting factors. A breaking constraint is a way of looking at spontaneous system change. "Limiting factor" in the old parlance is emphatically a part of nature that we discover, whereas a constraint is recognized as a conceptual device for adept prescription and description of the system. An adept prescription leads to prediction, it suggests a test with strong inference. A limiting factor gives prediction only after it is well understood.

Our general frame will require some definitions. We have stated firmly that definitions are arbitrary and are not right or wrong in any general sense. That needs to be said in ecological discussions because much ink has been spilled in pointless pronouncements over the true definitions of several ecological terms. The word "niche" is a case in point. It refers generally to the role an organism or species occupies in an ecological system. Niche is related to but is distinct from habitat, a term that emphasizes physical place rather than ecological role. Niche carries overtones of physical place more for plant ecologists than animal ecologists, but only overtones. The niche could be captured by the larger meaning of the word "place": as in place in society, but also with some hint of physical location. Note we give no firm restrictive definition to niche because we wish not to draw the fire of the adherents to any particular meaning; different schools of thought are vociferous on the subject.

The mistake with "niche" or any other term would be to insist on any definition being right in a general sense. Each definition has its purposes and can claim to be the right one for a given purpose, but that is different from it being correct in a general sense. Our definitions of the principal ecological structures are as arbitrary as those of anyone else. Nevertheless, they are important to our purpose, for without our definitions there would be no vehicle for moving the argument forward. However, our argument does not depend in principle on the details of our definitions. Indeed, anyone is invited to substitute their own for their own purposes. We do, however, insist that ecology has no general adequate frame at the moment, and offer one to remedy the situation.

The Criterion for Organisms. We have already identified organisms as arbitrary. Botanists are more comfortable with this notion than vertebrate biologists, because plants regularly reproduce vegetatively and gradually become autonomous and separate; however, vertebrates resemble ourselves in having an unambiguous physical boundary and obligate sexual reproduction in most species. Nevertheless, in humans, Siamese twins draw attention to the arbitrariness of even human organisms. Physical identity is an important point in the biology of placental mammals like humans, because confusion between self and not-self invites either spontaneous abortion or uncontrolled fetal parasitism. Both of these conditions reduce fitness.

Organisms are generally physically discrete, although members of a single tree species will commonly root graft so that a grove of trees may have a common root system. Organisms are generally of a single genetic stock, although fruit tree grafting can lead to branches where the outer layer of cells belongs to one partner of the graft while the core of the shoot belongs to the other. In botany, the formal definition of a new generation focuses on the reduction of the organism to a single cell, a fertilized egg, or a spore. The reduction to a single cell gives the genetic integrity of the individual. Even so, some population biologists view the parts of a plant, particularly grass shoots, as competing individuals. While the genetic and physical discreteness of organisms clearly fails in such conceptions, the individual grass shoot is adequately defined by its physiological autonomy.

Thus, to be an organism, a being should have: 1) genetic integrity, because it comes from a single germ line (egg or spore); 2) a discrete bodily form; and 3) physiological integrity within and physiological autonomy from other organisms. Normal humans are the archetypical organism; other beings pass for organisms even though they may be compromised on one or more of the above three general features.

The Population Criterion. Populations follow from individual organisms, for they are collections of individuals. Even so, there are individuals that could be defined as populations in their own right. Colonies of primitive animals like sponges and anemone relatives have all the critical characteristics that we used to define a single organism. For the most part, populations contain one species, but this is by no means a requirement.

Populations vary in size from a few individuals chosen ad hoc for a given purpose to millions of individuals spread over a large geographic area. While populations could be randomly culled from a matrix of individuals, there are several less capricious criteria that commonly apply in setting the bounds of populations. Spatial discontinuity between populations is often a helpful criterion, but discontinuity can be very much a matter of degree. Even the convenient tangible bounds of an island can be a matter of degree, in that influx from surrounding populations combined with significant emigration could make the population as transient as a collection of human beings in a crosswalk. The physical bounds of the island might be perfectly clear, while at the same time they correspond to no significant biological limit.

Often there are biological underpinnings to populations like a certain genetic homogeneity. This too is a matter of degree, in that pollen comes on the wind to insert genetic variation in plant populations. Animal groups may be readily invaded by newcomers. Although species are special populations of large size with a degree of infertility in crosses between groups, there is evidence of considerable permeability to the genetic boundaries of species. For example, woodland primroses and their close relatives in the adjacent pastures regularly produce hybrids of intermediate form. In animals, the circumpolar species of blackbirds blend into one another, with no regard for the ornithologists' boundaries. Even so, despite exceptions in populations of all sizes, populations can often be defined on grounds of relative genetic similarity within. Populations are collections of individuals that can be delimited by many criteria, depending on the question the ecologist chooses to ask.

The Community Criterion. Here we take a stance possibly at odds with convention. We do not insist on our point of view, and others are free to use their own definitions of communities. Research on the way ecologists conceive of communities shows us that ecologists do not, for the most part, work on communities using populations as the parts. An analysis of the proportion of research papers on communities involving particular organisms forces us to the conclusion that the majority of ecologists conceive of communities as consisting of individuals rather than populations. Community analysis disproportionately involves trees and their associates. Population studies focus on animals and small plants. Trees can be studied as populations but the fact that they are bigger than us presses their individuality upon us. That makes it hard to see populations in a forest. Animals can be readily seen to herd into populations and the spatial limits of the populations of small plants can be observed easily by standing and looking at them from above. That is why animals and small plants are the favorites for population but not community work.

Communities cannot be readily seen as multispecies populations because the concept of community focuses on biotic integration using many facets of biology. Populations that have more than one species are ad hoc entities conceived for a narrow purpose like interspecific disease transmission. Populations as broad-based conceptions rely heavily upon the integrity that comes from all members being from one species. The explanatory principles of communities may involve competition, but not the clean, clear competition measured in population experiments. Competition in communities is set in a variable environmental context that does not allow population competition to come to a simple resolution.

Communities are the integration of the complex behavior of the biota in a given area so as to produce a cohesive and multifaceted whole. This whole usually manifests properties of self-regulation and a self-assertiveness that often modify the physical environment. The forest is an archetypical community with its individual trees bound together as members of the community by a tangle of processes. Trees standing side by side represent an instant in a continuing complex process involving not only competition but also interference, accommodation, and mutualism amongst other factors. So fully integrated are communities that the neatly divided processes studied in populations are too deeply compromised to be seen as working separately. Our community conception only allows us to study competition, mutualism, and the many other population processes in aggregate.

The Ecosystem Criterion. The functional ecosystem is the conception where biota are explicitly linked to the abiotic world of their surroundings. System boundaries include the physical environment. Ecosystems can be large or small. Size is not the critical characteristic, rather the cycles and pathways of energy and matter in aggregate form the entire ecosystem. Clumps of floating carnivorous plants the size of a few handfuls in the Okefenokee Swamp of Georgia show all the properties of an ecosystem.

The origin of the term "ecosystem" goes back over fifty years to when Arthur Tansley recognized the need for an entity that blended biota with the physical environment. That original explication persists so that conventional accounts of the ecosystem include a large box, the eco-system, with four boxes inside labeled plants, animals, soil, and climate (figure 1.10). Unfortunately, such a characterization refers to the intellectual history of ecology in the 1930s more than to the powerful ideas that flowed from Tansley's brainchild. The box with its four parts is misleading, and is an example of ecologists' reluctance to abandon irrelevant tangibles. The ecosystem contains plants, animals, soil, and atmosphere, but those names are not helpful categories for seeing how ecosystem parts are put together. The functioning subunits in the eco-system consist not of plants, animals, soil, and atmosphere but of mixtures of them. Plants do not naturally separate from soil when the compartment in question is "below ground carbon."

Tansley's term met a need of his day, for ecosystem approaches did precede the name itself. A full decade before the term ecosystem was coined, Transeau measured the energy budget of a cornfield. He subtracted outputs from inputs to identify net gain. It would have been more productive had ecology taken that study as the archetype of an ecosystem. Transeau's implied ecosystem emphasized fluxes and pathways that are hard to address in an organism-centered conception that insists on preserving plants and animals as discrete entities.

The critical difference between an ecosystem and a community analysis is not the size of the study, but rather the difference in emphasis on the living material. In community work, the vegetation may influence the soil and other parts of the physical environment by the very processes that are important in ecosystems. However, in communities the soil is explicitly the environment of the plants. The biotic community is recognized as having integrity separate from the soil. The same vegeta-

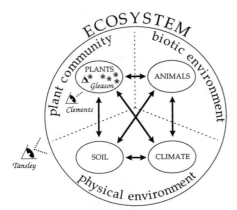

Figure 1.10. Gleason, Clements, and Tansley looked at vegetation in three different ways. Gleason saw the plant community as a collection of individuals filtered by environment; Clements saw the plant community as an integrated whole, set in a physical environment; Tansley saw the plants, their biotic environment and their physical environment as all components inside the ecosystem. It is the same material system seen from different perspectives.

tion on the same soil may be studied as a functional ecosystem or as a community, depending on how the scientist slices the ecological pie.

Ecosystems are intractable if the biota are identified as one of the distinct slices, particularly if separate organisms are allowed to be discrete parts. Take the case of the cycling of nutrients for repeated use. The leaf falls from the tree; worms eat the leaf; rainwater washes the nutrients into the soil directly from the leaf and from the feces of the worm; fungi absorb those nutrients and convey them to the root to which they are connected; the root dies leaving a frozen core of nutrients; in spring new roots grow down the old root hole, collecting the nutrients; the rest of the plant passes them up to the leaves (figure 1.11). In terms of a well-specified ecosystem the above is a simple, efficient nutrient pathway. In terms of a model that insists on emphasizing living material as different from nonliving material, the pathway is horrendously complicated. In

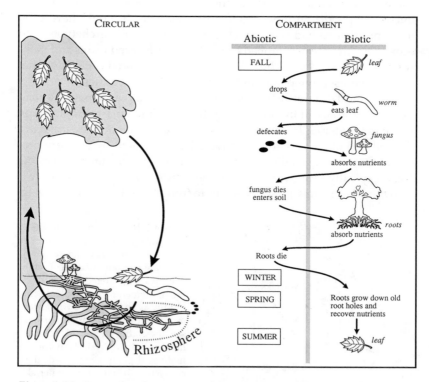

Figure 1.11. For a mineral nutrient to pass once round a nutrient cycle, it must pass in and out of the biotic component of the ecosystem several times in one year. Accordingly, biota are not often a simple subsystem of process/functional ecosystems.

just once around the cycle, the nutrients pass in or out of the biotic compartment at least four times.

The Landscape Criterion. Landscapes occupy a range of scales comparable to ecosystems and communities, but according to their own distinct criteria. Although landscapes are generally accessible from commonplace experience—and represent one of the earliest criteria for studying ecological systems—they have lain neglected for most of this century, ignored by almost all ecologists except those studying wildlife. The landscape ecology of the last century was supplanted by community conceptions. Community ecology has its origins in an abstraction of landscapes, one where the pattern of the patchwork on the ground is replaced by abstract community types defined by species lists and proportions of species abundances. Pursuit of the community abstraction left the landscape conception untended. When a new conception was deemed necessary in the middle of this century, the ecosystem and not the landscape prevailed. In a sense, Victorian country vicars did all that could be done with landscapes before the computer age. In the last decade, landscapes have again come to command the attention they deserve.

Spatial contiguity is the ordering principle for landscapes. As the dynamics of species replacement in community patches became a focus of study some twenty years ago, the scene was being set for a reawakening of interest in landscapes. More than in any other subdiscipline of ecology, landscape ecologists are buried in data. A casual glance across a vista reveals an unmanageable mass of detail. To achieve anything more than the obvious, the landscape ecologist must have cheap, fast computers. Recent years have seen the development of several measures of pattern that formalize and quantify what has been subjective, informal assessment. Remote sensing from satellites has allowed the landscape ecologist to move upscale. In many studies the student of landscapes has only to look; that which made landscapes so obvious as to be trivial by the end of the last century now makes them natural objects for study in a modern computer-assisted world. With tools to identify spatial structure, landscape ecologists now turn to fluxes of material in spatial contexts. Landscape ecology now investigates the consequences of spatial structure.

The Biome Criterion. By definition, biomes cover large areas. However, it is worth considering what is the essence of a biome aside from size. Biomes are defined by the dominant vegetation physiognomy which is not strictly scale-defined in itself. A biome should also have a critical climate such that the other characters are responses to some meteorological consideration. Often an assemblage of animals plays a central role in giving the biome its particular structure. Examples here would be the spruce-moose biome or the grassland biomes with their respective grazers.

As the name suggests, biomes are characterized principally by their biotic components, although soils and climate are important parts of the picture. Biomes, at first glance, are a hybrid of community and ecosystem with a strong landscape reference. The distinctive character of biomes is revealed when the concept is applied to situations scaled smaller than usual. Small systems that are simultaneously physiognomic, geographic, and process-oriented might prove very helpful. It avoids the confusion that arises when we try to use one of the other criteria to describe such a situation. Landscapes, communities, and ecosystems used separately or in tandem cannot do the biome concept justice. However, they are often pressed uncomfortably into service because we lack a term for small biomes. A frost pocket is a patch of treeless vegetation set in a forest. The absence of trees allows cold air to collect and kill any woody invaders. It is not adequately described as a community, for it has all biome properties except size: physiognomically recognizable, climate-determined, disturbance-created, and animal-groomed.

Even if we require that a biome should be large, a single large landscape could include more than one biome. Most ecosystems would, however, only belong to one biome. The same is true for plant communities, although animal communities could span more than one biome with ease. Under the conventional definition, a biome could be as small as a moderate-sized ecosystem or as large as the biosphere. In the biome scaled up to the entire biosphere, the atmospheric component gains the ascendancy and becomes the glue that holds the biosphere together as a functioning whole. In biomes of subcontinental size or smaller the glue seems to be an interaction of climate, wide-ranging animals, or catastrophic fire, all of which can waste tracts of vegetation.

The Biosphere. The biosphere is the one ecological system where the scale is simply defined. Its scope being the entire globe, it occupies a level that is unambiguously above all the other types of ecological system discussed so far. Stretching a point, the biosphere can be regarded as a large biome because of geographic considerations. Considering it a large landscape is possible but awkward. The biosphere is more often studied as a macro-level ecosystem. For example, students of the biosphere ask questions about global carbon balance, a problem involving the same sort of fluxes considered by an ecosystem scientist. There is, however, a subtle but unequivocal shift in the relative importance of Tansley's four parts to the ecosystem when we make a shift upscale to the biosphere. The atmosphere is definitely an overriding influence in the biosphere, whereas in subcontinental ecosystems soil interactions with biota through water are at least equal players with the meteorological components. The biosphere is not easily conceived as a large biotic community because there is not interaction of species on a global scale to hold the global community together as a working unit.

A FRAMEWORK FOR ECOLOGY

The critical distinction between the conceptual framework and the material system that ecology studies is the basis of all that follows. The material system appears to be strictly scale-dependent, but that dependence is only relevant to the extent that ecological happenings are bound to obey physical laws. However, ecology is more than complicated physics and chemistry. As ecologists we should be interested in the things which could occur given physical possibilities, but which do not occur for strictly ecological reasons.

The restrictions on the physical and chemical processes found in an ecological study generate the set of distinctly ecological structures (figure 1.12). Examples of distinctly ecological structures in an ecological system are organisms, communities, ecosystems, and the like. These are all either evolved entities, or at least structures whose functioning is based on evolved entities. Evolution gives ecology a certain epistemo-

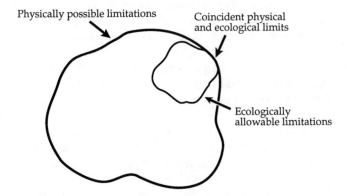

Figure 1.12. Only occasionally is what is observed in ecology directly the result of a physical limitation. More usually what is ecologically allowable is a subset inside what is physically possible.

logical style, one where neutral, valueless accounts of physical systems have limited utility.

The outcome of evolution has been constrained by what went before; organisms are endowed with a certain genetic potential the form of which came down from ancestors. Thus the systems that ecologists study in the contemporary biosphere have an inescapable biological history which can be expressed in physical terms only as surviving accidents. That history has involved successive organisms reading their respective environments and handing on some of their models for perception to their offspring. Evolution has endowed life with properties that can only be understood in terms of purpose. Looking at the world the same way as the study organism has great advantages. To study such things as organisms, communities, ecosystems, and other ecological entities we must take a point of view, preferably one that captures something of the points of view of the biological structure in question. Behind an adept biologist's perceptions are eons of evolved organisms taking their respective points of view on their environments.

In a sense, ecologists study biological subjectivity set in a scale-defined physical world. The subjective decisions we make in ascribing an organism to a species, or calling it an organism in the first place, capture the biological points of view that make the system ecological as opposed to only physical and chemical. Thus the subjectivity of naming things ecological is crucial to ecological understanding. Calling something an ecosystem involves taking a point of view on a given tract of land and emphasizing something different than if one called that same material system a community. The framework we erect is to take the subjectively defined entities of ecological systems and to embed them in a physical setting which is strictly scale-dependent. The distinctly ecological, as opposed to physical, aspects of structures are scale-independent.

Ecological systems are complex and require careful analysis if the student of ecology is to avoid being lost in the tiered labyrinth of his material. The most important general point covered thus far is the recognition of the role of the observer in the system. It is essential to resist the temptation to base ecological understanding on a belief that ecological systems are an ultimate reality beyond observation. Ignoring human subjectivity will not make it go away. Since one makes arbitrary decisions anyway, ignoring them abdicates responsibility needlessly. All decisions come at the price of not having made some other decision. By acknowledging subjectivity one can make it reasoned instead of capricious. A real danger in suppressing the ecologist in ecology is to be bound by unnecessarily costly decisions. These could be exchanged for a more cost-efficient intellectual device if only the subjectivity of the enterprise were acknowledged.

The most important benefit of consciousness of the observer is knowing the effects of grain and extent on observation. The formal use of grain and extent gives opportunities to avoid old pitfalls and opens up possibilities for valid comparison that have heretofore escaped our attention. Also of great importance is being cognizant of patterns of constraint. Constraint gives a general model for couching ecological problems in terms that generate predictions. At this point we have put most of the crucial tools in the box and are ready to go to work. It is time to erect a general frame; inside this frame we hope ecology will become a more predictive science.

Given the definitions of ecological criteria and the tools for linking them, the framework we suggest can be adequately stated in a short space. The temptation we resist is to stack types of ecological systems according to an approximation of their size. Yielding to the desire for tidiness in the conventional ecological hierarchy is one of the less cost-effective decisions to which we alluded above. Just because most populations have fewer members than most communities, we will not limit our arguments to only populations set in communities. We do not see that the context of a community is necessarily an ecosystem, nor are ecosystems always best seen as being set on landscapes. True, the term "organism" usually refers to relatively small entities that could be easily subsumed by most other ecological types. However, the floating carnivorous plants we mentioned above do function as the context of an entire small ecosystem, as does the rumen of a cow. The biosphere is unequivocally large and contains all earthly ecological considerations. Apart from organism and biosphere levels, there is plenty of room for entities from almost any type of ecological system to be contained within an entity belonging to any other class of systems.

Therefore, we will build a "layer cake" of ecological systems (figure 1.13) but not one stacked in the conventional order. The layers are not ordered by approximate size of representative organisms, populations, or ecosystems, but by strictly scale-defined criteria based on the principle of constraint, using grain and extent to define the scale of observation. Rather than putting types of ecological system on layers by themselves, we will put each type of ecological system on every layer: organism, population, community, ecosystem, landscape, and biome systems will sit side by side at every level. Across any given layer we can readily compare and contrast the types of ecological conceptions at the scale of that layer. The layers above and below will appear identical but will be stacked according to scale needed for observation. Thus, not only can we compare the ecosystem and community conceptions across a landscape of a given area, but we can do the same at larger and smaller scales.

We are not limited to making comparisons horizontally across a given spatiotemporally defined level. We can move up and down the cake and make comparisons between differently scaled entities of a single type of system. We could see how a given ecosystem contains smaller ecosystems, while being itself part of a larger ecosystem. As we move to levels below, we may find the mechanistic explanations of the behavior of the level in question. The levels above define the role of entities at the level in question in the functioning of larger systems. The upper levels define role, purpose, and boundary conditions. Predictions come most readily when the system to be predicted is up against a constraint imposed by the layer above.

Probably the most interesting—and certainly the most neglected—questions will involve slicing the cake diagonally. Here we change the type of system while also changing levels. Diagonal slices will allow us to see how community patterns at a given level are influenced by nutrient status maintained by a subsystem defined in ecosystem terms. With our ecological layer cake we can ask how low-level ecosystem function is constrained by population considerations of a dominant tree population. We need no longer see a difficulty when animal communities carry nutrients around an ecosystem that sits across the boundary between two biomes. The biome boundary is clear by its own criteria, but it is leaky with respect to nutrients involving animals and ecosystems. We do not recommend a compromise that forms a general-purpose system designed by a committee composed of one population biologist, a community ecologist, an ecosystem scientist, and biogeographer. Rather we suggest a formal change in the type of system description every time a new explicit question or explanation demands it.

We will use the above scheme to organize the material in the chapters that follow. We see it as both flexible and encouraging consistency.

Large Scale

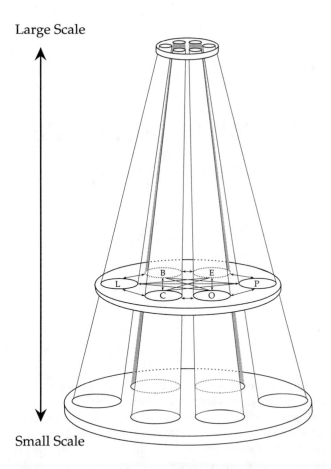

Small Scale

Figure 1.13. The layer cake metaphor for ecological criteria and ecological scale. The wide base indicates a large number of small entities; the narrow top indicates a small number of large entities. The cross-section across the entire cone represents one middle-level scale. Although there is only one here, any number of cross-sections could have been inserted, each at its own scaled level. Each letter indicates a different ecological criterion: O = organism; P = population; C = community; E = ecosystem;L = landscape; B = biome. The six lettered discs correspond to where the abstract, scale-independent criteria intersect with the scale to produce a given way of identifying an ecological entity at a given scale. Individually, the columns represent a criterion for looking at the material system, e.g., the abstract notion of community. The disc labeled C is an actual community with a particular scale. In the C column, larger-scale contextual communities occur above that community while smaller-scaled community subsystems occur below. A community context to an organism would be a C disc diagonally above a given O disc; the ecosystem that is a cow's rumen would be an E disc diagonally below an O disc representing the cow.

2. THE LANDSCAPE CRITERION

───────◆•◆───────

The previous chapter proposed that the major ecological criteria be seen as independent ways of viewing ecological systems, independent in that they need not be held in a rank order prescribed by the conventional biological hierarchy. The organism can be small- or large-scale, and so can the biome and the ecosystem. Thus there are two separate considerations for addressing any ecological system: scale and system type. Such a scheme opens up a formal approach to a whole complex of new relationships. There is something to be said for using this scheme to address first the most obvious and tractable ecological type. There are two candidates, landscapes and organisms; both are tangible. Because it is unequivocally ecological, we employ the landscape criterion as the point of departure.

For all the separate criteria, we have tried to follow the same plan. There is the option to hold the criterion constant and move up- and downscale, looking at smaller and larger versions of the same type. In terms of the previous chapter, this involves moving up and down one of the columns in the cone-shaped layer cake (figure 1.13). Alternatively, we can hold the scale constant and look at how the criterion under consideration is but one conception of a material ecological system at a given temporal and spatial extent. This cuts across the cake to expose one layer. Finally, it is possible to see how a given criterion offers either a context or an explanation for some other criterion. This involves simultaneous changes in both the scale and the criterion used to address the material system. Here we cut diagonal slices through the cake to cut the various columns at different layers.

Landscape ecology goes back to early expeditionary naturalists such as von Humboldt, but has lain quiet in America since the turn of the century. Only in wildlife ecology did an interest in landscapes persist, and only insofar as landscape applied to the particular purposes of that specialty. There has, however, been a resurgence of activity on landscapes through the 1980s which promises to persist. The switch away from landscapes seems to have occurred when the community concept came of age, husbanded by Frederic Clements. There was much debate about the proper meaning of community because there was conflict in the minds of early community ecologists between the community ideal versus actual stands. They gave the name "community" to both the collection of plants living on a particular site on the landscape as well as the abstract archetype community defined by associations of organisms over millennia. At first, the roots of communities in landscapes refused to die. The tension in the minds of early ecologists was between community concepts as they apply today and the patches on a landscape that were the origins of the community concept from the last century.

Landscape ecology has been neglected probably because landscapes were considered obvious. Certainly they are the most tangible of the large-scale ecological conceptions, for one only has to look at them to see them literally. The conventional wisdom is that landscapes are large, but some researchers in the spirit of this book are relaxing that consideration. Bruce Milne, for example, studies the landscape from a beetle's-eye view, so large physical size need not be a requirement. Unlike landscapes, ecosystems and communities have never been considered obvious, probably because they both require special measurements before they become apparent. However, times change and landscapes take their place as significant entities for study in ecology. Modern technologies have suddenly given us access to landscapes in ways that press them upon us as emphatically nontrivial systems. We can now see landscapes over enormous areas while preserving remarkable detail. It is possible to read license plates on cars from altitudes that allow mapping of large tracts. Remotely sensed images that distinguish between thirty-meter grid units using seven wavelengths are publicly available for ecological research. The satellite images make these fine distinctions while scanning county-wide swaths around the entire globe. Computer processing of such masses of data allows one to identify quantitative signatures of landscapes. Landscape ecology is coming of age. In fact, landscapes were never trivial, but contemporary image processing has brought that point home.

The Landscape Criterion Across Scales

Landscapes can be related to other ecological criteria for organization, such that the landscape becomes the spatial matrix in which organisms, populations, ecosystems, and the like are set. However, landscapes are meaningful in their own right and so it is possible to consider differently scaled systems while using only the landscape criterion. Within large areas there are smaller subsets which are mosaic pieces. Patterns may be extended so that they become segments of larger patterns. Smaller patterns may be just details from larger patterns or they could be autonomous with their own distinctive causes. Let us first consider the landscape criterion by itself, and look at landscapes nested in larger landscapes, or landscapes composed of smaller landscapes. Let us cut out and examine the landscape wedge of the cake with its differently scaled layers.

Over the last decade mathematicians have developed fractal geometry, a method for dealing with the shape of complex entities. These methods have been used to create intricate and strikingly realistic landscapes for science fiction films, and so have a place in popular culture. The fractal dimension of a pattern is a measure of its complexity. Fractals are particularly pertinent at this point in the discussion because they are derived from qualities of the pattern over different grain sizes. In this way a fractal dimension is an integration over levels while maintaining the landscape criterion.

Fractal dimensions can be calculated in several ways, but there is one that makes intuitive sense for the purpose at hand. It involves the length of the outline of entities forming the pattern. Once again, note that we avoid the ontological reality of the thing making the pattern, in that the length is a consequence of the scale at which it is measured. As one traces the outline of a natural irregular shape like a coastline, there is always a degree of smoothing of detail, no matter how carefully the measurement is made. If one were to follow the outline in infinite detail then the length would be infinitely long (figure 2.1). On crude world maps, the outline of Great Britain might be made as a simple long triangle: one point for Land's End, another for John O'Groats at the tip of Scotland, and a third for the tip of Kent in the southeast. Note that the straight lines connecting the points are long. If the distance between the reference points were shorter, then some other coarse features would emerge. With slightly shorter segments the top of Scotland could be seen as flat, not pointed, and Wales would appear as a rectangle stuck on the side of England. With still shorter segments, the outline would include a bulge for the Lake District, a rounded outline for East Anglia,

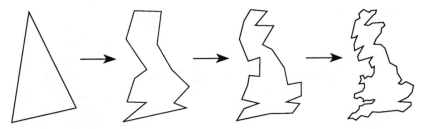

Figure 2.1. Maps of Britain capture more detail as the length segments used for the outline are shortened. With the longest straight line segments Britain appears only as a triangle. As the straight segments for the outline are made shorter, more and more detail emerges while the length of the coastline increases. For infinitely small straight line segments, the coastline of Britain is infinitely long. The length of the outline of Britain is unbounded.

and even a small notch for the Wash. As the outline is mapped with shorter segments, more details appear and the total length measured gets longer. The coastline of Britain measured so as to take into account the outline of every grain of sand would be of enormous length. The relationship between the length of the total outline to the length of segments used to make the measurement gives the fractal dimension of the shape.

Each shape has its own relationship of length of estimators (chords for a circle) and the total estimated perimeter. The more complicated is the shape to be estimated, the higher will be the fractal dimension. A simple shape has much of its detail captured when relatively long line segments are used as estimators (figure 2.2a). Therefore, shortening the length of the estimators will not add much length to the total length of the estimated perimeter. Complex shapes are poorly approximated by long line estimators, so shortening the estimators picks up much extra detail. This greatly increases the length of the total track indicating a complex shape and a high fractal dimension (figure 2.2b).

There are various protocols for estimating the fractal dimension of shapes. The equation for the relationship between length of the estimators and the estimate of total perimeter can be reorganized to emphasize the double logarithmic relationship involved. A log-log plot of length of the estimators and the estimate of total length gives an approximately straight descending line (figure 2.3). The line slopes down because the shorter the segments used in the estimation, the longer the estimate of the total perimeter. Complex forms respond more in total perimeter length to small changes in the length of the segments used to estimate the perimeter. Therefore, the steeper the slope of total length to seg-

A. Ratio
1 : 1.041

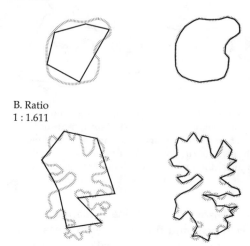

B. Ratio
1 : 1.611

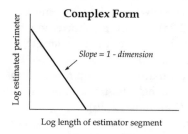

Figure 2.2. The more complicated is the outline of a shape,
the greater the total perimeter
will increase as the line segments
used to outline it are made shorter. The upper shape A is simple
and so has a small ratio estimated
perimeters with long and short
estimator segments (1.1041). The
lower shape is complicated so re-
vealing a greater increase in perimeter with shorter estimator
segments to yield a larger ratio
(1.1.611).

ment length, the greater the fractal dimension. The slope is one minus
the fractal dimension. Fractal dimensions can be estimated by regress-
ing log total perimeter length against log segment length.

The fractal dimension of a given pattern spans all scales from the

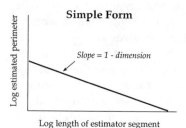

Figure 2.3. If a log-log plot of the length of
the estimators and the length of the total perimeter is a straight line, then the structure is
fractal.

length of the shortest estimators to the length of the longest. The implication of a straight line log-log plot of total length against estimator segments is that the complexity of the pattern is the same at all those scales. The simple way for this to be so is for small folds in the outline to aggregate to form larger versions of the same pattern of folding (figure 2.4). The pattern is then called self-similar. Very few patterns are strictly self-similar, but the straightness of the line on the log-log plot indicates a statistical self-similarity; patterns of similar complexity tend to repeat at successively larger scales, although they may not be strictly the same patterns.

It is, however, not uncommon for a pattern to have different degrees of complexity at different scales. This would be reflected in deviations of the log-log plot from a straight line. These deviations would be significant if, rather than noise about a straight line, the log-log plot bent at a particular scale. The indication would be that whatever family of processes was responsible for the slope of the line before the bend, a new set of pattern generators comes into play at scales beyond the bend (figures 2.5a, b). To identify different fractal dimensions at different scales, one simply does a regression to find the slope over several shorter sections of the abscissa of the log-log plot. The horizontal axis of the log-log plot goes from the shortest to the longest line segments used to estimate the perimeter, so the entire slope gives a fractal dimension that applies across a wide range of scales. Measuring the slope along only successive short sections of the abscissa gives a number of fractal dimensions. Each

Self similar pattern

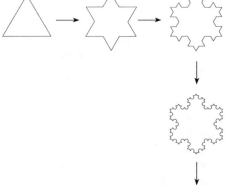

and so on...

Figure 2.4. In the Koch diagram, straight line segments are divided into four segments with a point emerging in the middle of each straight line. Subsequent straight lines of shorter lengths can then be further divided into lines with a point in the middle. This system is self-similar, in that successively finer-grained patterns follow smaller scale versions of the original pattern. The fractal dimension of a Koch diagram is 4/3.

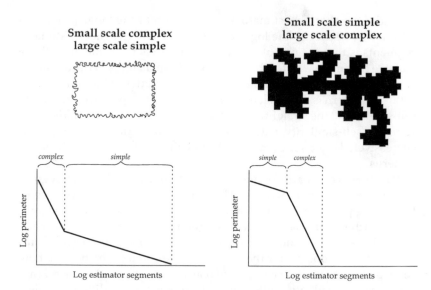

Figure 2.5. If the degree of complexity of the outline of a form is different at different scales, then the log-log plot of line segment against perimeter will bend at the scale at which the complexity of the pattern changes. At that point the fractal dimension itself changes. The outlines are merely to show simple and complex patterns at different scales; the differently scaled patterns are so specifically scaled that the shapes are not strictly fractal, but are to illustrate the point that large and small scale patterns can readily exhibit different degrees of complexity.

one of these fractal dimensions applies only to a limited range of scales. Each fractal dimension is scaled to the range of line lengths used to measure the outline for its respective short section of the slope. With a set of fractal dimensions, each with its local scale, it is possible to create a new plot of fractal dimension against scale.

Analyzing the fractal dimension of landscape patterns shows certain consistent behaviors of fractals. These have implications both for scaling the processes that generate patterns, and for the difficulties innate in the observation of ecological systems. The observations have been of forest islands in satellite images of the United States. Across a range of relatively small scales the fractal dimension is low. As the scale is increased the fractal dimension suddenly jumps to a high value and stays high for further increases in scale (figure 2.6). This result was consistent for many landscapes widely dispersed across the entire country. Two primary sets of forces seem to mold the landscape, one at small scales and the other at large scales. The explanation is that at small scales hu-

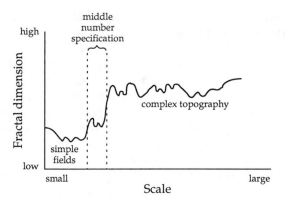

Figure 2.6. Krummel et al. (1987) discovered that the outline of forests in the United States have small fractal dimension at small scale and a larger fractal dimension at large scale. Small scale appears to indicate human activity while large scale seems under topographic control. At a point where the fractal dimension changes, the system has a middle number specification and is likely to be unpredictable (after Krummel et al.).

man influences are the principal cause of pattern while at larger scales topography is the constraining force.

Anthropogenic patterns include fields which are usually rectangular. Tracks and roadways simplify landscape patterns by straightening corrugated edges. Human actions impose simple patterns on the landscape with accordingly low fractal dimension. The reason that the pattern is simple is a matter of scaling.

When two waves of different periods are mixed together, sometimes both waves peak together. Together they make a big wave. Other times one wave is up and the other is down; then each cancels out the other. The aggregate wave is often complex and is the result of a pattern of interference between the component waves. Waves of the same length do not form complex patterns of interference when they are mixed.

Most human enterprises that mark the landscape operate on a time frame between a year and a few decades. Most things we do take about the same amount of time. It is therefore possible to think of human behavior as periodic waves of activity, most of which are approximately the same length. Processes with commensurate reaction rates are scaled the same and therefore do not produce an interference pattern. Although human endeavors are various they often share some common factor that causes the patterns left by humans to be mostly of one type. Property lines must be kept simple for recordkeeping purposes. Transportation is most sensibly conducted in straight lines. Because of this

similarity, property lines and thoroughfares coincide rather than interfere. Humans have made impacts on the surface of the planet at geological scales, but most human activity is relatively small-scale. The small-
scale pattern of forest mosaics in the United States has low fractal
dimension because it is human-dominated. Drury shows that as the English landscape moved from an open system through subsequent acts of
enclosure, successively larger-scale fields were enclosed, but still with
more or less the same straight line margins. At the scale of human enclosures in Britain, the system appears fractal.

By contrast, the pattern of topography is large-scale and complex.
The reason for the complexity of topographic patterns is the obverse of
the reason for the simplicity of anthropogenic patterns. The forces that
give topographic relief are many and each is remarkably independent
from the rest. The independence of the causes for topography comes
from their being disparately scaled. Glaciation, river scouring, rock
compression, and outcropping all contribute to topography, but do so
over very different time scales. They therefore exert their influences independent of each other except as each interferes with the work of the
others. These forces are not contained at the small time and space scales
of puny humans, and so topographic patterns are large. Because topography is the interference pattern between differently scaled processes, the form of topography on the landscape is irregular and has a
high fractal dimension. At larger scales the forest patterns are contained
by topography. This explains the sudden jump to high fractal dimension as topography takes over from humans as the constraint on forests.

There is other evidence of interference between human and natural
forces on the landscape, but over time instead of across space. A study
conducted by Monica Turner and her colleagues involved comparisons
of various aerial photographs from the 1930s, 1950s, and 1980s across
the major vegetational regions of Georgia. The areas studied were stratified random samples of the state, so the patterns can be taken as representative of more than just the places whose aerial photographs were
analyzed. With changes in the economy, much farmland has been abandoned and turned over to forest. As might be expected, the fractal dimension of the farmland was low. Interestingly, abandoned farmland
has a relatively high fractal dimension, indicating an interference between disparate causal factors. Poor land is likely abandoned first, and
that will often be the parts at the margin where the farmer had pushed
up to the practical limits for plowing and other agricultural activities.
These will have been the parts of the farm with irregular outline up
against the hillside, river, or some other natural form. The edges next to
the persisting fields would be straight enough, but the outer edge of

abandoned farmland would be irregularly corrugated leading to a high fractal dimension.

In the 1930s the fractal dimension of the forest was low. With successive decades the proportion of forest increased. The most recent aerial photographs have larger tracts of forest with higher fractal dimension. The explanation is that in the 1930s the forest outline was deeply constrained by the placement of fields. The first photographs were from the period just after the agricultural expansion of the 1920s. The low fractal dimension of forests then was because forest was merely the inverse of the cropland. By the 1980s the constraint of agriculture on forests was considerably relaxed. The forests have expanded in the last fifty years to fill the landscape and are now under a new set of constraints. Forest constraints have become topographical and the fractal dimension of forest has accordingly increased. Forests in Georgia are now larger and show the greater complexity of perimeter characteristic of landscape patterns that are large.

Predictability in Ecological Systems

The switch from low to high fractal dimension appearing with changes in scale has implications for predictability in landscapes. The sudden change in complexity indicates a new set of factors gaining control of the system. When there is a radical change in what controls the system, it is not possible to use behavior under the old regime to predict behavior in the new regime. Remember the constraint criterion for ordering hierarchical levels. If the constraint structure is stable, it is possible to make predictions about system behavior. These predictions amount to saying that the system will show nonrandom behavior in the vicinity of the constraints, and do so reliably. If there is a change in what constrains the system, then prediction is not possible.

Constraints only apply for a certain amount of time; for example, the constraints imposed by life processes only apply to the material of which an organism is made so long as the organism is alive. That is to say, constraints are scale-dependent. When a system is specified, that is, its parts are named and their significant relationships are asserted, a scale is imposed on the system at that point. The parts are defined at a given scale and this fixes the scale of operation of the system so described. Specifying the parts is above all else a scaling operation. More insidiously, the critical step of specification can be made unwittingly. Specification may be only implied in the form of the question that the scientist asks. Thus, the scaling of a system description is often hidden behind two veils: 1) the specification may be only implicit, and 2) the scaling of such a specification is only implicit. Therefore, the system is scaled as soon as it is recognized, even if all that has been done is to ask a question in all innocence.

The form of the question, with its implicit scaling operation, asserts a pattern of constraint. The question can only be answered to give predictions if the implied scale of system specification involves a stable ordering of the constraints. If the question asks what will happen after a certain amount of time, then what constrains the system at that time later will allow prediction, so long as those constraints are known. Those constraints must be at least implied in the form the question takes. However, since specification is often only tacit, the critical constraints may or may not be included. Problems arise when the implied constraint no longer applies. Some questions are difficult to answer because there is a lot of detail to be taken into account; that is not the problem here. Other questions cannot be answered at all because the constraints implied in the question are not stable over the period of the prediction. As we said above, if what constrains the system is not constant, then predictions are impossible. The question needs to be put in terms that involve a higher-level constraint that does indeed remain stable.

Let us show how instability in the ordering of constraint precludes the possibility of prediction. The scientist may not always be aware of the number of distinctive critical system parts that are implied by the question that has been asked. Answerable questions invoke predictable systems. Predictable systems are implicitly or explicitly specified so that they have a fairly small number of critical parts. A reliable equation can then be written for each part and its relationship to the others. Planetary systems would be an example here, where the respective masses and velocities of each planet are substituted one at a time into a single gravitational equation. These are called small number systems.

Sometimes the small number of parts involves using a small number of averages or representative values that subsume a very large number of parts. The number of entities contributing to the average must be large enough to make the average reliable. Equations are then written for the averages. These are called large number systems, but they work the same way as small number systems in that the number of reliable averages is small even though the reliability of the averages derives from a very large number of units contributing to the average. The gas pressure laws replace distinctive atoms or molecules with average "perfect" gas particles.

The problematic systems are called middle number systems. These have too many parts to model each one separately, but not enough to allow averages that fully subsume the individuality of all the parts.

Questions that cannot be answered imply a middle number system specification. The reason why middle number systems are unpredictable is that the constraint structure is unreliable. Any one of the too numerous to monitor parts can, at any time, take over the constraining role in the system. Any part that can gain the ascendancy cannot be disregarded as part of an average, and yet there are too many to treat them as individuals. Note that we are not asserting that nature is fundamentally unpredictable. It is the mode of system description that leads to the inability to predict.

In landscapes the change in the fractal dimension at middle scales causes the system to be unpredictable over those middle scales (figure 2.6). At small scales the landscape is reliably influenced by human endeavors. At larger scales the causes of pattern are reliably topographical. At scales in between, each part of the landscape has its own individual explanation. In some places topography might dominate, while in others the controlling factors might be distinctly human. In the middle scale, the landscape takes on middle number qualities, as the quirks of local considerations jostle to produce patterns inexplicable in more general terms. Thus a simple question, "What governs the form of abandoned agricultural land?" is in grave danger of invoking a middle number specification of the landscape. The answer is not very satisfying: "Lots of unrelated things." On one edge it is human endeavors, embodied in the land still under cultivation, while on the other it is the local pattern of glaciation or erosion that defines topography. Weather and economics also interact to change the interaction between these two disparate sets of constraints. The scale implied in the question is the problem here.

There are specific examples of middle number failure to predict in landscapes. Dean Urban found that when woodlots are small he can predict the number of birds with accuracy. If the patches have room only for one territory, then he only needs to count the woodlots and estimate percentage of occupancy to know the bird population. If the land is forested in large tracts, then he only needs to know the area forested and the size of the average territory. Dividing the former by the latter gives reliable estimates of bird populations. If, however, the woodlots have room for a small number of territories, say two and a half, then the estimates are unreliable. The reason is that the orientation, shape, and degree of isolation, among other things, influence the number of birds. The differences between the woodlots becomes a matter not of simple area but of many other considerations. The individuality of each woodlot makes a difference and so prediction is unreliable at best. Each woodlot makes a difference and so prediction is unreliable at best. Each woodlot constrains the outcome to an unmanageable degree. Thus an innocent question, "How many birds are there in my system?," is fraught with danger.

Repeated Patterns

Although there is a difference in complexity between small- and large-scaled patterns in human-dominated systems, it is remarkable that the fractal dimension remains constant for such wide ranges of scale. Indeed, if it did not, we could not say that the patterns were fractal. Clearly there are different factors governing the processes that give the patterns across the range of scales over which fractals are estimated. Therefore, the constancy of the fractal dimension is unexpected and requires an explanation. The answer is that there are remarkably few patterns used by nature. As Peter Stevens notes, "In matters of visual form we sense that nature plays favorites." He goes on to give us the reason "why nature uses only a few kindred forms in so many contexts."

The repeating patterns are spirals, meanders, branching systems, and explosive patterns. Additionally, many systems show 120-degree angles because they are both the consequence of space packing and what is left after stress release in a homogeneous medium. Cracks in soil and polygonal frost heaving patterns clearly fall into these classes of landscape. The requirement for homogeneity is often not met at larger landscape scales, so most 120-degree angles are at small landscape scales, with the one gigantic exception of stress release when Pangea, Laurasia, and Gondwana Land broke apart between 135 and 300 million years ago. Spirals, meanders, explosions, and branching patterns appear to apply across all landscape scales. Their ubiquity stems from four geometric attributes that have implications for processes: 1) uniformity; 2) space filling; 3) overall length of a linear system; and 4) directness of connection between parts of the system.

Let us connect the patterns to the geometric attributes. This will facilitate the linking of the processes common to landscapes (figure 2.7). Spiral patterns are beautifully uniform and fill space very effectively. They are also relatively short in total track, given the space they fill. Random meanders are very similar to spirals, although they are not as uniform. Like spirals they are not very direct in connecting arbitrary points. A river on a plain is a force that from time to time fills the space with water in very literal terms. Beyond the space filling of intermittent floods, the continuous processes of erosion and deposition tend to move the river across the space so as to fill it over time. Note that these processes and events apply to water courses over a wide range of scales. Rapid trans-

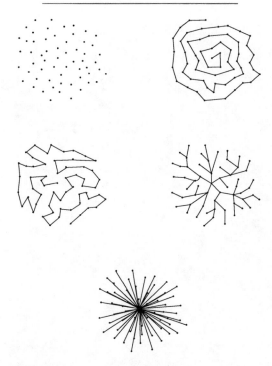

Figure 2.7. More or less any collection of points can be connected to show either a spiral, a meander, an explosion, or a branching pattern.

port of water from one place to another is not a critical factor on flat terrain, and so the lack of directness of meanders and spirals is not a problem.

Explosion patterns are very direct, particularly with respect to the center of the pattern. They overcrowd the center of the space and so do not fill space in the uniform manner of spirals. The angles between their tracks can, however, be very uniform. The total track of an explosive pattern is exceptionally long, and so in explosions conservation of construction material cannot be a consideration. When the rapid dispersion of material is critical, then the explosive star is the pattern to be expected (figure 2.8). Some landscape patterns are of explosive origin in the literal sense, volcanoes being a case in point.

While explosion has a literal meaning, that definition only applies to the narrow and arbitrary time frame of primary human experience. Something that takes a whole minute is stretching the literal meaning of the word "explosion." If, however, we recognize the scale-relative nature of explosions, then the notion can be applied to other scales. An

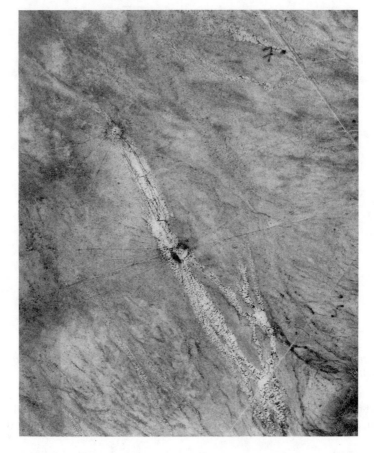

Figure 2.8. The pattern here from the Sevielleta LTER site shows an explosive pattern around water butts which cause grazing animals to move rapidly in toward the water resource (photo NASA).

explosion involves movement of material so fast, relative to normal system behavior, that it occurs as an event. The context has no time to exhibit behavior during an explosion and thus appears static. If we now take a system whose normal behavior is very slow, then something taking days or longer could amount to an instantaneous event. In this way patterns resulting from moderately paced dispersion of material can take on the explosion form, although there is not a literal explosion at the human scale; there is merely a functional explosion given the scale of the system at hand.

The form of a mountain changes over millennia except in cases of literal explosions. A literal explosion happens very fast relative to the time frame of the observer. Mountains erode very slowly, whereas most of the water from a thunder shower runs off the mountain in days. Therefore, at the time scale at which the mountain erodes away, the movement of water off a mountain is of explosive speed. Much of the appearance of explosion patterns on volcanoes on remotely sensed images is not due to the volcanic explosion over the time scale of seconds to minutes at the time of the literal volcanic explosion. Rather it comes from millennia of repeated removal of water from the steep sides in a matter of days. Given the size of a mountain, the removal of water from its steep sides happens at explosive speed. A concentration of water high on the mountain is dispersed on all sides a considerable distance. The only way to disperse water that far that fast is with a figurative explosion, in an explosive pattern.

Seasonal rain is removed in an instant by the temporal standards of the rock on the mountainside. The regularity of the arms of an explosion pattern comes from competition between the parts of the moving front of the explosion. A part of the mountainside that happens to receive more runoff has a shallow channel cut in it. In subsequent runoff events, that channel takes water from either side, so cutting deeper itself with the water from adjacent regions (figure 2.9). Deprived of their water, the regions on either side of the channel erode more slowly than the channel itself. Channels without competing neighbors cut faster than those that have to compete for water with other gullies. Thus, regularly spaced channels grow faster.

Competition is a positive feedback in which nothing succeeds like success. Positive feedbacks require a supply of resources that can drive the positive feedback and feed the growth of the winners. Note that ink dropped a short distance leaves only a generally circular blob, while ink dropped from a great height forms an explosive splat. The blob is not explosive because the force behind it was insufficient to feed the winners of the positive feedback over the time period required to manifest the pattern (figure 2.10). The prongs of the splat result from reduced surface tension in the areas of expansion, which causes further expansion, which starves the regions of the edge between the prongs. Thus the valleys on the side of a volcano compete for water and starve the ridges of the means of cutting their own channels, but for an explosion the sides of the mountain must be steep to provide the abundant energy required. The movement of water needs to be forceful. Gentle slopes or even equitable and modest rainfall on a steep slope blunt the force of the

Figure 2.9. The explosive pattern on a volcano is not always made by the explosion itself, commonly the explosion is a series of evenly spaced gullies moving away from the center created by rain runoff. Erosion on a California hillside shows how competition keeps channels apart (photo T. Allen).

positive feedback, and so competing linear channels fail to form. Sheet erosion leaves no explosive marks on a satellite image.

Explosive patterns also occur on urban ecological landscapes. Given the relatively long time it takes for an urban center to grow, the requirements of movement of commuters twice a day is a matter of explosive speed. That is the reason for the explosion pattern of roads from major cities. Roads are expensive but that cost is less important than the pressing need to move vast crowds in and out of the city quickly. Duplicating tracks of major arterial roads, each on a slightly different radius, is one of the costs of urbanization. For metropolitan road systems, the explosion is the pattern of choice. The Roman adage, "All roads lead to Rome," was true. A satellite image of the Iberian Peninsula from the time of the Roman Empire would have shown the explosion pattern of Roman roads we might expect. The flow that caused that star was slower than the flux of modern commuters, but it was faster and larger than any other flow at the time. Note that the interstate system of high-speed freeways that connect the major cities of the United States does not form

Figure 2.10. When ink is dropped onto paper it does not leave the characteristic splat pattern unless it is dropped from a significant height. With a short drop the ink only forms a blob because there is not enough energy to drive the positive feedback of the winning sectors of the perimeter.

explosion patterns even when it serves New York City or Chicago. The interstate system is a network because it functions as a national transport system. It belongs to a higher level of organization than mere metropolitan transport.

Branching systems are also common on landscapes. The geometric properties of branching networks are a compromise between the single circuitous route of the spiral and the heavily duplicated tracks of the explosion. In a branching system the total track of the system connecting all points is exceptionally short. They achieve this economy of track material at the expense of some small indirectness of connection of individual points here and there. Branching systems are less uniform than either the even, constant curving pattern of the spiral, or the uniform angles of the explosion. There is irregularity of detail in branching systems, although they do fill the space. When economy of track and relatively direct connection of all points to some central point are the important factors, then branching systems are the pattern of choice.

Note that the explosion of arterial roads from a city gives way to a branching system at a distance from the city center. This is because the urgency of moving people at rush hour is less, but the cost of building roads remains high. While duplication of major highways on adjacent radii is a necessary expense at the city center, in outlying districts costs are minimized by serving adjacent small communities with one road that eventually branches to serve both. Similarly, the explosion of water courses from peaks soon forms a branching network of rivers once the slope eases. Animals like sheep free ranging on the Welsh Mountains need to move to all parts of a space eventually so as to exploit resources. There is not an explosive flow because neither the sheep nor their resource base are centralized, and so traffic does not justify duplication of pathways. Therefore, grazing animals leave branching networks of trails on the landscape.

Through Stevens' analysis of the ubiquitous processes that underlie spatial pattern, we see that the distinctive patterns on a landscape are composed of a small set of component patterns. Movement of material at various pressures is responsible for pattern, but the options for regular pattern are few. This is the reason for the unlikely self-similarity of landscapes. Very few ecologically pertinent factors apply across many scales. In fact, as a rule of thumb, when different explanatory principles are suddenly needed, it is because there has been a change in scale. Given that there are so many different types of entities in an ecological system, it is remarkable that we find self-similarity to any degree, and very unexpected that we find a unifying set of ideas that offers so many explanations from disparate scales. The same general class of causes underlie branching systems of mighty rivers and the trails left by ants on a tree trunk. The explosion of Mount St. Helens is matched by the tracks of livestock to water buttes in the desert grasslands of New Mexico (figure 2.8).

Linking to Other Criteria

In the opening section of this book, we referred to the layer cake of ecological levels. So far in this chapter we have moved up and down the landscape wedge. Now we will look at landscape in relation to the other ecological criteria; we will slice across to other parts of the cake.

The notion of constraint becomes important again here. The patterns Stevens recognizes emerge as necessary constraints on system structure. It is characteristic for landscape patterns to have an underlying fine-grain process. This process would continue indefinitely but it eventually encounters an upper-level constraint. The marks on the landscape represent limiting factors imposed on some lower-level process (figure 2.7).

Accordingly, landscapes approached at the level of those constraining boundaries should be predictable because systems with reliable constraints allow predictions in the vicinity of those constraints. In the middle of a homogeneous patch, prediction as to the fine distinctions of the continuing pattern is not possible without a change of scale, and maybe not even then. However, at edges where the low-level process encounters the constraint, many predictions are possible. It is difficult to predict the position of the grains of light and dark in a textured expanse of a field, but it is easy to predict that when the slope reaches a certain angle, the farmer will stop plowing and the field will give way to another habitat. At a larger scale, it is hard to predict the variations of fields and fence rows on a quilted agricultural plain, but it is easy to predict that a mountain or a desert will constrain agricultural activity.

To test hypotheses about landscapes we need to know the characteristics of a neutral landscape, one that does not reflect the constraining effect of topography, aggregations, disturbance history, and similar ecological influences. Such landscapes would need to be formed by a random process that propagated pattern in the absence of constraints. Such neutral landscapes constructed by Robert Gardner and his colleagues have distinctive characteristics. If we consider a random landscape in terms of the percentage of the surface occupied by the habitat of interest, then the largest number of patches occurs when the habitat occupies 30 percent of the total space. At lower values there are a few well-spaced patches. Above 30 percent the number of separate patches decreases because the addition of more occupied habitat tends to connect existing patches more often than it creates new ones. The ecological processes on real landscapes seem to generate fewer and larger clusters than expected from the neutral random model with no constraints.

It emerges that changes in total habitat area do not have the same effect at all percentages of land occupancy. This is illustrated clearly in the case of the neutral landscape. There a 10 percent decrease in land covered below the critical 30 percent level will have negligible effects on the number, size, and shape of patches. Above 30 percent occupancy the same 10 percent reduction will have dramatic effects. Thus, the effect of disturbance that removes habitats of a given type are not simple. The effects will be greatly influenced by the original quantity of the habitat in question, independent of the size of the disturbance.

In random landscapes, at values greater than 59.28 percent of the landscape occupied, the network of habitat interconnects so that there is a pathway from one side of the map to the other without leaving the habitat in question. This number is well known from a body of theory concerned with the percolation of liquids or passage of electricity through aggregates of material. For ecological landscapes the capacity of the system to allow percolation is important. It couples landscapes to other criteria for ordering ecological systems. Communities and eco-systems are not strictly spatial but involve integration across space. For a community or ecosystem to have meaning the parts must be able to in-terrelate, that is, energy, biota, and abiotic matter must be able to perco-late across the landscape involved. The ability to allow percolation re-flects the capacity of the system to permit indefinite spread of organisms including disease, or physical factors such as fire. Real landscapes, full of pattern-generating ecological constraints, usually allow percolation at a lower threshold of area occupied by the habitat in question than do random landscapes. In the next section below we consider the special parts of real landscapes that encourage percolation.

The Importance of Fluxes on Landscapes

At the beginning of the rekindled interest in landscape ecology in America, there was a phase of naming and classifying landscape pat-terns. This resembled the first stages in community ecology when, at the turn of the century, Frederic Clements collected old terms and coined terms for different aspects of community form and process. For example, Clements listed many words for nuances in plant establish-ment and names for different means of doing that in different places. Such a large ecological vocabulary with so many fine distinctions may seem superfluous now, but underlying each word was a new concep-tion. Most of Clements' community terminology has fallen from use now that the underlying concepts are fully integrated into the modern conception of communities. The same may come to pass with the bur-geoning vocabulary in landscape ecology, but for the moment the words draw attention to the richness of pattern and the meaning that the vari-ous forms might have.

In landscape ecology, the terms imply underlying processes. In some cases these words are for structures and processes that link to other cri-teria for organizing ecological systems. A case in point is the term "cor-ridor." A corridor is a linear form on the landscape, but the choice of that word, as opposed to any other thing that is long and skinny, invokes a metaphor involving movement. Corridors have dramatic effects on the capacity of the system to allow percolation of ecological material across landscapes. They are places of flux; they are communication channels. This has importance not only for the workings of communities and eco-systems, and the movement of organisms and populations, but also has particular significance for the scale-oriented approach that we employ throughout this volume. Let us weave together the more abstract scaling implications and the use of the simple tangibility of corridors to relate landscapes to other organizing criteria.

We define levels in a hierarchy by the characteristics of the bounded entities occupying that level. There are strong connections inside the surfaces that bound the entities occupying the level in question. Although there is stronger bonding of the parts inside the surfaces of the lower-level entities, for upper levels to emerge there must be some exchange, albeit weak, between the low-level entities. These weak exchanges are what allow the low-level entities to become the parts of the upper-level entity. The next level up depends on interactions of the low-level entities immediately below.

Thus, surfaces that separate the low-level entities become involved in communication channels for the next level up. Some examples might help here. The skin of an animal contains the single organism, but it acts as a communication channel in blushing as the animal plays its role in a social structure, or in communication of diseases. Surfaces filter material and energy; in disease transmission, the pathogen exists in relatively high concentration in the animal already infected but only relatively few pathogens pass through the organism's surface to achieve a new infection. The course of the disease reflects strong interactions within the animal's surfaces, but the small number of disease agents involved in infecting a new individual reflects weak interactions across the surfaces of organisms. Communication channels have properties opposite to surfaces; they allow the passage of signal unimpeded. Accordingly, corridors are the opposite of landscape boundaries and pertain to higher levels of organization than the patches which they connect.

Corridors are the connectors across boundaries. They allow interaction of the parts of large-scale communities or ecosystems. If communities and ecosystems are worth studying, they cannot be happenstance collections of biotic and abiotic material. Communities and ecosystems have integrity coming from a binding together of their parts. Those interactions are mediated by communication channels like corridors. Some of the exchanges between system parts can be by means of short-distance communication not along corridors, but by diffuse interaction. However, over longer distances corridors are often involved because they facilitate flow of organisms and abiotic material.

In traditional Euclidean diffusion, the mean square displacement varies as kt, where t is time. Bruce Milne, of the University of New Mexico, has pointed out that in fractal environments, mean square displacement is proportional to $kt^{.7}$ (for random walks on a percolation cluster).

Therefore, by virtue of their linearity, corridors transform a fractal landscape into a Euclidean landscape. Corridors speed up flow across the landscape by a factor of $t^{.3}$.

By channeling material, corridors allow long-range movement of ecological signals with minimal attenuation. Thus corridors are an important part of the means whereby larger-scale ecological systems form and persist. A system with significant corridors is larger scaled over space than one with few corridors. The relatively rapid movement down corridors allows parts at considerable distance to interact. Rapid flux down corridors allows communication of distant parts, leading to a system that can persist in spite of destabilizing forces that operate transiently and locally in space. Corridors allow the long-term survival of the system through reinvasion from a distant unaffected region within the large-scale system. Long-term survival indicates a high-level system. The short time for transit down a corridor should not be mistaken for high-frequency behavior of a low level of organization. The rapid movement down landscape corridors is no more low-level than is the telephone system that allows cohesive behavior in large metropolitan centers like New York City. Corridors move the system upscale by allowing large-scale integrity.

This integrating role of corridors has an intuitive appeal. However, the functioning of corridors as the communication channels of communities and ecosystems is not a simple matter. Many types of organism encounter linear structures on the landscape differently. Only some types of organism will use a given type of linear structure as a corridor. Large mammals might use the open strips under power lines or along logging roads as corridors for rapid movement. Using these mammals for dispersal, weedy plants with burrs may also move down these same corridors. On the other hand, birds from the forest center may avoid these corridors. Accordingly, fruits whose seeds are dispersed by such birds will not move around on the logging trails.

The fact that different organisms respond to potential corridors in different ways has important consequences for the scaling complexity of communities and ecosystems. When we assign a different upper level, say a man as a factory worker instead of a father, the difference is a matter of recognizing alternative communication channels as significant: the connection to coworkers and workplace instead of family members. In the same way that we can emphasize alternative connections to arrive at some other upper level (factory versus family), organisms on a landscape are emphasizing one connection over another when they do or do not use a linear form as a corridor. When organisms respond to their distinctive perceptions of linear landscape structures, they produce distinctive upper-level systems, one for each type of reaction to the critical corridors. To the extent that corridors are the connectors that give community integrity and character, organisms living at the same spot but with contrary perceptions of a corridor can be functional members of different communities. This is not a matter of us choosing to recognize this as opposed to that ecological system. It is a matter of functional difference for the organisms, whether we do ecology on them or not. A deeply integrated system from one organism's biological point of reference is a capricious happenstance assemblage from another's experience.

In a similar vein, abiotic components are transported to various degrees down corridors and so could belong to alternative ecosystems depending on which types of corridor move them and which do not. For example, wind and water move different sized particles, wind often uphill but water always downhill. Water-soluble materials will move down water courses, but anything insoluble will not. Different degrees of solubility change the rate at which water can transport deposits of various materials. The reason why lime is so important in natural and agroecosystems is that it changes solubility of minerals and so changes the way those ecosystem components relate to corridors of water. For both communities and ecosystems, inconsistent uses of corridors by different organisms and abiotic materials may not be the rule but neither are they the exception; this inconsistency confounds simple conceptions of communities and ecosystems as places on the landscape.

The problems raised for the ecologist by organisms' different perceptions of the landscape are very real. The major difficulty is that differently scaled entities may be forced together in our conceptions of the relationship of organisms to communities. We have already emphasized that changes in scale radically change apparent system behavior. The difficulty with these alternative responses of organisms to given landscape linearities is that each new type of response rescales the environment in an important functional way. Anything that responds to a linearity as a corridor is a functional member of a relatively large system. Anything which does not move along a corridor belongs to a local system, unless it does respond to some other sort of corridor of about the same length.

This raises the question of whether a corridor for one purpose is not a boundary for another. While the Alaskan pipeline is a conduit for human purposes, there was deep concern that it might be a barrier for animal migrations. Many deep forest birds have an aversion to open spaces such that a logging road is a considerable boundary. In primeval forest, these birds belonged to a system that was larger, occupying great tracts that were more or less continuous. The rescaling that has occurred through human intervention could have large consequences for genetic diversity and population persistence. By decoupling routes of reinvasion, narrow logging trail corridors do not allow local extinction to be reversed in small sections of forest. Over time this leads to global extinction. Meanwhile, birds that have more neutral responses to small open spaces are not rescaled at all by the corridors. Yet other species, those inhabiting forest edges, suddenly find themselves thrust into a much expanded and linearly connected network of forest edges. They become part of a new large-scale system. If such opposite effects can pertain within one order of animals, it is clear that difficulties in prediction across orders or kingdoms are to be expected. Conservation is suddenly a very complicated business.

The differences in response to potential corridors is no more marked than in the plant-animal kingdom distinction. Animal movement scales fauna in a way that has little relevance to plants. Animals move fast down corridors, and across all but the most difficult spaces they pass quickly by plant standards. In a search and statistical analysis of the literature, we have found that animals are used principally for studying ecological conceptions that overlap very little with those studied through plant-centered investigations. Plants move slowly across landscapes, almost all large movements requiring a reproductive event. Hedgerows are barriers, not corridors, for some animals but they are not often barriers for plants. Even when hedgerows are considered as corridors for both plants and some animals like birds and rodents, there are great differences in the temporal aspects of corridor use. Corridors usually rescale the context for animals in a way that is different from the way they rescale it for plants. Therefore, the questions that plant ecologists ask are often very different from those of animal ecologists. For all the lip service paid to the unity of ecology across the botanical and zoological divide, there are very real differences in the conceptions of plant and animal ecologists, and for sound reasons of differences in scale. Plant-animal interactions are well worth studying, but they can be fairly described as zoological studies with some substrate for consumption and perhaps trampling.

There is some distinction in the literature of landscape ecology between corridors which are wide enough to have interiors as opposed to those which are so narrow that they only have edge. Corridors with interiors can function as discrete entities in their own right. For all their fully apparent flux, stream corridors have permanent residents of their own. When corridors intersect they often show unusually high diversity of organisms. In landscape networks the flux is high and the alternative tracks at each intersection change the scale of the system.

Topography and Distance as Surrogates for Interconnection

The linking of the landscape criterion to the other ecological criteria often involves flux, but not always along corridors. Sometimes ecological material passes across the landscape diffusely. Landscape topography may slow down that diffusion. In this case topography operates as a filter on ecological signal. The wider the expanse of unfavorable terrain, the more the high-frequency signal from low levels of organization is filtered. Landscape topography changes the functional scale of the communities on the ground.

Bruce McCune's study in the Bitterroot Mountains reveals how topography can constrain communities. The Bitterroot Mountains run north and south on the Montana-Idaho border for some fifty miles. At regular intervals there are valleys that cut into the range from the east. Since they are so close together and are geologically homogeneous, one might expect them to be replicate systems. If communities are the result of filtering of biota by competition mediated by the environment, then the communities in the valleys should be the same in composition. The unexpected observation is that they are not the same. Furthermore, the differences do not align themselves with any environmental factor, not even a spatial ranking from north to south. After years of tedious measurement of the environment, McCune was forced to presume some other cause (figure 2.11a-c).

The valleys are completely separated from each other by the mountain ridges. The ridges filter out propagules so that only what is in the valley already is available to colonize. In a more open system, we might expect the constant pressure of invasion to produce a full suite of competitors for the environment to select. Selection over millennia should produce an environmentally determined outcome, probably one with the same vegetation patterns in all the valleys. However, the mountains constrain the valleys so much that they are too small to reach any sort of unique equilibrium between vegetation and environment.

Ilya Prigogine, Nobel laureate for chemistry, studied the emergence of higher-level order in chemical systems far from equilibrium. The particulars of that high-level order are the consequence of the particular configuration of the system as lower levels of order went unstable. For example, the details of form of the nucleus that starts a snowflake determines the particular form of the high-order symmetry in the finished

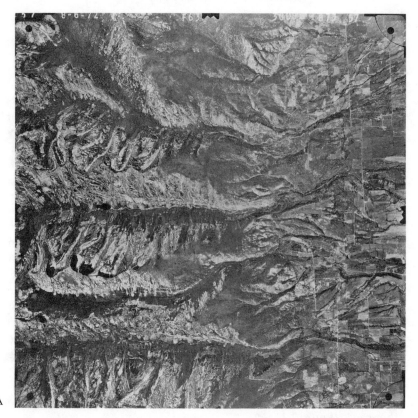

A

Figure 2.11. The canyons of the Bitterroot Mountains are isolated and, being small, each has its own vegetation arising from a unique disturbance history. A. shows Fred Burr and Bear Canyons. *Larix* does not occur south of Fred Burr. Fred Burr and Bear lack *Thuja* but Mill Creek Canyon, on the bottom margin of the photograph, has it. Bear has *Pinus monticola* but the other canyons in this image do not. Note how the Bitterroot Valley running north-south up the right side of the image isolates the canyons from each other vegetationally (photo US Forest Service). B. A map of the entire range showing McCune's study sites. C. is a view from the trail to St. Mary's Peak looking south down the Bitterroot Valley with the ridges between the canyons marching into the distance (photo B. McCune).

snowflake; the instability is of liquid water at temperatures below freezing. Complex systems are formed by successive reorganizations where a series of instabilities cause the emergence of a series of higher levels. That is why complex systems require several levels of organization for their adequate description.

Prigogine has stated that complex systems contain past disturbances

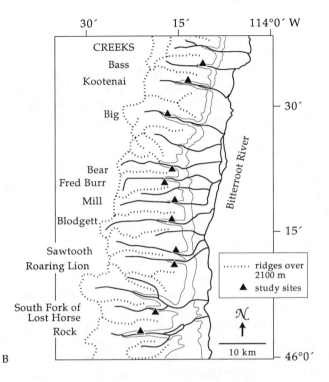

B

C

Figure 2.11. (*Continued*)

81

in their lower levels. A low level is disturbed and collapses up against a new, more global constraint of the new higher level. What makes the system unstable becomes an integral part of the working of the new higher level, for example the nucleus of a crystal determining crystal form. An ecological case in point is fire-adapted vegetation (high level) where individual fires destroy and disturb the susceptible community (low level). Fires change the vegetation over time until it is fire-adapted. At that point, long-term survival of the community requires fire to remove invaders that are not fire-adapted. Fire-adapted communities have incorporated fire as part of the system. Fire is a friend of the emergent higher level, but its destruction of biomass indicates its historical role as disturber of the primitive low-level system.

The emergence of higher levels is dependent on the system being large enough in temporal and spatial terms so that the disturbance can be put comfortably inside. In the case of the Bitterroot valleys, they are too small to incorporate local disturbances. With the mountain ridges filtering out the propagules of many species that are in the Bitterroot Range at large, each valley is dominated by the happenstance of what was available at crucial times of establishment. Each valley is controlled not by a process of orderly maturation of community structure through environmentally monitored competition, but by historical events.

Note here that what might be historical accident at a large scale could be seen as environmental context at a smaller scale. Being somewhat isolated from seed sources from other valleys, the processes at work in each valley are isolated and see the local single valley system as global. Within that narrower purview, what occurs across one valley is environmental context, even though seen in the context of the mountain range at large, what occurs across one valley is historical accident.

Consider how forest herbs respond to the local forest as an environmental constraint, although the local forest trees may be the result of some accident of past fire which allowed some trees to survive and seed into the open area after the burn. If one were by chance to be present at an important historical event like a fire, the particular movement of the fire would be perfectly understandable in terms of the interaction between environmental factors such as slope, wind direction, and state of the tinder. Nevertheless, the fire would be a historical accident over the long years that it leaves a mark on the tree populations at the site.

At the large scale in which species composition is the result of long-term filtering of a complete species set, the vegetation in each of the Bitterroot valleys is governed by accident. These accidents occur in such isolated conditions that the vegetation as a unit is too local to be able to incorporate the disturbance in some higher-level context. Thus, the factors responsible for what is present in each valley are very local in time and space. Wind direction on a given afternoon is the explanation for what one sees perhaps centuries later. In a larger system, the constant invasion from undisturbed places at a moderate distance would soon remove the effect of the fire that happened on that particular afternoon. The fire would be incorporated and disappear in the larger system in a way that does not happen in the local isolated valleys of the Bitterroot Range.

Landscape Signatures Across Biomes

While there is much to be said for putting effort into linking particular patches on remotely sensed images to ground truth, the power of remote sensing for broad comparisons needs to involve a single all-encompassing signature for each entire landscape in the comparison. The generalized unified patterns can be compared as wholes rather than as a set of local patterns with local causes. Nontrivial comparison at large scales has to be conducted at a level of great generality, otherwise the sites are merely completely different.

It is possible to generate a single number that describes one site such that comparison between several sites across several biomes is possible. That number appears to be a quantitative estimate of the general significance of biological processes in generating pattern. The argument is as follows. Pattern on remotely sensed data can be helpfully divided into just two types: 1) fossil pattern which is a scar left by some ancient event, and 2) pattern that is reinforced by a continuing process. The difference between reinforced pattern and fossil pattern is a matter of temporal scale. Anthropogenic and biologically significant patterns both fall into the self-reinforcing category for the most part. For example, a cornfield generates a crop which is sold, and the money is used to put the field back the next year. Geological pattern is of the ancient scar variety. The relatively large amount of water in eastern United States ecosystems makes them biologically driven. Therefore, the eastern United States is given to self-reinforcing pattern over scar patterns. The opposite pertains to dry western sites. Since biology involves growth about a point in space, biologically driven pattern should be isodiametric and should smear the edges of fossil patterns. Fossil pattern will be relatively sinuous and have sharp edges.

There are methods of analysis that classify the texture of an image. These analyses take each pixel in the image and compare it for brightness with its neighbors. The computer can deal with forty different tones on the image and so it is possible to generate a forty-by-forty matrix that summarizes the chances of all brightnesses being next to all other brightnesses. It is possible to operate on this matrix to obtain measures of heterogeneity, contrast, and other pattern qualities. This can be done for vertical, horizontal, and both diagonal pixel comparisons sepa-

Figure 2.12. A. The Okefenokee Swamp in Georgia (photo B. Patten), B. the Northern Lake site in Wisconsin (photo Northern Lakes LTER), and C. the Jornada Long Term Ecological Research site in New Mexico (photo Jornada LTER) represent a successively drier series. D. Konza Prairie LTER also studied for texture by Musick and Grover (1991) (photo Konza LTER). With a drier landscape there is less biological pressure on the system, less continually reinforced pattern, and more linearity to the pattern on the ground.

rately. If the pattern is biological, the signature should be isodiametric. That is to say, the different directions yield the same degrees of contrast giving a low coefficient of variation across all directions. All directions should appear the same.

The Okefenokee Swamp is a system driven by biological processes. There is essentially no topography so the pattern should be blotchy, not streaky. Computer processing of the image indicates generally radial patterns because of a low coefficient of variation across direction with respect to pattern contrast. The Jornada research site in the deserts of New Mexico is very different. It has a sinuous pattern and gave high coefficients of variation indicative of a fossil pattern. Elsewhere in the desert, the marks of General Patton's World War II tank training maneuvers in the desert west are still fully apparent and will probably remain for centuries. To an extent, fossil patterns fall prey to the forces that drive reinforced pattern. Biological processes soften fossil patterns and so lower the contrast in sinuous patterns from steep topography. If biological processes became significant in the desert because of a change in climate, the reworking of material in biological processes would fidget Patton's tank tracks to oblivion in a few years.

The more radially symmetrical a pattern, the more important is the biology as the principal constraint. Since biological processes are constantly reworking material, isodiametric pattern indicates a landscape dominated by high-frequency constraints. This is certainly true for the Okefenokee Swamp which is as dynamic as one could expect a large landscape to be. Decomposition bubbles up the swamp bottom to make islands. Trees grow on these. Meanwhile, the alligators worry the sides of the island until it breaks away, floating around with its trees in place. Thus, relatively large landscape units in Okefenokee Swamp are uncommonly mobile. Not all biologically driven systems show this extreme high frequency of movement but in general, the more biology constrains the system, the shorter is the significance of historical events, because rapid biological processes destroy landscape memory (figure 2.12a-d).

Conclusion

Landscapes are the most tangible of the ecological criteria. Therefore, we tend to study them at conveniently human scales. There are, however, small and large scales at which we can profitably study landscapes. Despite the wide range of scales of ecologically interesting landscapes, there is a remarkable unity to the landscape criterion. The reason is that many of the patterns at the scale of an unaided human experience, say in a landscape painting, are remarkably universal. There are indications that the fractal dimension of the landscape experienced by insects is lower than the environment in which we function. However, many processes like surface compression and tension, explosions, or meandering flow recur at scales from the landscape of a leaf surface as seen by a mite, all the way up to remotely sensed images of continents taken from 22,000 miles above the earth. The reason is that Euclidean space has the same geometric properties no matter what the scale. Much of the world rests in whatever state that requires minimum energy for persistence, for example tight packing to minimize stress. These minimum energy states are generally limited to certain geometric configurations, for example meandering when there is no energy to ride over obstructions or branching systems when material for track is expensive. These and a few other patterns we see repeating over again at all scales.

The power of the landscape approach is in its intuitive appeal. We can use our unaided senses to great effect on landscapes. The familiarity of the landscape criterion makes it a mode where interesting hypotheses are most often generated, even hypotheses that pertain to other criteria like populations, communities, or ecosystems. For example, the pulsing epidemics of spruce budworm are population and ecosystem problems, but the manifestation of the event on the landscape is what first grabs our attention.

Although many interesting phenomena rightfully belonging to other ecological criteria do appear on landscapes, the relationship of landscapes to other ecological conceptions is far from simple. The problem is that the other conceptions appear to be based on many more primitives. The essence of landscapes can be captured by only a few patterns, but communities, for example, contain many species. Worse than that, the primary units of communities are not scale-independent as are the units in landscapes. While big landscapes can be self-similar, employing the same few patterns as small landscapes, there are great differences between large and small populations, communities, or ecosystems.

The organism criterion seems the only governed by a set of constraints about as small in number as the set for landscapes. Large organisms are stockier than small ones, but big organisms take the form they do for the same reasons that explain the form of small organisms. It is again a matter of minimal energy. Note also that the organism criterion is as deeply anthropocentric as the landscape criterion. We have a vested interest in recognizing organisms, for our very own level of organization is full of them. What we recognize easily is a reflection of what is important for our own biological survival.

As with organisms, human perception of landscapes is probably the result of selective pressures. It is reasonable to suppose we have been selected to perceive the world in a way that allows prediction. Prediction comes easier in familiar circumstances. Since changes in scale change perception radically, it would be of advantage to perceive in a way that recognizes patterns that occur at multiple scales; then the world remains familiar even under changes in scale. These scale-independent patterns are meanders, spirals, explosions, and branching systems. Landscapes and organisms appear to be very special criteria for organizing across hierarchical levels, special because of the peculiarities of human perception.

Mapping communities, populations, and ecosystems onto landscapes is not simple. This is because the tangible organism criterion and the tangible landscape criterion form a complex interference pattern which is the community or ecosystem. Not all organisms read landscape pattern in the same biological terms, and so some respond to certain landscape features and some do not. The problem is that corridors and other landscape features rescale the context, depending on how the organism responds. If the landscape is taken as the frame of reference, then finding one type of organism in a place has different scaling consequences than finding another type of organism in the same place, even though they apparently live there together.

If organisms of different species on the same spot of ground respond to differently scaled patches of landscape, then the conception "organism" becomes inconsistent as a scaled unit. Alternatively, if we use organism as the normalizing frame, then the landscape becomes a curved space because different organisms view it at their own scale of reproductive dispersal, movement, or home range. The curvature will be different, depending on which organism is used as the reference. Non-Euclidean spaces can be seen as locally Euclidean (the perception of the single organism or species) but are globally curved as the metric for distance changes across the space (the alternative perceptions of other organisms of other species). If there were no accommodation between species, then the different local spaces could be studied separately and we could treat the space of ecology as Euclidean. However, that is not the case, for there is a rapprochement between different species that presses together their differently scaled, personal realities. It appears worthwhile to look at, compare, and try to integrate the differently scaled spaces that make ecological spaces non-Euclidean for many considerations. This forces us to view the community as a wave interference pattern.

Community ecology with population biology exist as devices for unraveling these scaling problems. Some methods of community ecology (e.g., principal components analysis) press species performances into a Euclidean space and display the ecological space as curved. Other methods (e.g., species mode ordinations) allow the individual species each to define the space, and these display the ecological space without curvature, although the relationship to the individual species is complicated by an unseen curvature. But more of this in later chapters. Because of the relative scaling involved, communities are very scale-dependent conceptions in themselves, and lack the simple internal consistency of either landscapes or organisms. Ecosystem conceptions have the same problems as communities, but are very explicit about abiotic material as it moves across landscapes at various and confounding rates. Although this issue is raised here in the landscape chapter, it receives a fuller treatment in the community and ecosystem chapters. It is no accident that we have started our comparison and contrast of differ ent ecological conceptions with the tangible landscape.

3. THE ECOSYSTEM
CRITERION

———◆◆◆———

Because landscapes are tangible, they are particularly useful for putting into practice our scheme for comparing ecological observational criteria. The scheme laid out in abstract in chapter 1, and then implemented in chapter 2, is: keep the criterion constant change the scale; hold the scale constant and change the criterion using that same tract of land; change the scale and the criterion together so that landscapes can become the context above or the mechanism below ecosystems, communities, organisms, or any other type of ecological structure. Now we are in a position to test the general usefulness of our scheme with something more challenging, the intangible ecosystem.

But first we must deal with a potential problem of terminology. Robert O'Neill and his colleagues, in "A Hierarchical Concept of Ecosystems," mean something broader by the term "ecosystem" than we have in mind. Their process-functional type of ecosystem is what we mean by the term "ecosystem." They also include in their definition of ecosystem "population-community" types of ecosystem. We reserve the term "community" for that type of system and will give it status equal with "ecosystem" rather than make it a type of ecosystem. O'Neill and his colleagues were, like ourselves, at pains to discourage ecologists from requiring that communities be contained within process-functional ecosystems. That is why they included community as a type of ecosystem, so it could not be taken as a process-functional ecosystem component.

However, O'Neill and colleagues, in their broad definition of ecosystem, depart significantly from conventional usage. Ecologists who call themselves ecosystem scientists would cleave themselves away from ecologists who study "population-community" types of ecosystems in the parlance of O'Neill et al. Conversely, students of "population-community" types of ecosystems would not often call themselves ecosystem scientists. Rather they would answer to the name "community ecologist" or "population biologist," depending on how they prefer to approach multiple species assemblages. We have invested so much effort elsewhere in this book to make sure that communities are not seen as necessary parts of ecosystems that the terminological device of O'Neill and his colleagues is superfluous for our treatment. We do not need to pay the price of using an unconventional vocabulary on this point, so we choose not to use O'Neill et al.'s terminological device. Nevertheless, there is an exact correspondence across the two vocabularies: organisms are often powerful explanations and predictors of many properties of communities, by our definition of "community," and of the population-community type of ecosystem of O'Neill and his colleagues.

The Organism in the Intangible Ecosystem

There are several meanings for the word "ecosystem" in the ecological vernacular. We do not insist that ours is in any way the right one, for all definitions are arbitrary. Our definition compromises the integrity of organisms in a fashion which may at first be counterintuitive.

Organisms are as tangible a set of entities as one could want. The essential intangibility of ecosystems can be brought home by watching organisms melt into the intangible pathways of the ecosystem. We define the parts and explanatory principles of ecosystems as pathways of processes and fluxes between organisms and their environment. Note that the critical parts are the pathways that may involve organisms, not the organisms themselves. Using the appropriate observational protocol, we could find organisms inside the boundary of an ecosystem, even one defined in our terms. Also, ecosystems would certainly be dysfunctional if their biota were removed. Nevertheless, a reduction of ecosystem phenomena does not, for the most part, lead to organisms as the explanatory lower-level entities. The failure of organisms to offer ecosystem explanations and predictions comes from their lack of discreteness in ecosystem function; organisms do not represent the functional parts. The pathways in which organisms are subsumed are the functional parts.

There is more to being a system part than merely being present within the sector of the universe contained by an ecosystem; there is an added requirement beyond just being inside the ecosystem rag bag. The critical feature of a scientifically relevant system part is integrity of that part as it plays its role in the functioning of the larger system. Note that here again we are concerned with organization of perception rather than what may or may not be real above and beyond scientific observation. It is in recognizing a set of interesting behaviors and ascribing them to ecosystems that we delimit what are the system parts. Contained by the same general boundaries of an ecosystem there may well be other entities which are irrelevant to an ecosystem conception. These irrelevant entities contained in ecosystems are pertinent to other types of system behavior, but that does not make them so for ecosystem types of behavior. Ecosystems may or may not be out there in the real world. What is important is that they appear to be helpful conceptions that lend predictive power. If the ontological status of ecosystems is beside the point, then it hardly makes sense to insist that the ontological reality of organisms makes them ecosystem parts. Organisms need be considered a part of our ecosystems only in the sense that they, like pebbles, can be found by rude disaggregation of most ecosystems. Only occasionally does a discrete organismal entity autonomously play a role in ecosystem function; if it does so, then it is an ecosystem part by any standards.

Science is about explanation, not finding collections of unrelated, arbitrarily asserted parts. A philatelist is free to add to a collection any postage stamp on arbitrary criteria. Not so the scientist adding parts to an object of investigation, for only those that work predictively are acceptable. At the start of a scientific investigation, we are free to choose a certain limited set of system behaviors as interesting, and indeed we had better make such choices or otherwise be swept away by a flood of options. Once we have made the unavoidable decision to focus only on a certain set of system behaviors, then the set of acceptable explanatory devices is fixed; the trick is finding them and rejecting anything which is irrelevant. For the most part, explanations of behaviors that we characterize as ecosystem do not reduce to the behaviors of organisms as discrete explanatory entities.

Organisms for Ecosystem and Community Scientists

Certain special approaches to animals necessitate balancing calories in, say, locomotion physiology. However, most organism-centered biologists of a taxonomic ilk spend little time dealing with fluxes of matter and energy, and for the most part are unconcerned with keeping track of the mass balance in the system. The notion of mass balance focuses on conservation of matter and energy; certain quantities enter the system and must remain if they do not come out. It is not that communities violate conservation principles, it is rather that such principles do not predict community structure or behavior. Evolution by natural selection works through principles like competition, mutualism, and predation. Communities are ordered on evolved organisms, and the above principles are used to explain the workings of the accommodation between community members. Evolution does not violate conservation of energy or matter, nor does it violate the thermodynamic principles of increasing global entropy. However, insights into relative fitness are not often gained from knowing that organisms respire or otherwise expel material and energy at a rate commensurate to their consumption, minus their growth. Conservation of matter and energy is, for the most part, irrelevant to the community ecologist. By contrast, the ecosystem scientist could not do the most elementary bookkeeping in ecosystems unless he invoked conservation and principles of mass balance.

Thus, important predictors in one type of system are of little use in the other. For example, the community structure of forests in the southeastern United States was radically altered by the blight that removed the American chestnut as a critical component of the canopy of the eastern deciduous biome and its communities. Meanwhile, the contemporary record at the end of the last century gives no indication that the ecosystem function in those same places was altered one jot, even at the height of the epidemic. The chestnut, as indicated by simulation studies, seems to have been merely one of many equally workable alternatives for primary production and energy capture. This notion that something can matter a lot in one framework but not in another will be an important consideration when, at the end of this book, we turn to strategies for basic research as well as to management issues.

One part of an ecosystem, say a pathway, corresponds to many parts in a community, because each of the organisms involved in the pathway can be an autonomous community member in its own right. This applies to most ecosystem parts. No matter what scale of community is employed, each of the many parts of an ecosystem corresponds to many parts of the community. There is no rescaling operation that can make the relationship simple. Accordingly, the relationship of ecosystems to communities is called a many-to-many mapping. The mapping of communities back to ecosystems is also a many-to-many mapping. That is to say, neither conception invokes with any regularity the entities that pertain to the functioning of the other. A community factor on one side of the comparison relates to, and is controlled by, many factors belonging to the ecosystem conception.

To see the consequences of many-to-many mapping, consider ecological succession where a dominant tree species is replaced by another. Expressed thus in community terms, one sentence has laid out the situation, and the consequences of that successional event for the community can be readily described. For the ecosystem there may or may not be a change in the rate of recycling, the carbon budget, or mycorrhizal efficiency, to name just three uncertainties. Worse than that, take any one of those many ecosystem factors related to the one community factor, and we find that it relates back, in turn, to many community factors; for example, mycorrhizae could have much the same ecosystem consequences while representing a whole suite of different fungal community assemblages. Only occasionally does the same chunk of the world relate one to one across the community and ecosystem conceptions in a one-to-one mapping.

One notable exception is the case of Pacific salmon in the rivers of the western United States and Canada where we find that the fish belongs as a discrete entity to both types of system. Salmon are distinctive in that they spend the bulk of their adult life cruising the ocean, but return for reproduction with astonishing accuracy to exactly the same place they were spawned. One of the reasons salmon are such a sporting challenge for the dry fly fisherman is their lack of interest in feeding during their spawning run. Salmon eat almost nothing as they return to spawn. Both the male and female fish have all the energy they need for breeding, and so their guts are superfluous and are disintegrating. Once having laid or fertilized the eggs, they die. Heavy fishing over recent decades has caused the salmon run to fail. The first impression was that not enough salmon had survived the catch to return and lay an adequate number of eggs. It seemed that the small number of fertilized eggs had not offered a large enough population of hatchlings. This deficient year class supposedly manifested itself as the missing adults of years later. The remedy was thought to be restocking the rivers with the missing hatchlings. Unfortunately, this was only moderately successful. The problem was apparently not missing hatchlings seen as community members, but an ecosystem variable that mapped onto the adult salmon.

The error had been easy to make, for it mistook salmon for components of the fish community. Of course they are members of the fish community, but that was not the role which they played in the system's sickness. What was missing was not live hatchlings but dead adults who had just spawned. Rivers constantly flush away nutrients, an ecosystem consideration. Dead adults rot, release their nutrients, and stimulate algal growth. The algae are food for microscopic invertebrate animals, the food of the hatchlings. The adult salmon were indeed sources of eggs, but the adults were more significantly sources of nutrients. The fish were a critical ecosystem property, a nutrient pump upstream, as well as members of the fish community. The solution to the fishery management problem was to put mineral nutrients into the headwaters of the rivers at the critical time.

The Size of Ecosystems

Having shown that the organism is only occasionally a discrete part of an ecosystem, now let us see how the ecosystem criterion relates to the other tangible criterion, the landscape. Area is a landscape criterion, but ecosystems are not readily defined by spatial criteria. Ecosystems are more easily conceived as a set of interlinked, differently scaled processes that may be diffuse in space but easily defined in turnover times. Processes pertaining to very differently scaled areas encounter each other in the full functioning web of the ecosystem. Thus a single ecosystem is itself a hierarchy of differently scaled processes. This should not be confused with the hierarchy of differently scaled ecosystems where bigger airsheds belong to larger-scale ecosystems, and where higher-level ecosystems have longer-term soil-based processes. There are differently scaled processes inside a single ecosystem, as well as sets of differently scaled, more inclusive, and less inclusive ecosystems.

Attempts to specify a particular area for an ecosystem raises difficulties. The problems stand out clearly for the meteorological part of ecosystem pathways, but they are by no means limited to that part of the system. A place on the ground does not adequately delimit the climatic aspects of ecosystems. It rains most afternoons in the Great Smoky Mountains in the summer; plants cause the precipitation through their transpiration. Thus the rain is not ecosystem context, it is part of a pathway that is itself a critical component inside the ecosystem. There is an appropriate airshed that is partner to a given watershed, but it is much bigger than the watershed. With the atmosphere as part of the ecosystem, the spatial boundaries of the ecosystem move every time a new weather system passes through the region. Area is not a helpful general attribute of ecosystems. It is not that ecosystems lack a boundary in space, it is that such a dynamic boundary is impractical for most investigations. There have been attempts to use averages of the position of air masses but that is different from defining the explicit boundary of the ecosystem at an instant (figure 3.1a, b).

As we will show at length in the community chapter of this book, using a spatial criterion for communities is no easier than it is for ecosystems. However, difficulties of defining the areal extent of a community has little to do with indistinct atmospheric boundaries. This is because meteorological influences are the environment rather than part of the community; therefore, being outside the community system, atmospherics can be left unbounded. The issue is further complicated here because, even though the ecosystem does contain a climatic component, it still has a climatic context; however, that climatic context belongs to higher-level, larger-scale ecosystems of which the local ecosystem is part.

Having emphasized the spatial intractability of entire ecosystems, conversely we should not forget that some ecosystem parts may be attributable to a place. Some ecosystem parts like surface water flows can be mapped to a restricted, stable site. Thus nutrient retention processes of a particular ecosystem could be associated with a particular watershed. Even so, on criteria other than nutrient retention, the entire ecosystem regularly violates the watershed boundary. Such violations come, for example, from the ecosystem's internal airflow and form endogenous processes associated with animal movement.

If an entire ecosystem cannot be said to be in any given spatially defined place, then we need some other way of telling ecosystems apart from each other, or a single ecosystem from its context. Put another way, how big is an ecosystem? Entire ecosystems vary in size not by area but by the scale of the pathways that comprise them. Some of those pathways may be confined to a given area, but most will not correspond to a place on a landscape. To gauge the size of ecosystems we must look for complete pathways contained within natural surfaces. The search is for natural surfaces, ones that coincide with a large number of limits. Natural surfaces are not necessarily tangible, they may not occur as separations of place from place. They can be separations of diffuse process from diffuse process. Later we will show how to identify and specify such intangible surfaces of ecosystems using some of the devices that biochemists use to separate their intangible pathways. For the moment we will deal with limits associated with tangible surfaces of ecosystems, like watershed boundaries. We choose tangible surfaces without prejudice against diffuse surfaces that give the intangible bounds of an ecosystem. However, the preliminary discussion of ecosystem boundaries can move forward more easily using tangible surfaces. Even so, we still do not surrender to the notion that entire ecosystems can be mapped onto places on the landscape for a workable duration of time.

The watershed makes a robust natural boundary because a number

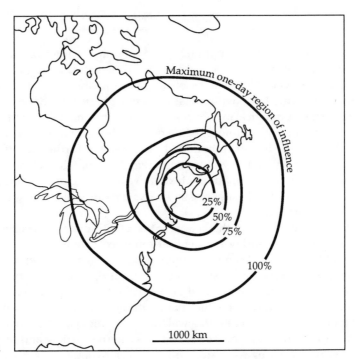

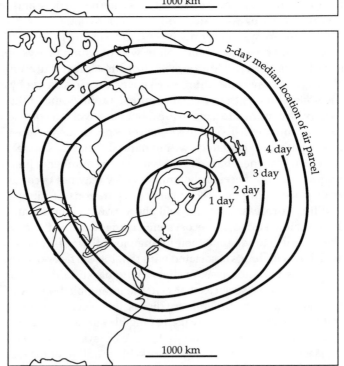

of critical processes reach their limit at the edge of the catchment area. Nutrient deposition input to the ground and surface water flow patterns are both held within the drainage basin. The coincidence of a boundary with many processes indicates a linkage between those processes, and that linkage embodies the system itself. The system connects nutrients arriving through deposition to all parts of the ecosystem by the flow of water. That is why both nutrient input and water flow coincide at the edge of the watershed.

It is important to work across the entirety of the system, so that we encounter all the pathways that give critical system behavior. We need to know the size of the system so we can study it all and then go on to find how it is different from its surroundings. The boundary of an ecosystem encloses the parts exactly, although that enclosure is not unambiguously in space. The size of an ecosystem is given by the largest extent that only just contains the definitive pathways of the system. The ecosystem is relatively homogeneous inside its boundary. As a result, there is a big difference between increasing the size of the study site while remaining inside the system boundary, and increasing the extent so as to cross the system boundary. The relative homogeneity of the ecosystem means that there is great similarity across regions, big or small, within the system boundary. The difference between a greater part of a watershed and an entire watershed is small because the change in area only includes more of the same. The same cannot be said for a larger area which crosses the natural boundary, because suddenly a whole set of pathways is involved and the integrity of the local entity is compromised. Rain falling on just the other side of the hill runs away from the local system center instead of toward it. Including water just over the hill forces us to take into account an altogether larger system with less homogeneity and integrity. At points where there is a break in the scale of the ecosystem we could recognize the upper level on some other criterion. We are free to abandon upper-level ecosystem criteria in favor of recognizing the journey over the hill as something on a landscape. However, let us continue up the column of ecosystems within ecosystems; let us hold the criterion constant and increase the scale.

At the exact scale of the entire ecosystem, there is an asymmetry between studying an area a little bigger as opposed to one a little smaller.

←────────────────────────────────────

Figure 3.1. Not only is it possible to identify the watershed of a given ecosystem, but it is also possible by calculations of average wind speeds and directions to identify airsheds for a given site in North America. The two maps show: A. the percentage of the time a one-day airshed is contained within the circumscribed area; and B. the median airshed for 1 to 5 days for Kejimkujik, Nova Scotia (after Summers 1988).

This is because of critical changes in the patterns of connections at the limit of the system. The ecosystem is held together by strong connections within. The weaker connections are to the world outside, to other local ecosystems. Go beyond the bounds of the watershed and one of two things happens. Either the ecologist leaves one watershed ecosystem and enters another, or he encounters a new and much larger system, of which the original watershed is but a part. The choice of scale at which the situation is viewed determines which event pertains. In the first case, the bounds of the pathways of water and nutrient flow have been crossed, and the ecologist has entered a new and perhaps equivalently scaled adjacent ecosystem. In the second case, the new larger system is large enough to encompass many more streams and rivers, including drainage from the smaller ecosystems whose boundaries were just crossed. If we see the change as entering a larger system, suddenly all critical parts change because of the shift in scale (figure 3.2a–c). In the larger system, local microclimate is no longer a predictive system part, and the scientist must consider larger air movements pertinent to precipitation in the new larger basin. A change in scale of soil-based parts of the system changes the scale of all other pathways, including those of the meteorological components.

Much in the same way that one can move over the hill in space and change the scale required for system description, one can study a system for too long and cross a sort of temporal watershed boundary. Entities appear to have a certain time span for their identity, after which they lose their identity and become only a part of a larger system. An organism lives for only a certain time. Ecosystems are no exception; they too have a certain lifetime, so to speak. The degree to which processes of different types express themselves and the length of time they do so, are both ways of describing the uniqueness of particular ecosystems. Much of what we observe in ecosystems is better set in time rather than space.

By tracking an ecosystem over successively longer periods of time, we can find the temporal limits to the critical processes. Just as crossing local spatial boundaries in ecosystems can require a rescaling of system parts, crossing a temporal boundary imposes the same requirements. Some processes only hold for certain periods of time, after which they attenuate or change. Trees only grow for so long. The influence of a fire in making nutrients available only lasts a definite time. Over a long enough period, individual fires become irrelevant and fire frequency becomes the pertinent measure. The processes of an ecosystem operate together over limited areas and beyond those limits we must rescale to other more widespread specifications of the processes. Similarly, pro-

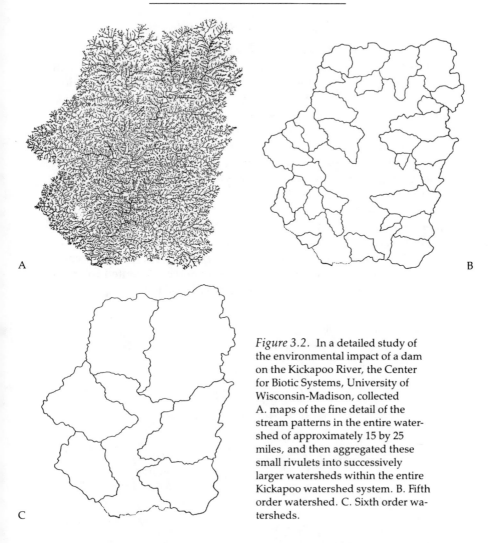

Figure 3.2. In a detailed study of the environmental impact of a dam on the Kickapoo River, the Center for Biotic Systems, University of Wisconsin-Madison, collected A. maps of the fine detail of the stream patterns in the entire watershed of approximately 15 by 25 miles, and then aggregated these small rivulets into successively larger watersheds within the entire Kickapoo watershed system. B. Fifth order watershed. C. Sixth order watersheds.

cesses only operate over certain time spans, after which we again need to respecify if we are to predict ecosystem function.

The watershed example above has an intuitive appeal, but the same arguments apply to less tangible temporal aspects of ecosystems. We are so insistent on resisting a spatial definition of ecosystems because that definition excludes many aspects of ecosystems that simply do not fit the landscape neatly. The ecosystem is a much richer concept than just some meteorology, soil, and animals tacked onto patches of vegetation.

The Critical Role of Cycles

An ecosystem resembles a complex of biochemical pathways in that cell metabolism and ecosystem function both mix and interdigitate with neighboring systems. They share their respective spaces. The various pathways shunt nutrients and energy around the ecosystem very like the way that biochemical pathways shunt molecules and chemical energy. The Krebs or citric acid cycle does not occupy a particular place within the cell, for it is better considered as an intangible but discrete part that is defined by its role in cell functioning. It keeps its identity by maintaining strong connections between its parts dispersed across the cell (figure 3.3). Despite its intangibility, this biochemical pathway has very real consequences. The same is true for ecosystems; they are consequential but are for the most part intangible. Ecosystems can be seen more powerfully as sequences of events rather than as things in a place. These events are the transformations of matter and energy that occur as the ecosystem does its work. Ecosystems are process-oriented and more easily seen as temporally rather than spatially ordered.

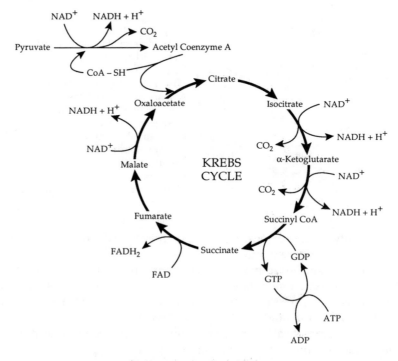

Figure 3.3. The Krebs cycle.

Consider the way biochemists plot the course of biochemical cycles. They add radioactive material in a chemical form which belongs early in the pathway, and then let the system run for successively longer periods of time. Molecules that occur early in the cycle show radioactivity quickly. Later parts of the pathway take successively longer to pick up the radioactivity. In doing this, biochemists engage in a temporal version of our looking for the areal extent of the watershed. They look for the sequence of events that takes the longest time to occur while still reflecting an orderly pathway; they seek the extent of their system to find the limits of the cycles they investigate. They take note of pathways that appear to turn around and eat their own tails, for it is in recurrence that the pathway achieves closure and so a discrete identity.

In the end the biochemists lose track of their pathway as radioactivity becomes dispersed throughout the whole cell. After a long enough period their pathway of interest leaks radioactivity into related cycles. At that point the biochemical experiment has crossed into another biochemical "watershed." By extending the time of the experiment they eventually address a new, higher level of organization, namely integrated cell function, where the individual pathways disappear in a sea of biochemical exchange. Instead of finding carbon atoms shunted into the mainstream of the pathway (cf. watershed) in question, they have thrust upon them the fact that all parts of cell function are linked. Those links are made as the many pathways cross by sharing common organic compounds (figure 3.4). Thus, larger-scale ecosystems arise from the sharing of water, minerals, and organic material by collections of small ecosystems.

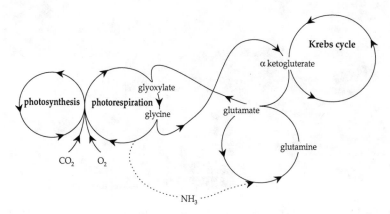

Figure 3.4. The Krebs cycle is but one cycle in a complex web of metabolic pathways. Radioactivity fed into the Krebs cycle eventually leaks into other parts of cell function.

Note that the smaller ecosystems may be as distinctive as the individual biochemical pathways in a cell. Therefore, the move to larger ecosystems is no more a simple summation than is the move to a cell from its biochemistry. The manner of integration of cell parts to make a cell need not be more complicated than the manner of integration of small ecosystems to make large ones. Big ecosystems can be as different from small ecosystems as cells can be from biochemical cycles.

The methods of ecosystem ecologists resemble those of biochemists. Ecosystem experimentalists commonly pulse their systems and wait for the signal to pass through the entire network. Sometimes they even use radioactive tracers like biochemists; therein lies the history whereby Oak Ridge National Laboratory has become one of the world-class centers for ecosystem research. Radioactive tracers tell the ecologist what are the important connections in the system.

Many experimental pulses in ecosystems do not involve radioactive tracers and might take the form of a clearcut of the forest with or without removal of the downed timber. This was the approach of Gene Likens and the team of researchers studying the Hubbard Brook watershed (figure 3.5). It is fortunate that at least some aspects of the diffuse ecosystem do map onto the ground because that allows the experimentalist a means of pulsing the ecosystem in a meaningful but tractable way. The

Figure 3.5. The Hubbard Brook site where some of the first ecosystem clearcutting work and nutrient cycling calibration was performed. The weir in the foreground allowed measurement of water and nutrient fluxes (photo Gene E. Likens).

clearcut action is unambiguous but still relevant to the entire ecosystem whose intangibility does leave room for ambiguity. Clearcuts performed across entire watersheds facilitate the mass balance calculations. Clearcuts on slightly larger or smaller areas are difficult to calibrate.

One often measures materials that flow around inside the system in a complex fashion. However, relatively few measurements may be needed if one is satisfied to let the ecosystem integrate the signal. This is done by making the measurements at the ecosystem outflow in the stream. The inputs can often be calculated by measuring precipitation or deposition at a few local sites in the ecosystem and making the reasonable assumption that input is diffuse and relatively even. The Hubbard Brook study found that the system leaks mineral nutrients when the trees are removed, but reseals itself very quickly as soon as the woody vegetation achieves significant cover again. By measuring system loss and input through deposition, ecosystem scientists calculated how long it would take the system to achieve the same nutrient capital as before the clearcut. The clearcut allowed the ecologists to assess the role of woody vegetation in recycling resources. Woody vegetation in combination with soil and its microbes comprised the nutrient retention part of the system. As with many critical ecosystem functions, nutrient retention is performed by a cycle.

Like biochemists, ecosystem scientists are very interested in cycles. If ecosystems were only conduits through which material and energy passed, then they would not be very profound conceptions. It emerges that ecosystems cycle material to a considerable degree. Cycling causes the ecosystem to achieve an interesting identity. Without cycles the system would quickly run out of resources and be very directly constrained by the physical environment. Living systems, including ecosystems, are characteristically open systems, but that openness is not necessarily the best characterization. A river is a hydrologically open system where most of the substance of the river is water passing straight through. Living systems are open but are more conservative of their resources than a river is of water. If we think of rivers as ecosystems, the least interesting aspect of them is the flux of water through them. Their distinctly ecosystemic properties come from the way mineral and organic materials spiral downstream, passing through the biota many times before leaving the system. What makes living systems such as ecosystems distinctive is the way they escape the constraints of the physical world by recycling resources.

Recycling does not come free; it demands dissipation of energy. The order of life comes from energy dissipation in driving the critical cycles. Ecosystems, like organisms, expend much energy in repair and use of

their cyclical pathways. In holding nutrients in the system, the ecosystem invests energy in microbial growth in the summer and in root
growth in the autumn and again in the spring. The fungi die wholesale
once a year, while fine roots die and must be replaced twice. In organisms and ecosystems the cycles are integrated to become the system itself.

The ecosystem has as much identity as does an organism. In fact, the
various facets of an organism have some equivalence in the various
ways that ecologists study their material. There are equivalent camps of
scientists in organismal biology and superorganismal ecology. Physiologists correspond to ecosystem scientists; geneticists correspond to
evolutionary biologists; perhaps anatomists correspond to community
ecologists. In organisms the material substance of the body turns over
several times in the life span. Living systems, including ecosystems, are
very like the woodsman's ax; it is still his ax after he has replaced the
handle four times and the head twice. It would become a different ax if
he was forced to replace all parts at once.

As often as not, the faster cycles are the more conservative. The mean
residence time of carbon in a deciduous forest is of the order of 200
years. Although carbon moves rapidly in and out of leaves through photosynthesis and respiration, some of it becomes locked into tree trunks
for centuries. For most of the 200 years the carbon does nothing and is
not in the process of cycling. Nitrogen, on the other hand, moves
around the system from biota to soil and back again on an annual basis.
Nevertheless, the mean residence time of a nitrogen atom in the forest is
of the order of 1,800 years. There is not much carbon by percent in the
atmosphere, but it is reliably there in workable amounts and so the ecosystem keeps only a 200-year inventory. Most ecosystems are nitrogen-
starved and so the inventory for nitrogen is about ten times longer than
for carbon. Nitrogen and carbon, having different turnover times, respond in opposite ways to grazing pressure on grasslands. The difference in turnover time decouples these critical elements.

Biota in Ecosystems: Plants and Primary Production

We have been at pains, at the beginning of this chapter, to point out that
the explanations of most ecosystem behaviors are not organisms as discrete entities. Nevertheless, the activities of organisms are crucial in
ecosystems, even if it is not often profitable to emphasize the discreteness of one organism from another in ecosystem studies. Certain biotic
processes are crucial to the working of ecosystems; we now turn to the
role of the biota. Various classes of organism play particular roles in eco-

systems, depending upon their special biology. Plants and animals relate to a generally different set of ecosystem functions. We will now contrast the different roles of plants and animals in ecosystems. First we consider plants.

Often as an organism performs some biotic function in an ecosystem, its special ecosystemic role is only one facet of its biology. For example, there is more to an animal than eating, but degradation of biotic material through consumption is its principal role in an ecosystem. Furthermore, that part of the organism's biology becomes a minor subsystem three levels down, a part deep inside an ecosystem part. For example, the roots of trees are crucial parts of the rhizosphere, but it is the rhizosphere, not the root, that is the functional unit in the nutrient retention subsystem. There are other biota, the fungi, that also allow the rhizosphere to do its job. Note that the fungi play their part not as discrete entire organisms, but rather as members of a guild of fungi whose members individually work like interchangeable movie extras.

Along with Curtis Flather, we analyzed the ecological literature to see which organisms were used for work on which major concepts. We performed a computer search of BIOSIS, a literature-searching data base, to retrieve papers by their paired use of types of organism in the study and the ecological concept investigated. Literally hundreds of thousands of paired hits were found and tabulated. The numbers of each organism/topic pair were expressed as proportions of total effort devoted both to each type of study and to each type of organism. Knowing the amount of research on a given class of organism and the amount of research on a given concept, we could calculate how much research would be expected to be devoted to both at the same time. The actual amount of research in a given area was sometimes more and sometimes less than we would have expected. Those items in the matrix scoring above expectation indicate where there is a natural match between organism and concept. The results gave insights into the way ecologists think about all sorts of ecological entities, particularly ecosystems.

We found that the archetypical ecosystem organism has the life form of a tree. Trees are also the prime entities for community and succession studies. There are, of course, ecosystems and communities that are much smaller than human size, but the intuitive feel we have for a community or ecosystem is that they are both typically large systems. We are talking here of the gestalt, not the fact of ecosystem size. When something is larger than ourselves we cannot readily see it all at once. Because in a forest the dominant organisms are larger than us, the intangibility of the upper-level ecological entity is not only acceptable, but expected. In ecosystem and community concepts there would seem to be an unspoken integrity of an unspecified whole; there are unifying processes but ecologists do not feel confident to tease them apart in communities and ecosystems. Trees comprise the primary production compartment as well as being the site of carbon storage. They live a long time and are, therefore, helpful long-term capacitors that smooth variations in fluxes. By behaving slowly, they constrain other ecosystem parts and so lend the ecosystem stability. Their behavior is so slow that it passes unnoticed at the scale of human primary perception. This slowness fits in with the intangibility of function that is implicit in the ecosystem concept.

Our study of BIOSIS showed that herbaceous plants are not generally considered as ecosystem parts. Plants smaller than ourselves can be seen easily as collections in which the individual maintains its identity. This is unhelpful for an ecosystem conception where the biota are not usually discrete subsystems, but are parts of pathways. The principal exception are grasses, which appear readily to lose their identity as discrete biotic entities. The reason for this is that grasses all look the same to the untrained eye and so can be happily integrated into the primary production compartment. They are also often seen playing their ecosystem role of supporting secondary production in grazed systems. Organisms that can play a role unambiguously in an ecosystem function are favorites for ecosystem research.

Biota in Ecosystems: Animals and Primary Consumption

An animal eating means different things at different levels. Becky Brown has considered primary consumption in ecosystems in a way that leads to insights about the importance of scaling in measurement protocols. These scaling effects determine which processes in herbivory emerge as explanations.

There is a school of ecologists, including Sam McNaughton, interested in the way that plants grow more vigorously after grazing than had they not been grazed at all. The term used for the process whereby plants grow to make up the loss due to grazing is "compensation." When the plant more than fully recovers the loss, it is called "overcompensation." The literature is, however, confused because the scale at which the compensation to grazing occurs is often unstated. As a result, the theoretical point of exact compensation, against which observed conditions are compared, becomes equivocal.

Although herbivory is a process, it is often measured as the percentage of leaf damage at a given moment. Clearly there is no time for compensation to occur in such data. Measuring the status of damage to leaves relates two processes: animals eating leaf material and leaf growth. The two processes occur at very different rates; damage to individual leaves is a fast process, much faster than leaf growth. At the level of the individual leaf, compensation is not possible. The leaf is the constant against which changes of state are measured.

Some plants, like grasses and their monocotyledonous relatives, do not lend themselves to estimate of leaf damage at an instant in time because they keep growing their leaves from the base, below the point where the animal nibbled. In these plants there is the possibility of compensation at the level of the leaf. Note that grasses give their name to the process of animal primary consumption, grazing. Grasses appear to have accommodated to grazing more than other plants by rescaling their parts so as to make the individual leaf the unit of response. This takes the natural unit of ecosystem primary production below the level of the whole plant and fixes it at the leaf. This biological rescaling appears to give grasses great selective advantage in that they are the dominant plants in ecosystems where animal pressure on plants is greatest.

The problem for other herbaceous plants, forbs, is that not only are the leaves damaged by primary consumption, but so are the growing points of the plant. Leaf capital is destroyed and the means of replacing the capital is disorganized. The unit for compensation to grazing in plants other than grasses is the entire shoot, plant, or clone, not the leaf. The growing point being nipped off, a new growing tip must become active to take over replacement of losses.

Let us move upscale and measure removal over time. With removal seen as a continuous dynamic instead of an event, there is time for the plant to respond to grazing, and we can now investigate compensation. If we assert that the compensation can be measured by comparing the size between grazed and ungrazed plants at a given time after grazing, then there are several hidden assumptions. First, the growth of the biomass of the grazed and ungrazed plants increases linearly. Second, only one defoliation occurred. Both assumptions would only be sufficiently close to the truth for relatively short periods.

Extending the period between defoliation and measuring plant recovery allows time for 1) further grazing to occur, and 2) growth patterns to depart from linearity. Recovery could easily involve a lag period in which the plant reorganizes its resources, followed by a period of rapid growth which eventually asymptotes to normal growth rates. A sigmoid growth curve is reasonable and would give different accounts of compensation depending on when the recovering biomass is measured. Therefore, many protocols for measuring recovery from herbivory suffer from one of two limitations. If they attempt to measure long-term compensation, they are based on unwarranted assumptions. Otherwise, they pertain only to short-lived aspects of ecosystems and are likely to miss some of the higher-level aspects of compensation.

Only through sophisticated data collecting protocols is it possible to move upscale and leave time for some of the more interesting patterns of compensation to emerge. When we leave time for compensation, there is also an increase in the spatial aspects of the system, since we now have to take into account below-ground parts of the plant. After defoliation the plant can compensate the leaf compartment by transferring material from roots. Only after the plant has had time to use that new leaf tissue is there a possibility that the whole plant can compensate or overcompensate for the lost tissue. At the level of the whole plant, the response of the roots is what an economist would recognize as deficit spending to stimulate the plant's economy. It is a form of compensation. Deficit spending takes time to have the desired effect, so there is a considerable increase in temporal scale if we wish to measure compensation at the whole plant level. That increase in temporal scale is large enough to violate the assumptions of a single grazing event and linear growth of the experimental plant and the control.

Let us move higher in the stack of nested ecosystems. Over still longer time periods, changes occurring because of grazing operate at a level higher than the plant itself. The plant may grow faster than it otherwise would have done because it lives in a grazing regime. The critical change wrought by grazing over the long term is not on the plant directly but on the context of the plant. If grazing opens space by damaging other plants, grasses that can recover faster do so. They now live in a context altered by grazing to favor their survival and growth. It is over this long term that grazing impacts ecosystems at the high level and large scale at which we normally apply the ecosystem concept.

Biota in Ecosystems: Plants, Animals, and Nutrients

There are many examples where herbivory has changed nutrient cycling rates. Such an increase in nutrient cycling can in turn increase primary production. Steven Carpenter and Jim Kitchell showed that phytoplankton are negatively correlated with zooplankton when observations are separated by two days. This is because the higher numbers of zooplankton crop down the floating plants. Nevertheless, Carpenter and Kitchell found that phytoplankton are positively correlated with zooplankton over a ten-day window. Ten days is long enough to allow the nutrients consumed in the cropped population of phytoplankton to reemerge in the water and be expressed in increased production. Correlation changes completely with the time frame of the context. The extra days allow a larger-scale process to manifest itself, a nutrient-based phenomenon that is only tangentially related to the original grazing phenomenon.

Animals often play a role in complicating phenomena related to nutrients. As they cross the system boundary, animals play an important role in the nutrient budget. Their rapid movement sometimes allows them to be nutrient importers, as in the case of salmon returning to the stream of their birth. They are particularly important in systems of that sort, where nutrient loss is a critical factor. Insects are important parts of the nitrogen import into bogs where carnivorous plants trap them and digest the insect bodies on sticky leaves. Pitcher plants trap insects in chambers in their leaves. Bogs are particularly dependent on these nutrient imports because their acid water keeps nutrients in solution, so allowing the nutrients to be exported with water. In unequivocally terrestrial systems, A. D. Bradshaw has pictures showing how dogs are crucial in bringing nitrogenous waste into reclaimed urban landscapes. The explanation for patches of bright green, vigorously growing grass in parks created from leveled slums is the spots marked by dogs in their territorial ritual urination (figure 3.6a, b). There are not enough dogs to support the nutrient budget of the reclamation for the missing component was a critical mass of nitrogen-fixing plants. Despite the input of nutrients from insects, bogs remain nutrient-poor. Thus, unequivocal input of nutrients by animals with local effects may not always change the nutrient status of the system at large.

Plants do not often play the role of nutrient import to ecosystems because they simply move too slowly. Plants are more often critical components of nutrient cycling. Animals also play an important role in cycling nutrients inside ecosystems, as in the case of the zooplankton cited above. Also, soil arthropods break down leaves in a timely fashion so that the nutrients can return to the plants. Most ecosystems receive only small amounts of nutrient input. The slowest part of the system constrains all the others, and so painfully slow nutrient import might be expected to constrain most ecosystem components deeply. However, many ecosystems escape this nutrient constraint by cycling and building up significant nutrient capital. On the shortgrass prairies of Colorado, Bill Parton and his colleagues sensibly divide the nutrient budget into a fast, a middle, and a slow nutrient cycle. They calculate that the system is very resilient to even long-term abuse, because only the capital changes as the slower pathways continue to deliver nutrients for the system primary production.

In summary, biota in ecosystems may not be readily identified as dis-

Figure 3.6. Patches of bright green healthy grass on this restored urban site are attributable to the ritual urination of dogs bringing nitrogen into the system. Although the restoration used topsoil, the essential absence of nitrogen-fixing plants led to low nutrient status, allowing the signal from the dogs to stand out clearly. According to A. D. Bradshaw, the smaller clumps of green are "one dog once," while the most vigorous patches are many dogs many times (photo A. D. Bradshaw).

crete organisms, so the naturalist may not have an intuitive understanding of the role of biota in ecosystems. Nevertheless, biota do play a crucial and readily identifiable set of roles in ecosystem function, particularly with respect to nutrients. The ecosystem opens a whole new window on the activities of organisms.

The Special Case of Aquatic Systems

Returning to our BIOSIS investigation on the proportional use of various types of organisms for various types of study, fish were the only exception to the rule of underutilization of animals in ecosystem work (table 3.1; figure 3.7a, b). They are the only animal taxon used in ecosystem research more often than expected. There are two reasons for this. First, fish live in an environment alien to us. Without scuba equipment, we cannot see them clearly in their native habitat. Out of sight leads to out of mind, as fish become a component inside a black box. Black boxes are studied from outside; input to and output from the box are compared and the state of the system is calculated from conservation principles. This is the method of ecosystem science, where fluxes and storage compartments, not the individual organisms, are the entities of interest inside the system. Being unseen as autonomous organisms in their natural habitat, the image of a fish with head, tail, and fins does not interfere, in the minds of ecologists, with the ecosystem functional role that fish play.

The second reason for fish being preferred ecosystem attributes is that fish play some of the same roles in lakes that trees play in forest ecosystems. They are nutrient storage units that bridge across times of paucity or destruction. Fish integrate nutrient input to the system as they grow larger and persist for years at a time. Lacustrine ecosystems appear to operate faster than terrestrial ecosystems, so years in a lake are probably equivalent to centuries in a forest. A fish holding nutrients in its body for a few years plays the same role as a tree locking up carbon and minerals in its trunk for centuries. In springtime it is fish-nitrogenous waste that keeps the phytoplankton primed, ready to lock onto the nutrient inputs when the ice on the lake melts.

Algae are favorites for ecosystem work. This is because, like fish, algae perform discrete ecosystem functions. They are the primary producers and nutrient-capture compartments of the lake. In terrestrial systems, many types of plants are primary producers, and their role of nutrient capture is less conspicuous than that of algae in aquatic ecosystems. Also their size makes unicellular algae invisible to the naked eye. Accordingly, they sit easily, unseen in the ecosystem black box.

Table 3.1

Binary matrix when 0 means less than and 1 means more than expected effort in research on that concept for that organism. Column and rows are ordered to show blocks of concept/organism concurrence.

Taxon	Compe-tition	Distur-bance	Population	Symbiosis	Evolution	Resource Capture	Island Biogeog-raphy	Niche	Habitat	Ecosystem	Community	Succession
LICHEN	0	0	0	1	0	0	1	0	1	1	1	1
ALGAE	0	0	0	1	0	1	0	0	0	1	1	1
ERICACEAE	0	1	0	0	0	0	0	0	1	1	1	1
TREES	0	0	0	0	0	0	0	1	1	1	1	1
BRYOPHYTE	0	0	0	0	0	0	1	0	1	1	1	1
CONIFER	0	0	0	0	0	0	1	1	1	1	1	1
GYMNOSPERM	0	0	0	0	0	0	1	1	0	1	1	1
PTERIDOPHYTE	0	0	0	1	1	0	0	1	1	0	1	1
BASIDIOMYCETE	1	0	1	1	1	0	0	1	1	0	0	1
ASCOMOMYCETE	1	1	1	1	1	0	0	0	0	0	0	0
ROSACEAE	1	0	1	0	0	0	0	0	0	0	1	1
GRAMINAE	1	1	1	0	0	0	0	0	0	1	0	0
COMPOSITE	1	1	1	0	0	1	0	1	1	0	1	1
ANGIOSPERM	1	1	1	1	1	0	0	0	0	0	0	0
MAMMAL	1	1	1	0	0	0	0	0	0	0	0	0
FISH	0	0	1	0	1	1	0	0	0	1	0	0
INSECT	0	0	1	1	0	1	1	1	0	0	0	0
BIRD	1	0	1	0	0	1	1	1	1	0	0	0
AMPHIBIAN	1	0	1	0	1	1	1	1	0	0	0	0
REPTILE	0	0	0	0	1	1	1	1	0	0	0	0

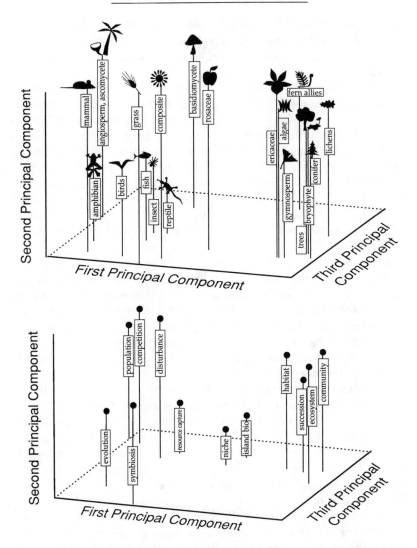

Figure 3.7. Principal component analyses of the relative use of organisms as research objects for certain ecological concepts, and the inverse analysis where ecological concepts are placed relative to each other depending on the organisms which are used to study them (after Hoekstra et al. 1991).

It is probably no accident that much of the pioneering work on ecosystems was performed on aquatic ecosystems. Lakes represent one of the few examples where a landscape boundary clearly coincides with most aspects of an ecosystem boundary. Also the change across the system boundary is such that the edge of a community coincides with both a natural ecosystem and landscape boundary. Given this coincidence, aquatic ecosystems make an interesting case study for the size of ecosystems.

Water operates as a vehicle for exchange in aquatic systems, and so aquatic ecosystems are particularly well integrated. This means that the physical size of a body of water determines the type of system. The size of a lake is more influential than the size of a tract of land in determining the type of ecosystem that can be supported. The difference in size between lacustrine and oceanic systems leads the systems of each type to be driven by different factors. This weaves together notions of size, patterns of constraint, and prediction. John Magnuson proposes that it is useful to think of lake fish assemblages as being related to extinction while the species list in an equivalent volume of the ocean is driven by invasion.

Magnuson has shown that factors related to extinction are better predictors of small lake fish assemblages than are factors associated with invasion. Some temperate lakes are small enough so that water freezing in the winter can eliminate all fish. If a lake is subject to winter kill every five years or so, a happenstance fish invasion might do well for a few years. The invasion is unpredictable. However, we can predict with confidence that after only a few years the whole population will be killed. Most of the time, the state of that lake can be reliably predicted on the likelihood of extinction driven by lake form and chemistry. Extinction follows a regular pattern over the passage of decades. However, over those same decades, lake invasion invokes the wrong time scale for predicting fish assemblages and consequent ecosystem properties. If an invasion were to occur by means of fish eggs arriving on a duck's feet, the limiting factor could be as trivial as the duck turning left instead of right. Lake invasion is not a predictor because it scales to the level of historical accident.

The size of oceans changes the scale of the causal factors and reverses the patterns of prediction. In oceans, invasion is a reliable driving force. Being emphatically open systems for the biotic components of local ecosystems, oceans allow invasion to override the happenstance of local extinction. Apart from extinction of an entire species at a global scale, extinction is always a local event relative to invasion. This is because invasion necessarily involves not only the place invaded, but also the place whence the invasion came. For a given scope of extinction, relevant invasion must involve that scope plus somewhere else. Lakes are small and so can be predicted if we scale to local extinction. Oceans are large and so can be predicted from invoking the larger-scaled process of invasion. Lakes experience invasion and parts of oceans experience local extinction, but neither is predictive. Invasion in lakes and extinction in oceans are too ephemeral in their effects to be constraining forces; they are not limiting factors. Prediction is scale-dependent in ecosystems as in systems in general.

Complexity and Stability Wars

One of the patterns that has caught the attention of ecologists has been the trends in species diversity across different biomes from the pole to the equator. There are fewer species found in the arctic tundra than in the boreal forests to the south. As one moves south through the temperate deciduous forests, diversity increases further. In the high tropical moist forests the diversity is very high indeed, reaching levels of hundreds of tree species per hectare. From this pattern has come assertions that increased diversity increases stability. Robert May has disparagingly called this notion part of the "folk wisdom" of ecology. One of the arguments is that the multiplicity of species provides alternative pathways should any community member suffer a local extinction.

Theorists have often shone light on the controversy over ecological stability. This is because the mathematical formulation of ecological systems fixes the level of discourse and avoids confusion over definitions. The difficulty with instability is that aspects of the system that were taken as constraints, namely system structure, suddenly change radically when the system becomes unstable. Change in system structure is of a different logical type and belongs to a higher level of organization than change that is normal system behavior. There is ample room for confusion in an unwitting change in level. Change in system behavior involving the old structure can be confused with change in system structure itself.

The difference between behavior and structure is fundamental. Behavior is rate-dependent, while structure is symbolic and rate-independent. Once the system is structurally defined and observation protocol is fixed, behavior follows independent of the observer. On the other hand, structure is always observer-dependent, and a change in structure follows from the observer finding that the old system specification is untenable or contradictory. After specification through mathematical formalism, constants and the manner of change of what can change are fixed.

An ecological system becoming unstable is such a dramatic event that it is hard to remember that stability is a relative matter. Stability comes from the observer's specification of the system as much as it comes from the system under observation. Like complexity, instability is not a property of the system itself, but an aspect of the mode of system description. Consider a system showing unstable behavior. If the system is specified to be more general, then what was unstable behavior becomes subsumed by a coarser system specification. For any given system, a set of potential states and changes of state define the range over which the system is asserted to be stable. The wider the range across which the system is said to be stable, the larger is the implied system, and the higher is its level of organization. Only when the observer identifies change beyond those bounds does the system manifest instability. A tree crashing to the forest floor can either be seen as the local tree exhibiting instability or as a healthy, normal process of replacement on a forested landscape. What is unstable over a narrow range of tolerances becomes normal system behavior if the observer uses a broader system specification.

The level of system description is not always transparent. If it were then there would be no need to consider scaling problems. Accordingly, aspects of system description that might not appear to have anything to do with stability can respecify the system so that it stays within or travels beyond the bounds of stable behavior. The length of time a system is studied determines whether it passes through unusual states. Study it for a long time and it passes through unusual states often enough for them to be considered normal; this implies a relatively large system. Study it for a short time and the unusual state may arise only once and be seen as a system aberration; this implies a smaller system. In a sense, the time period of the study fixes the level of the implied system. If the observer asserts that the unusual transition is within the bounds of stability, then the implied system is large enough to encompass whatever causes the unusual condition. If the transition is a manifestation of instability, then the cause of the transition state is a disturbance from outside the system.

In the same way that the length of time the system is studied affects the very identity of the system and the bounds of stability, so does the level at which the parts are specified. If the parts are small and ephemeral, then critical parts could be lost in a given period of observation, causing the system to appear unstable. In diversity studies, the arbitrary level of the system parts is often at the species level. Other levels such as ecologically functional groups of species or guilds of species are available as attributes of more broadly specified systems. In ecosystem studies, rather than species represented by organisms themselves, the functional parts will be pathways of connection wherein organisms play a subsidiary role. These pathways can be specified with various levels of generality, much as taxa can be specified at various taxonomic levels.

There is a hidden assumption in relating diversity to stability, which is that the new species entering to increase diversity are connected to the rest of the system. This assumption comes from the nineteenth-century idea of the economy of nature, that everything is connected to everything else. In fact, that view is mistaken, in that most things are not connected to any but a few others. True, there is a gravitational interaction between all things, but that is practically immeasurable as well as irrelevant at ecological scales. Ecological theorists build matrices of interaction to describe many ecological situations and in these are inserted terms for the strength of interaction between the column and row entities (figure 3.8). Most of the values in these matrices are zero except in restricted cases. Most pairs of species chosen at random do not interact in any meaningful way except for highly focused matrices (figure 3.9a, b). Several theorists in the 1970s, including Robert May, Robert Mac-Arthur, and Richard Levins, pointed out that if indeed there is significant interaction between all parts of an ecological system, or even if the connections are random, then diversity would be held low. This is because destabilizing positive feedback is bound to arise with increasing diversity. The theorists' point was not that diversity decreases stability, but that ecological systems are connected relatively weakly and very nonrandomly.

The critical feature of stabilizing elements added to a system is that they be only locally connected, that is, to only part of the system. In large stable systems, there are local knots of connection that amount to

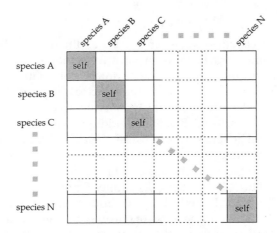

Figure 3.8. Interaction matrices have columns of species often arbitrarily arranged, but the rows are in the same order. Thus the diagonal trace refers to self-interaction while other values away from the trace are filled with values reflecting the direction and strength of the interaction between species.

	Fox	Rabbit	Lettuce	Farmer
Fox		eats		competes for rabbits
Rabbit	is eaten by		eats	competes for lettuce
Lettuce		is eaten by		is eaten by
Farmer	shoots	shoots and eats	tends and eats	

A

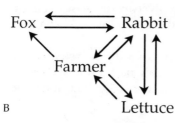

B

Figure 3.9. A. A simple interaction matrix in an agroecosystem. B. The corresponding species interaction diagram.

relatively discrete subsystems, and stable additions must be to members of only one subsystem, or at most to members of adjacent subsystems. In an appropriately ranked interaction matrix, these subsystems occupy local blocks astride the principal diagonal of the matrix. The principal diagonal contains the terms whereby every column or row element influences itself. Stabilizing additions are connected inside these blocks and generally nowhere else (figure 3.10a, b). Destabilizing additions to the system form bridges between otherwise disconnected major sectors or blocks in the system. In the interaction matrix, these entities would have large interaction values far off the principal diagonal (figure 3.11a, b). The effect of such bridges between quasidiscrete subsystems is to create long loops of effects in the system. The delay in these long loops often causes the signal to return at a time when it amplifies system change, a destabilizing positive feedback.

In a stable system, the blocks of interaction terms on the main diago-

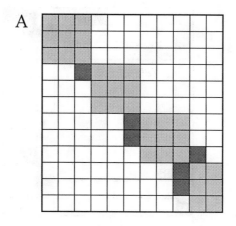

A

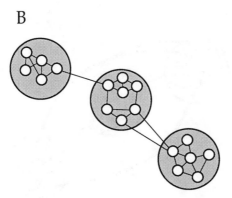

B

Figure 3.10. A. In a matrix of interaction, it is possible to rearrange rows and columns so that successive rows and columns reflect tight interrelationships in the system. When this is achieved, most of the large values of interaction are close to the principal diagonal of the matrix, the trace. B. To the right is a diagrammatic representation of this type of species interaction. Values close to the trace define subsystems.

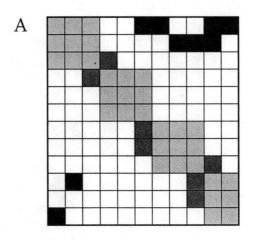

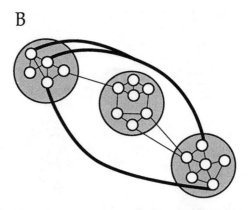

Figure 3.11. A. When values are inserted away from the trace of an interaction matrix, this amounts to inserting long loops that are destabilizing. B. The diagram that corresponds to that in figure 3.10 but with the destabilizing links introduced.

nal of the matrix that make up the principal subsystems are significantly smaller than the whole. Therefore, they are effectively and reliably constrained by the total system. Systems with orderly, unambiguous constraint are generally stable. The difficulty with the long loops formed by making connections between subsystems is that the long loops exhibit slow behavior which is only marginally faster than the behavior of the total system. With only a small difference in reaction time of the long loops and the total system, constraint becomes conditional. In the right conditions, where an external influence causes the whole system to show a fast reaction, the long loop behaves more slowly than the whole system, and so the loop escapes constraint. The long loop returns with a signal that causes a correction that has already been made by the total system. The loop returns a signal for overcorrection that causes more problems than the original disturbing force. Error then amplifies through the pathological relationship between the long loop and the total system. The whole system exhibits unstable behavior because system structure has become a variable.

Random connections in the system are only close to the principal diagonal by chance, and that is why randomly connected systems become readily unstable. Randomly connected systems are complex because there are many long loops in the system, most taking their own unique period of time to return a signal fed into the loop. The differences in periodicity put each long loop at its own level of organization as defined by return time. Remember that complexity has been defined elsewhere in this book as belonging to situations that require many widely spaced levels of organization for adequate description. It is hard to control randomly connected systems because a correction of one interaction term is likely to have unexpected effects on some of the long-term loops of connection at some other level of organization. This means that small changes often amplify to give large changes in system behavior. In complex systems, there is more to go wrong than in simple systems; there are many small components in a complex system and any one of them is connected fairly directly to the behavior of the entire system under study.

The Empirical Strikes Back

Complexity is a matter of the number of system parts and the strength and patterns of connection. Diversity of species by itself appears not to have a simple relationship to complexity. There is a large literature on diversity and stability, but the striking feature of it is the paucity of empirical data to support the myriad hypotheses. A notable exception is the empirical test of complexity against stability of Peter Van Voris and his associates. They used microcosm ecosystems, made an empirical observation that implied complexity, and related complexity to system stability.

They took eleven plugs out of an old field and enclosed them each in its own chamber. Ten plugs were used for the test and the eleventh was the control. Although they were from a single community and had levels of diversity that were in the same range, the microcosms were not replicate sample communities. This is because they were too small to contain the same taxa. The differences in the particulars of the flora and fauna were large enough to be consequential. For example, some had ants in them while others did not.

Despite the clear differences in sample communities, it did emerge that the microcosms were replicate ecosystems (figure 3.12). To investigate the ecosystemic aspects of these microcosms, Van Voris and his associates turned away from species composition and toward measurements of fluxes.

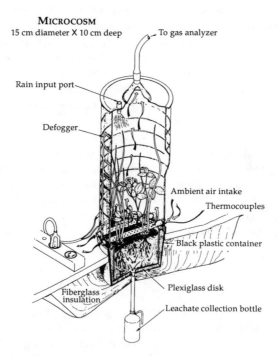

Figure 3.12. The small plugs of soil and vegetation from amonitored output ports are shown (after Van Voris et al. 1980).

Each microcosm was watered once a week and carbon dioxide was sampled hourly for six months. The measures of stability they chose both related to calcium leached from the system after the watering. One was relative system resistance, the capacity of the microcosm to retain calcium after an insult of cadmium. The second was a measure of relative system resilience, the speed with which the microcosm bounced back to patterns of calcium leaching that existed before the poisoning with cadmium.

The relative system complexity was assessed by looking at the patterns of carbon dioxide released from the microcosms. Complexity is an attribute of the total system and so it does little good to look at a collection of the parts as does diversity. Carbon dioxide considered in the appropriate way is an integrator of the system, much in the way that water quality coming out of a watershed is a good indicator of many aspects of a total watershed system. Van Voris and colleagues took the time series carbon dioxide data and transformed them into a power spectrum. To obtain a power spectrum, a sine wave is run through the data and the amount of variance in the data given account by that wave is recorded. If the data show a periodicity at the length of the wave, then the wave will account for much variation in the raw data. If there is no periodicity in the data at the length of the wave, then when the data show a high value, the sine wave may be high or low or in the middle. The wave would show no relationship to the data and therefore it would give little account of variance in the time series data of carbon dioxide. After one wave has been considered in this way, a slightly shorter sine wave is passed through the data. The variance which that wave extracts is noted. The process is repeated until a graph can be plotted of length of the successive sine waves against the variance extracted from the time series data.

The peaks in the graph indicate periodic cycles in the carbon dioxide measurements. Each peak in the power spectrum indicates at least one loop of connection, which in turn indicates one of the working components contributing significantly to total system behavior. Remember that the microcosms were too incomplete and capricious in species compositions to be replicate sample communities. The fact that they did appear to be replicate ecosystems was indicated by the way that several peaks were exactly replicated across the microcosms. This would suggest that the systems shared a basic set of processes that made them replicates in ecosystem terms.

Despite the replication of peaks across all the microcosms, some microcosms had significantly more peaks than others. These microcosms were dominated by a series of low-frequency cycles. The presumption was that power spectra with more peaks indicated a more complex system. Each peak would correspond to a critical subsystem in the microcosm (figure 3.13). Although Van Voris and his colleagues were particularly concerned with low-frequency peaks and therefore longer loops in the microcosms, we are not dealing with loops here that are so long that they would necessarily correspond to the pathological loops in the interaction matrices. This is because even the longest loops detected by the power series would have to have cycled several times during the experiment in order to be detected. Any loop that has time to cycle numerous times through the course of the experiment is not close enough to the frequency of the total microcosm to escape constraint on grounds of relative frequency alone. We are therefore dealing with a fairly narrow universe of relative complexity in the microcosms. The more complex a system, the more long loops of connection it would have. These long loops would each cause periodicity. Most of them would do so at different frequencies, and so it was the relatively larger array of frequencies that was taken to indicate greater relative complexity.

There was general concurrence between the two measures of stability. The ranking of the microcosms in terms of stability did not corre-

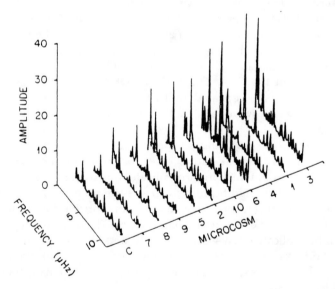

Figure 3.13. When the Van Voris et al. microcosms were ranked according to the number of peaks in their power spectra, those to the right with the largest number of peaks appeared the most resilient and resistant to disturbance. C was the control (Van Voris et al. 1980).

late with diversity or any other microcosm measurement, except for the number of peaks on the power spectra. The more peaks the more stable was the microcosm. This is one of the few empirical tests of complexity and its relationship to stability in an ocean of speculation and clever reasoning. The little empirical information we have indicates that complexity, at least up to a point, is positively correlated with stability in real biological, albeit experimental, systems.

Conclusion

The ecosystem criterion is very distinct from all others. Like communities, it bears a complex relationship to landscapes. For the most part it is unworkable and certainly depauperates the concept to think of an ecosystem as a place on a landscape. Although it is more inclusive than the community, in that ecosystems contain meteorological and geological components, it is not a higher level than community. The description of a single ecological place can be couched in either community or ecosystem terms. They are merely alternative specifications. The reduction of an ecosystem to its functional parts does not usually lead to organisms or populations. Only by reducing on very specific phenomena does one find plants and animals as discrete entities inside ecosystems.

The attributes of communities are the end products of biological evolution. Ecosystems do depend on evolved entities for some of their functions like primary production, but any one of a large number of separately evolved plants can do the job. Therefore, evolution is only tenuously connected to ecosystems. Ecosystems are not evolved in any conventional meaning of that term. Nevertheless, some of the same principles hold for ecosystem development as apply to evolution. The particular ecosystems we find take the form they do because those patterns are stable and therefore hold the material of the world in those configurations long enough for us to observe them. We find stable configurations regularly. Given a world where there is a scarcity of biologically available nitrogen, ecosystems develop organization that cycles nutrients. It is stretching Darwinian evolution too far to suggest that there is a group selection explanation for the sharing and cycling of nutrients. The ecosystem is therefore a parallel development that is highly structured like strictly living systems, but with a different cause than natural selection in the conventional sense.

4. THE COMMUNITY
CRITERION

———◆•◆———

Like landscape and ecosystem, the community criterion is not a level in
and of itself. Rather, it is a way of describing ecological systems by their
floristic and faunal similarity. On the figure of the layer cake in chapter 1
(figure 1.13), the community is but one of the columns. There are various conceptions of community that cut into the community column at
different scales: movement up and down the community column. The
contentious debate between Clements and Gleason, upon which we
will expand later, is a matter of changing the scale while keeping the
criterion constant. In the manner prescribed in the first chapter, we
compare the community criterion with other related ordering principles
caught in cross-section at a given scale. After contrasting communities
with ecosystems and landscapes expressed at comparable scales, we
will look at differently scaled community conceptions inside the community criterion.

Relating communities to ecosystems, we note that like ecosystems,
communities have been mistaken for landscape entities. Both have been
conceptualized as places. Using watershed boundaries as the limits of
an ecosystem puts unwarranted primacy on soil and water in the ecosystem over atmospheric components. The essence of ecosystem is flux
of energy and matter of all sorts, including atmospheric and animal matter. Defining the ecosystem as a place on the ground generally rips
whole sections of the pathways and cycles out of the system. Like ecosystems, communities do have some aspects that can map onto a spatial

matrix. Nevertheless, the spatially defined community is as inadequate as the spatially defined ecosystem.

We can think of communities as collections of types of organisms on a landscape, but that depauperates the concept of community. The community as a place refers to a happenstance collection of plants or animals, rather than a functional whole with interrelated and interdependent parts. The relationships embodied in the community transcend the places where particular individual plants grow and animals live. The difference between a community and a mere collection of organisms is the accommodation that the different species make for each other. The community is not the presence of a particular set of organisms, it is the difference in the organisms because the other community members can be expected to be present.

There is nothing tangible about the community defined thus, for it invokes a potential condition that is not observed, the state of organisms in the absence of the community. Even more intangible, the community itself is the difference between that unobserved condition and what is observed, the states and behaviors of the organisms in the presence of community constraints. The community is therefore a hypothetical differential, a very intangible entity.

Esoteric as such a definition of a community might appear, it is only a formal expression of some commonly held views of what constitutes a community. A community without accommodation between members would only be a collection and not worth studying. A community equated with the observed state of the organisms could imply accommodation but would be a confusing structure. It is also a very static conception. We would ask how much of what one sees is attributable to the organisms independent of their community membership, and how much is attributable to community constraints? The community seen as the collection of observed community members compromises those two considerations irreconcilably.

A community at an instant is the embodiment of prior processes of accommodation, for example character displacement where a species has been selected to avoid direct confrontation with another species. The history of give and take between species allows the particular organisms at a site to coexist as community members. Some accommodations may be to site conditions ameliorated by former occupants. Thus, adjustments may be made between community members that do not even occupy sites at the same time. Therefore, there is a distinctly temporal component to communities that extends beyond just the place itself at a moment in time. Anthony Bradshaw showed that plants living together on the spoil of lead mines in the Snowdonia Mountains were distinct ecotypes. The mine spoil was only decades and not centuries old, indicating that building a community of short-lived species may not take long, but it still takes time.

One might argue that the past determines landscapes as well as communities. However, that is beside the point. While there are processes like erosion that change a landscape over time, these pertain to only the state in which the landscape is left. They are not what makes it a piece of landscape in the first place. We would argue that the past processes of accommodation in communities actually make it a community, rather than determine the state of the community. The past processes that built a community have become part of community structure. A quite separate set of historical accidents are the determinants of the state in which we happen to find a community. These local events are the equivalent of past erosion on a landscape. They are differently scaled from the ancient accommodations between species that have given the community its status as a community.

The temporal aspects of community can fail to map onto the ground in any concrete fashion. For example, the accommodation between organisms that makes them community members could refer to a history of encounters between individuals scattered along time and across space. On a given tract of land two species may have an intimate relationship wherein the poisonous allelopathic potential of one party fails to act on the other. At that point, the allelopathy binds the two species together as community members, since both parties benefit from the exclusion of occasional species that have not been selected to overcome the allelopathic substance. Allelopathy which no longer works because there has been an accommodation to it has thus become a community attribute. Even so, when a naive target population, an outsider to the community, succumbs to the poison, then the allelopathy can be seen as the possession of the individual allelopathic species. Allelopathy is a potential for the community but is an actuality for individual species that possesses it.

This raises an interesting difference between plant and animal communities. As described above, time is the binder for plant communities, while space is the separator. Patches of similar vegetation are separated in space but are bound together over evolutionary time for genetic selection. Over ecological time, the community parts are bonded by both the time for the working of ecesis (establishment of arriving propagules in the local environment) and the time to play out the progress of succession.

On the other hand, place binds the animal community together, while time tears it apart. Over the normal time period for field observations, members of the animal community come and go. Therefore, time gives a disjointed appearance to animal communities, much in the way that space separates islands of plant communities. The plant community ecologist can return to the site some time later and still find the community. Such a point of purchase is not readily available to the animal ecologist. The reference for the animal community is the place where the observer works for the brief time that the inventory is taken.

We do not wish to overstate the time/space dichotomy of plant and animal communities, for it is only a matter of scale relative to the human observer. Certainly we do not want to deny accommodation over evolutionary time between animal community members. Also, animals do show a sort of succession, albeit over shorter time frames. For example, the grazers of the Serengeti Plains pass through a given area in a given growing season in a recognizable sequence. Zebras mow the grass to a height that is workable for a series of cloven-hoofed ungulates that appear in a sequence of decreasing body size. Animal communities can be bound together over something of the same order of time frame that holds the identity of plant communities in succession. Even so, the capacity of animals to move during the period of data collection does give plant and animal communities contrasting textures.

The Development of the Plant Community Concept

The history of the development of the community concept has been a tug-of-war between landscape and organism as the points of reference: the community as a collection of organisms versus the landscape with certain communities scattered across it. The modern view of community denies victory to both sides and requires complementarity. The tension arises from different species of organisms failing to read the landscape in the same terms because each species occupies the landscape at its own scale. A complex system requires many scales and therefore levels of organization for its adequate description. Since the parts of a community, organisms assigned to species, are variously scaled in their use of the landscape, communities are complex. The early literature on communities has an intellectual polarity that reflects the dilemma of organism (part) and landscape (context). In this section we will explore that intellectual tension that spans the community concept.

The organismal prong of communities has its origin in evolution. Important evolutionary notions precede Darwin by a full half-century. Furthermore, one of the founders of ecology as a discrete discipline was Frederic Clements, who died in the middle of this century a Lamarkian.

Nevertheless, it is Darwinian evolution that generated the organismal side of the community origins. The organismal pole of community concepts originated with German efforts in plant physiology applied to organismal adaptations. Therefore, the source of the organismal end of the community is found in the middle of the nineteenth century. The landscape half of the community concept arose a little earlier, through biogeographic exploration in the first half of the nineteenth century. The impetus behind these biogeographic expeditions was a need to record the patterns of life form found in the different zones of the globe. Vegetation physiognomy was an important part of the descriptions brought back to Europe.

Both the biogeographic and the physiological, organism-centered approaches to organisms in the field related to nineteenth-century European imperialism. It is easy to forget that the voyage of the *Beagle* was primarily to chart South America for British imperial interests, with Darwin's presence as almost an afterthought. German expansion was the setting for various German-speaking schools of biogeography, leading to Warming's text on biogeography. Following the biogeographers, other young Germans went to the tropics with the notion of adaptation and its physiological basis in plants. Schimper's text covers much of this work; its physiological focus places an emphasis on the organism.

Although the pertinent studies on landscapes and organisms arose in Europe, the critical melding of these two separate endeavors to generate the new concept of community happened in the United States. Warming, Schimper, and other European writers greatly influenced the young American ecologists involved in state land surveys at the end of the last century. These midwestern graduate students were to play a crucial role in establishing, at the turn of the century, a self-conscious discipline that called itself "oecology." These Americans pressed the notion of community beyond its biogeographical, physiological, and adaptation-centered origins in Germany. The Americans gave it an intellectual autonomy that was distinctly ecological.

The Botanical Seminar in Nebraska was singularly important. This school produced the plant or vegetation "formation" as a new concept in the 1890s. MacMillan's 1893 thesis on vegetation formations of the Minnesota Valley was the first with a distinctly community ring to it. It is interesting that MacMillan himself retreated in later studies to the physiological origins of the community. He failed to follow through and fully establish intellectual autonomy for the "formation." What we now call communities were called formations and associations by Clements and Cowles, but that is really only a shift in parlance. The term "formation" had already been in use with plants for some fifty years, but at the turn of the century it took on an independence from landscapes, with the works of the young Nebraskan, Clements, and Cowles from Chicago.

These workers—and particularly Clements—took the idea of the association of plants and gave it a set of distinctly ecological explanations. Until they did this, vegetation ecology only extended plant physiology into the inconvenient setting of the field. New fields, like ecology at the turn of the century, characteristically start as a cumbersome groping upscale using established ideas. Furthermore, they generally do not make significant contributions to knowledge until they abandon the explanatory principles of the parent discipline. The principles of the parent disciplines of biogeography and physiology became overextended when applied to species accommodations in the field. Then an intellectual collapse gave autonomy to the new entities, plant communities with their organism/landscape tension.

By the early part of this century in America the association was the entity that organized thinking in plant ecology. Associations were recognized as having integrity over time such that, in a given association, species composition could change as the processes of environmental amelioration and competition allowed succession. The processes of invasion and ecesis were recognized and investigated. Note how these explanations of community behavior are neither physiological nor spatial but are identifiably ecological.

As often occurs in ecological advances, a new protocol or method allowed a break with the past. New technology often gives new vision in science in general, as with Galileo and the telescope. The critical step in the separation of plant formation and community from landscape principles was the development of the quadrat. The quadrat is a sampling plot, say a square meter, in which the investigator conducts a plant census. The quadrat census started as a means whereby the Americans corrected the vegetational assessments of traveling European biogeographers. By addressing large-scale biogeographic questions through counting individual organisms, the quadrat drew together the two poles. In the end, it gave autonomy to species associations as a means of identifying communities. Places that gave similar associations viewed through a set of quadrats could be considered the same, whether they were contiguous or not. Thus began the move away from the general appearance of vegetation on the landscape as the dominant organizer of vegetation studies. It was a move toward the floristic criterion of the community. Nevertheless, the shift from landscape to association and formation as the dominant organizing principles for vegetation was gradual, and it is significant that on the continent of Europe the association has remained an important tool for mapping vegetation by Braun-Blanquet and his associates.

The community of Frederic Clements was different from the landscape notions that preceded his work. The distinctions can be enumerated: he saw the community in floristic not physiognomic terms; he emphasized sites with vegetational homogeneity, but was not a landscape ecologist; he was cognizant of the dynamic nature of the community at a site; he was aware that each species requires particular conditions to become established in the community. Although emphases within community conceptions have changed somewhat through this century, Clements' conception of the community is really very close to our notion of the community as a tussle between organismal and landscape-referenced views.

In this chapter we shall develop the idea that the community is the accommodation between species with different periodicities as organisms from each species occupy the landscape at different scales. We use the metaphor of a wave interference pattern. The modern conception of a community gives about equal weight to both organisms and landscapes as players in the emergence of communities. We can see that a Clementsian community definitely comes from the landscape end of things, although it by no means neglects the organism side of the community dilemma.

Clements' view was very much a reflection of the landscape in which he was raised. The opens plains of Nebraska had vast tracts of similar vegetation, with much of the variation attributable to differences in stages of recovery from disturbance. Accordingly, Clements proposed the monoclimax which suggests that over large tracts of land, the vegetation can be characterized by a single climate-determined mature phase. All other floristic compositions reflect immature stages of the self-sustaining climatic climax vegetation. The vastness of the plains and the long temporal framework of recovery to climax gave Clements a large-scale view of community, much larger than the individual organismal considerations that were to challenge the Clementsian concept of community.

Although the individual organisms in Clements' quadrats played an important role in prying communities away from a strictly landscape conception, individual organisms were quickly subsumed in the "superorganism" of the Clementsian community. However, as early as the first decade of this century, efforts were made to redress the imbalance; the organism as an organizing principle offered a challenge to the landscape as a point of departure into community conceptions. For the first half of this century, academic wars were fought over the nature of the community: is it an association on a landscape or is it a collection of organisms? We now see that it can be profitably considered as both simultaneously. Henry Gleason championed the organism as the basis for communities in his "Individualistic Concept of the Plant Association." Gleason was raised in the prairie-forest border region of northern Illinois. There the vegetation on the landscape is dynamic with prairie and forest waging war on each other with fire and shade, respectively. The monolithic single climax for a whole region does not fit the observed facts.

Actually there was remarkably little difference between Clements' and Gleason's considerations of the critical community processes. Both recognized the importance of invasion from local seed sources and the way invasion feeds into ecesis as an accommodation is reached between flora and environment. Thus, Gleason's recognition of spatial proximity of seed source, as well as the habitat condition, show him to be cognizant of the landscape origins of the association that is central to the Clementsian community. On the other side of the argument, Clements' extended consideration of ecesis shows him to be aware of the importance of the performance of individual organisms at a site, a cornerstone of the Gleasonian conception of communities. Nevertheless, the difference in emphasis on tracts of land, as opposed to individual organisms, drew the battle lines for a quarter of a century of academic acrimony.

Gleason's thesis was that the vegetation at a place at an instant in time was the product of the happenstance of the local flora available for invasion, and the selection of individuals by the site-specific environment. A careful reading of Gleason's original statements shows that he did not deny the integrity of the larger-scaled system implied by the Clementsian community concept. However, both the defenders of the Clementsian faith at the time, and some of Gleason's supporters in later decades, thought that he was completely at odds with Clements. Gleason preferred to focus on the underlying processes that select individual organisms for a given site; Clements emphasized community integrity across wide regions. However, the difference between the Gleasonian and Clementsian concepts of community is not just a matter of scale, where Gleason sees small-scale relationships and mechanisms while Clements focuses on the coarse-grain constraints imposed by the whole community. The difference is also one of the criteria used as the point of departure: organism for Gleason and landscape for Clements.

Although it is the interference between two tangibles, landscape and organisms, the community is not readily part of commonplace experience (figure 4.1). This fact led ecologists to use metaphor freely as they

Figure 4.1. An interference pattern painted on a clay egg. The Acoma of New Mexico call this the sunburst pattern. Note that the lines do not relate simply to the emerging pattern (photo C. Lipke).

tried to wrestle the beast down. Cooper's analogy of a braided stream of development of communities over eons attempts to reach a compromise between the distinctness of individual patches of vegetation and the obvious continuity over time across vast tracts of landscape. Eventually the clear modern view of community emerged through advances in analytical techniques.

Much in the way that the quadrat allowed Clements to achieve a view of community that went far beyond its landscape precursors, the ordination techniques developed by John Curtis and his students at Wisconsin and Robert Whittaker in the Great Smoky Mountains both gave the distinctly modern view of community. Ordination is a family of data manipulations that place sites with similar vegetation close to one another on a gradient. Dissimilar vegetations place sites at opposite ends of gradients. At the same time, David Goodall developed classification devices that place sites with similar vegetation in clusters. Classification is just a discrete version of ordination. Goodall's classifications worked alongside his own and other ordination techniques all emerging at the same time. These advances allowed the description of community structure to be isolated on flora and fauna, independent of the landscape position. These multivariate methods made it possible to rearrange descriptions of patches of vegetation so that they would be displayed next to their closest floristic or faunal relatives, whether or not these were neighbors on the landscape. Thus, species representation in a stand of vegetation could assert itself as the organizing principle, unfettered by landscape questions of contiguity.

Analysis of Vegetation

The multivariate methods of Curtis, Whittaker, and Goodall in the 1950s wove together the parts of the modern community concept. The woof of the fabric of the community is made of strands in time, temporal gradients of succession as the community recovers from disturbance. The weave is made of spatial gradients in environmental factors. The result is a patchwork on the landscape. Let us explain.

Clements and Cowles at the turn of the century both focused on the way that the community can change over time, while still maintaining its identity. This is a move upscale from the community as a snapshot that only captures the instantaneous flora or fauna of a particular patch of ground. By emphasizing changes over time, early community ecologists allowed time for community processes to exert their influence. This temporal aspect of communities, in particular, did not fit into the landscape conception. Thus a gradient in time was introduced.

Gleason emphasized local invasion filtered through ecesis at the site. This perspective brought to bear the spatial gradients from, for instance, wet to dry with movement up a hillside. It was probably important that the first community ecologists broke with landscape by introducing time. The spatial emphasis of Gleason probably could not have struggled free from the landscape conception in the way that Cowles' and Clements' communities in succession over time achieved their independence.

Immediately before development of the explicit gradient analyses, soil scientists contributed by talking of changes down hillsides of soil moisture, texture, and nutrient content. These gradients, called catenas, provide continuous change of substrate for the plants. These catenas give gradients of local conditions that amount to gradients in the factors controlling ecesis. Of course, not all changes in soil arise as smooth monotonic change across space, but the obvious hillside examples give an intellectual point of purchase for considering more convoluted or dissected spatial gradients in the physical environment. Vegetation does not, however, change in species composition in smooth, continuous floristic gradients. The reasons are twofold. First, the soil gradients may not be in one direction, and contiguous plots could have radically different substrates. Second, even if the substrate does change simply and slowly across space, disturbance factors like fire would leave the continuous hillside covered with patches in various stages of recovery. The landscape is covered with a patchwork mosaic that is the interference of spatial and temporal gradients. What was needed by the 1950s was a way of giving the patches floristic autonomy from where they happened to be in space.

The two solutions were classification and ordination. It is no accident that the two approaches emerged at the same time in ecology. Ordination techniques fall into two major categories. One type arrayed vegetation along continuous gradients according to their vegetational composition. The other type ordered sites according to the physical environments at each place. Both of these gradient criteria are independent of the landscape criterion of spatial contiguity. With both ordination and classification, the folded gradients in space and the interference from different stages of recovery can be accommodated in simply ordered community representatives. The fact that time since disturbance can mimic movement on soil gradients toward the mesic zone presents no problem. The individual patches of vegetation are free to order themselves on strictly community criteria, namely biotic composition. Thus, a technical advance in methodology moved us to our modern conception.

Vegetation and Environmental Space

Implicit in both the classification and ordination approaches is the notion of species space. As a starting point in a discussion of ecological spaces, consider a site with many species present, each with its own number of variously vigorous individuals. There are many ways to represent species quantitatively: numbers of individuals; amount of area covered; biomass; and number of subsample units occupied. Any one of those or any other sampling method used consistently will serve here.

Given a quantification of the vegetation by species, we can set up the first species as an axis and place the site on that axis according to the quantity of that species at the site. We can set up the next species as another axis at a right angle to the first. The site can then be positioned on that axis in the same way. The site is thus placed on a plane defined by the first two species (figure 4.2). When we introduce a third species with its axis at a right angle to the first two, then the site can be positioned in a volume which is the space of the first three species. A second site will have different quantities for our first three species and will be positioned by three different coordinates in the species space. The distance

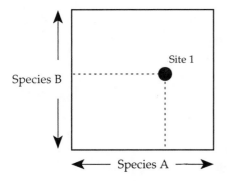

Figure 4.2. A site may be defined by the quantity of its various species. In this case it is a point on a two-dimensional species plane.

between the first and second site in the space is a reflection of their differences in composition in terms of the first three species (figure 4.3).

That distance is the hypotenuse of a right triangle, as we shall show. The distance between the sites on the plane of the first two species is the hypotenuse of a different right triangle. The sides of that right triangle are the differences between each site in the quantities of the first two species. From Pythagoras' theorem it simply follows that the distance between the two sites on the plane is the square root of the sum of the squared differences in scores for the first and second species (figure 4.4). We can easily extend this to three dimensions by considering the third species. The third species contributes the distance that one site is above the plane. We now have a new right triangle standing upright, with its

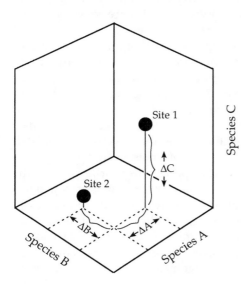

Figure 4.3. As in a two-dimensional species space, sites may be separated from one another in three dimensions according to differences of occurrence of three species at the two sites.

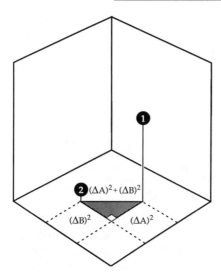

Figure 4.4. The distance between two sites in a species space is the hypotenuse of a right triangle where the base and the side are the differences on the first two species axes.

base being the hypotenuse of the triangle on the plane of the first two species. The hypotenuse of the upright triangle is the difference between the sites according to the first three species. The square of the base of the upright triangle is the sum of the squares of the sides of the triangle on the plane. Take that square of the base and sum it with the square of the distance that one site is above the plane, and we have the square of the grand hypotenuse in the three-dimensional space (figure 4.5). The square root of that quantity is the distance between the two sites in the space defined by the first three species.

What we have done above is a special case that can be generalized. Take the sum of the squared differences accumulated so far between any two sites. This is the square of the grand hypotenuse of a hypertriangle that lies diagonally in the space of all the species so far included. To include the difference contributed by a new species, add the square of that difference to the sum of squared differences for all species included so far. Each time a new species is included, we are taking the accumulated distance so far and treating it as the base of a new, even larger right triangle. In the same way that the third species axis is orthogonal to the axes of the two preceding species, each new species axis is orthogonal to all previous species axes, no matter how many of them there are. We can keep accumulating the difference contributed by a fourth, fifth, or even a hundred and fifth species as the grand hypotenuse that cuts diagonally across spaces with successively more dimensions. Thus, there is a generalized Euclidean distance that measures difference between

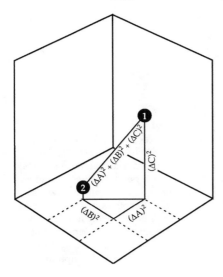

Figure 4.5. The generalized Euclidean distance is the square root of the sum of the squared differences on each dimension of the space.

sites in a species space with dimensionality equal to the number of species in the study. The sites are therefore points inside a hypervolume space, positioned according to their species compositions. We have been at pains to be explicit about the species space because it is central to many of the ideas and controversies in modern community theory.

There is another vocabulary for describing the distance between two sites in species space with which the reader may be familiar. The critical term there is "vector." The distance between two sites in species space is the length of a vector from one site to the other. By subtracting all the species scores for one site from the scores of the other (or indeed any other) site, we place the first site at the origin of a coordinate system. After the subtraction of the first site's scores from those of the second, the values left to the second site constitute a vector that passes from the origin (the first site) to the second site. If there were a large number of sites, one could perform the same operation on them (subtracting the values of the first site) (figure 4.6). That would generate a cloud of points, each one being a site at the end of a vector from the origin, that is, from the first site. The calculations in all the various multivariate techniques are operations performed in vector notation. Vector notation is a powerful protocol, but for many people it has less intuitive meaning than the geometry that underlies it. Accordingly, we will generally use the intuitively appealing geometric expression of multivariate methods of vegetation, rather than the equivalent expressions in matrix algebra and vector notation. However, in the final analysis, the serious student

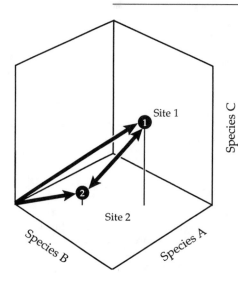

Figure 4.6. As opposed to using the coordinate frame directly, it is possible to use vector notation which can consider distances between points as direct straight lines, or as the angle between lines to both points from a reference point, taking into account the length of the arrows pointing to each site. This is only an alternative mode of description.

of vegetation analysis will need to have at least a passing familiarity with matrix algebra.

The clustering techniques for community analysis work implicitly in species space. Sites with similar vegetation are put into a class because they are close together in the species space; indeed, they are the closest according to the clustering criterion used by the particular method being employed. It is also possible to perform inverse analyses wherein species are clustered according to similarities of occurrence, rather than sites clustered according to vegetational similarities. Gradient analyses, also called ordinations, use the proximity of sites on axes as approximations of site similarity.

The data that underlie these analyses are matrices of species scores across sites. The rows of the matrix are sites while the columns are species. In "normal" analyses the site entities are clustered or arranged on the axes of gradient analyses. Since a matrix can be transposed so that the species columns become rows and the site rows become columns, there is a converse set of analyses. The points scattered across a gradient analysis do not have to be sites. After matrix transposition the species rather than the sites appear as points on the axes of gradient analyses. Also it is possible to cluster species according to similar occurrences at sites rather than sites according to similarities in vegetation. These alternative analyses are called inverse analyses. While normal analyses reveal a description of relative vegetation, inverse analyses display the relative ecology of species.

Data Analysis by Point Projection

A silhouette is a projection of a three-dimensional object onto a two-dimensional area. A rotation of the object in the light gives a different silhouette. Some silhouettes give more information about the original object than others. Similarly, the gradient analysis techniques of community ordination rotate the point cluster in the hypervolume so as to cast an informative shadow of the whole. The criteria for rotation of the point cluster are different between specific ordination methods. Some techniques explicitly project sites from the full dimensional space of the full data matrix to a smaller dimensional space that is tangible. Other methods are not strictly geometric, but can still be profitably seen as approximations to projections of points to a smaller dimensional space. Poor projections foreshorten distances which are long in the original large species space. Effective projections minimize such foreshortening; that is to say, the points in the lower dimension space are well scattered with high variance. Conversely, the sum of squares of the lengths of the projections is minimized in an effective projection (figure 4.7a, b).

In both normal site analyses and inverse species analyses, and in ei-

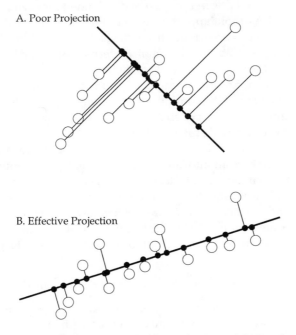

Figure 4.7. An efficient projection of a cluster of points minimizes the sum of the squared lengths of the projection to the summary axis.

ther cluster or gradient analysis, association of species or underlying species correlation structure orders the outcome. In gradient analysis, by projecting to an axis that aligns with the correlation between species, the patterns of several species can be well represented by a single line drawn through the total species space. The gradients that summarize the smaller space after the analysis are long lines that pass diagonally through the original space aligned with the principal pattern of species association.

A great contribution of Curtis and his students was to recognize explicitly the importance of site selection. In retrospect, we can see significant geometric implications to the randomization of sites characteristic of the Wisconsin school of vegetation analysis. The sites have to be selected to be independent of each other in some sense, so that they are all at right angles in the vegetation space. If the sites are not randomly chosen, they are not orthogonal to each other. If the sites are not all at right angles to each other, then there is no reliable reference frame against which we can assess correlations of the species associations. Perhaps two species occur together more often than we expect in the data because the bias of the sampling regime puts them together, even though there is no basis in their actual distribution for saying that they are associated. Without appropriate randomization, the ecologist cannot tell artifactual co-occurrence from that based on a pattern in nature.

By sampling with stratified random patterns, Curtis and his students could see whether it was justifiable to divide communities into discrete types or to accept an underlying continuum of change in vegetation structure. They found the continuum which one might expect if indeed the community did consist of species all with different tolerances, each site being invaded by a distinct local flora.

Our geometric analysis of vegetation leads us back to the tension between the Clementsian and the Gleasonian view of community. The continuum is a vindication of Gleason's perspective and is often seen as a refutation of the Clementsian community concept. However, because Gleason's and Clements' views are not actually competitive, Gleason can be vindicated without Clements being wrong. Clements' view is not confounded by the existence of continua of vegetation on the ground.

Gradient Analyses and the Gleasonian Community

The different views of Clements and Gleason can be seen in terms of our grand organizing scheme. Both are concerned with organizing perception in terms of the community criterion. However, each cuts into the community column at a different layer of the cake. Thus, Gleason's conception of the community is set in the context of Clements'. Gleason's community offers the mechanisms for putting the Clementsian community in a particular state. The community as described by Gleason is constrained by the processes to which Clements gives primacy. By coming at communities from the organismal side of the dilemma, Gleason considers a smaller-scale community. Clements would come at the same set of species and organismal interactions from the landscape end, and so would deal with a larger community. Of course, it is possible to have a Clementsian community that is local in time and space, involving relatively few species. Meanwhile, inclusion of more species and a wider spatiotemporal extent could make a Gleasonian community that is in fact larger scale. Clements only thinks larger-scale than Gleason so long as all else but their different emphases are constant.

At the time of the altercation between Gleason and the supporters of Clements, both sides agreed about the facts concerning the patterns of vegetation on the ground. Nevertheless, both sides were steadfast in their difference of interpretation and insisted that if they were right, the other side must be wrong. Actually, the two views are barely in conflict, for they are more different specifications than competitive theories. Each view expects to find particular patterns.

The Gleasonian view would expect to find each site with its own particular species composition. The Clementsian view would expect there to be upper-level community constraints restricting the constellations of species associations. Some combinations of species would occur much more commonly than others, and some may not occur at all. Careful consideration of these different expectations recognizes that although different, the two worldviews are not mutually exclusive. Gleason pays attention to the continuum in the occupied species space, whereas Clements notes that most of the species space is unoccupied. Gleason emphasizes the processes whereby continuity of composition occurs, whereas Clements focuses on the forces that deny the existence of most constellations of species concurrences.

Gleason focuses at a small scale that recognizes individual organisms. His specification of the system is middle number; any one of an unmanageable number of causes could produce the particular species composition one finds at a site. Gleason works with system parts, not with upper-level constraints that embody the community as an entity. The continuum of vegetational composition across species space reflects the continuity of the endogenous processes that sift through the species to give community behavior. Clements works at a large scale, that of the community context. He comes at communities from the contextual landscape criterion. At that large scale, the Clementsian observer is closer to the level of the community constraints. Consideration of system constraints allows one to predict what vegetation will never or hardly ever occur. The basis of these predictions is the instability of certain vegetational configurations that make such community exemplars transient. There is, nevertheless, plenty of room inside the constraint envelope of the community for there to be a continuum of species abundance. If a caricature of the Gleasonian community were completely right and Clements were exactly wrong, then we would have no business studying communities; communities would be associations through happenstance, a mere collection of species with arbitrary relative abundances.

The classification techniques sit more comfortably with the Clementsian perspective on communities. Early gradient analyses were an extension of primitive classification devices. The creators of these ordination techniques were forced into gradient analysis because they realized that there appeared to be continuity between vegetation classes. It became clear very early on that there was a deal of continuity in the occupancy of species space. With each species exhibiting its own environmental tolerances, and each site being uniquely situated in between variously occupied pools of potential invaders, continuity of species composition is to be expected, just as Gleason hypothesized.

The Dynamic Synthesis Theory of Vegetation

Community ecology appears less experimental and predictive than other ecological fields such as population ecology. John Harper once joked that there seems to be a touch of voodoo magic to community ecology. Indeed, predicting community structure and behavior is difficult because there exists no one-to-one mapping of community composition onto environment. From the environment, it is not possible to predict community composition or vice versa.

Recognizing this problem, David Roberts drew upon the writing of Sukachev, a Russian ecologist whose work became accessible to English-speaking ecologists in 1964. The cornerstone of the Russian work is the recognition that plants interact with each other only through their environment. Therefore, an adequate description of vegetation must equally involve both vegetation and environmental space. Relying upon the gradients in only vegetation space, as does the Curtis school of ecology, is inadequate. Curtis thought that plants are better indicators of the pertinent aspects of the environment than any measure that an ecologist could devise. That may be true, but it does not allow for the two-way interaction between vegetation and its environment. On the other hand, the direct environmental gradients of Whittaker miss the vegetational side of the interaction, and so are no better. For predictive community ecology, we need a description of the interaction between happenings in vegetation and environmental space.

Accordingly, Roberts developed a relation between the environmental and species spaces that is flexible enough to link the two in an orderly fashion. He maps events and continuous changes reciprocally between the spaces. A change in one space often produces no commensurate change in the other. While some environmental factors may respond to shifts in vegetation, others remain unchanged, in the short term, by even dramatic shifts in vegetation. Roberts uses the term "elastic" to describe site factors that respond immediately to changes in vegetation. In a forest, light could be one such factor. Tall trees and thick understory shade the ground. However, no matter how little light reaches the floor of the mature forest, clearcutting the forest or a crown fire causes light levels to snap back to full insolation immediately. Light levels are easily changed by the growing vegetation, but are always ready to snap back to unvegetated brightness. Other factors, like soil pH, change slowly, and even then only if there is unremitting pressure from the vegetation. It takes several decades for spruce trees to acidify the soil, but acid soil persists long after the spruce trees have been harvested. Factors of this type Roberts calls plastic.

Elastic environmental factors exhibit short memory of the influence of vegetation, while plastic factors can reflect changes wrought by vegetation long since removed from the site. This is exactly the sort of situation that generates complex behavior. The state of site factors at any time is the result of the interference between processes with very different reaction times. As a result, a myriad of ways can lead to any given state of affairs. That explains the difficulty we experience in mapping vegetation to environment with any reliability.

The situation is further complicated by unidirectional changes in one space being related to tortuous change in the other space. Consider a two-dimensional species space, the plane of species A and B. Consider a related environmental space with two factors, say, light and soil moisture. The unvegetated site is high in light and soil water. A site with no vegetation is at the origin in the vegetation space, while the same site in environmental space is far from the origin, showing high values for both factors (figure 4.8a, b). It is possible to identify separate places in the environmental space where species A and B find their respective optimal conditions. Some of the work of Ellenberg in Europe involved experimental gradients upon which the pH optima were found for the species in his community. If A is a pioneer species and B is a mature for-

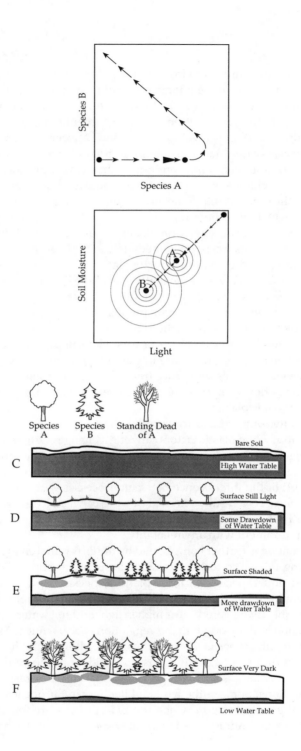

Species B

Species A

Soil Moisture

Light

Species A

Species B

Standing Dead of A

Bare Soil

C

High Water Table

Surface Still Light

D

Some Drawdown of Water Table

Surface Shaded

E

More drawdown of Water Table

Surface Very Dark

F

Low Water Table

est species, then A is likely to be the first to establish and grow (figure 4.8c). As A grows in stature and numbers, the trajectory of the vegetation in species space is from the origin along the A axis. As species A grows, the ground receives less light. Furthermore, the increasing vegetational cover transpires larger amounts of water and so draws down soil water (figure 4.8d). In environmental space, the site begins to move down both axes together as it moves toward the origin. If species A does best in conditions between the starting point and the origin, then its growth generates a positive feedback where the plants ameliorate conditions and enhance the growth of their own species.

After a period of this sort of growth, species A has modified conditions so that they are optimal for itself. During this period, species B finds conditions less than optimal, but better than the conditions of the open site (figure 4.8d). Species B has an optimum closer to the origin than species A. Accordingly, species B enters the stand, at first in small numbers. The trajectory of the site in vegetation space starts to turn up the B axis while continuing down the A axis.

The growth patterns that A used to improve the site for itself may well continue to change site factors beyond the optimum for A. The increasing stature of A further darkens and dries the site. Not only does this take the environment closer to the origin than the optimum of A, but it moves the site in environmental space further toward the optimum for species B. The effect of this in vegetation space is to stop increases down the A axis and accelerate the turn up the B axis. At this point, species B is in positive feedback and takes the site factors further beyond the optimum of A, toward its own optimum (figure 4.8e, f). This final development reduces the contribution of A to the vegetation. In vegetation space the trajectory turns toward high values of B and away from high values of species A. Aspen is a transient forest species and behaves like species A. Its shade suppresses its own offspring but also makes conditions conducive to growth of pine, hemlock, and sugar maple, which in turn behave like species B.

Think of the environmental space in Roberts' model as an undulating surface (figure 4.9). Hollows on the surface are domains of positive feedback where success of a species leads to further success. For transient species, the surface slopes in one direction or another in the vicinity of

Figure 4.8 A. A site at a given time occupies a position in space. Here are the environmental optima of species A and B. C., D. Early in succession there is high light and high water table. E., F. As stand development proceeds the system moves closer to the optimum of species B as the water table drops and the light intensity at the ground diminishes.

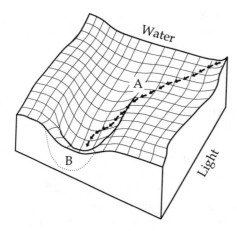

Figure 4.9. The environmental space of the example in the text expressed as an undulating surface where the site behaves like a ball rolling from high ground through the optimum of species *A* to the hollow of the optimum of species *B*..

their optima. If the site is a ball on this surface, it will roll into optimal regions for transient species' optima, but roll on past into deeper hollows in the region of the optimum of the species that replace the transients. Some species have optima at the bottom of local hollows such that the ball cannot escape without application of some external force. For example, heaths in positive feedback change the soil to an acid condition in which they do well. Other species then find it difficult to colonize and establish on such poor soil. The heaths thus create a condition and then perpetuate it, thereby creating tracts of very similar vegetation. Similarly, yew in England can form canopies so dense that the shrub holds the site against all invasion of trees for centuries. The same may be true for rhododendron thickets in the Smoky Mountains of Tennessee, called heath balds, although the Appalachians have a shorter historical record than England and so we are less sure of stasis for balds (figure 4.10). Thus, the environmental surface in the vicinity of the optimum of, say, aspen would be gently sloping (figure 4.9, point A), which is conducive to continuous and relatively unguided change in vegetation and site factors. Alternatively, the surface could have local deep hollows which hold the vegetation and environment in stasis, as in heathland (figure 4.9, point B).

There is no firm rule as to which type of pattern will hold sway in a given vegetation study. Furthermore, there is no reason why the environmental space may not be dissected by strong positive feedback in some parts of the space, while in other parts of the space the surface could be relatively flat with little to retard or direct vegetational change. Different degrees of sculpturing on the environmental surface would clearly lead to different patterns of vegetation typing. A relatively flat

Figure 4.10. Heath balds in the Smoky Mountains appear to be a terminal vegetation type which creates its own optimal environment. High on the ridge the bald looks deceptively open, but it is an impassable thicket about 5 meters tall (photo T. Allen).

surface would lead to continuous change giving the continuum that the devotees of gradient analysis expect. Alternatively, an environmental surface dominated by a few deep hollows holding the vegetation in certain states would give a landscape that yields reliable maps of discrete types of vegetation. Such maps are the aim of the European phytosociologists who study discrete vegetation types.

There is still argument as to whether vegetation composition is continuous or discrete; considering site dynamics in both vegetation and environmental space at the same time appears to resolve this question. Recovery from disturbance is clearly continuous, but the fact that recovering vegetation often seems to get stuck at certain stages introduces a quasi-discreteness in vegetation. Rapid transient stages will be underrepresented relative to the widespread occurrence of the stages that change slowly. The vegetational sticking places give the identifiable associations of Clements. Meanwhile, the transient vegetation straddles the boundaries between the associations of Clements and gives the continuity that Gleason would expect.

Predictions Relating Vegetation and Environment

We see how straightforward movement in environment space is related to complex trajectories in species space. Both positive and negative feedback play important roles in this pattern. Add to this the complications of differences between elastic and plastic site factors. It then becomes clear how the same vegetation can occur in different physical environments, and how different vegetation can occur in separate sites that replicate site factors. Therefore, vegetation is not a function of the environment, rather it has a relation to its environment.

A function is only a special case of a relation. If X is a monotonic function of Y, then only one value of X corresponds to each value of Y. A relation does not require only one value of X for every value of Y. For a mathematician or statistician, saying that environment is linked to vegetation as a relation rather than a function is unremarkable. In purely mathematical terms, little appears to be gained by relaxing the special condition that comes with a function. Nevertheless, the difference between a function and a relation for mapping vegetation onto the environment is important for the ecologist. The difference between a mathematician's view and that of an ecologist pertains to the models that brought community ecology to its modern state, namely the vegetation space model of Curtis and the environmental space model of Whittaker. Both of these models, particularly that of Whittaker, implicitly assume that a function links vegetation to environment. Roberts insists that the relation is not a function and so opens the door for prediction on more workable grounds. What the conventional view might call noise in the relationship between environment and vegetation can be recognized as slack within constraints. Thus, Roberts is conceptually in line with notions of hierarchy theory in a way that Curtis and Whittaker are not.

As we have said several times before, we cannot make predictions successfully until the question implied by the prediction is posed against a constraint. Roberts has devised a method that discovers the ranking of constraints, a basic requirement for posing community questions with effect. The published details are less important than the argument behind them. He starts with vegetation data from sites and groups them according to vegetational similarity. The technical name of the groups is "cliques," groups where all members are connected to all others by a particular level of vegetational similarity. The cliques were maximal cliques, a special case where only the largest groups that meet the condition are considered. Thus it is possible for a given site to belong to several cliques, but no clique can be a strict subset of any other. The method proceeds to work out the sets of maximal cliques in the data for various thresholds of vegetational similarity.

Remember that very similar vegetation can occur on sites with different environmental conditions. Some cliques would by chance contain only members with more or less identical environments; these are not interesting, since they represent no limit and reflect no constraint. What we want to know is how much variation in environment could be tolerated in cliques of a given level of vegetational similarity. In a plot of clique width against widest environmental range within cliques, Roberts found that very strict cliques with very similar within-clique vegetation had relatively narrow ranges of environment between clique members (figure 4.11). Very particular vegetation could be generally related to en-

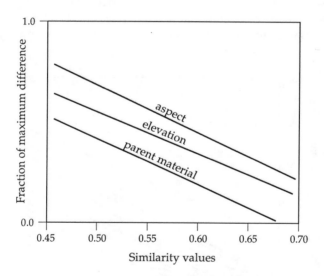

Figure 4.11. The regression lines plotted refer to average clique width with respect to the factor in question. As clique width increases in the Little Rockies, aspect as an environmental factor is quickly unable to distinguish between wider cliques. However, altitude and particularly parent material can distinguish between wider cliques than aspects, indicating that they are the primary constraints. Note that at only moderate similarity of vegetation within cliques (0.45) almost any aspect could appear within a clique, giving such cliques poor predictive power for aspect. However, those same cliques still possess some predictive power as to altitude and parent material. For narrow cliques (0.70) with very similar vegetation inside the cliques, all three environmental factors could be predicted from a knowledge of vegetation (after Roberts 1984).

vironment. However, the weak relationship to environment breaks down completely if even moderate amounts of vegetational variation are introduced. Even cliques defined by only moderate differences in vegetation across clique members were not restricted to any environment. Knowing a site was from a clique of only moderate vegetational variation, and knowing the environment of the other clique members, gave no indication about the site factors for the site in question. The site factors lost all predictive power as to what vegetation type would be found within only moderately different patches of vegetation. In a way, the environment of the wider cliques had become saturated as a measure of vegetational similarity. The environment's capacity to exert constraint was exceeded by the variation of vegetation in the wider cliques.

The tighter the environmental constraint, the tighter the relation between environment and vegetation. The tighter the environment constraint, the less slack there is in the system. The tighter the constraint, the greater will be the capacity of that factor to resist becoming saturated by wider ranges of vegetational difference within the cliques. Various site factors were tested against cliques with successively larger vegeta-

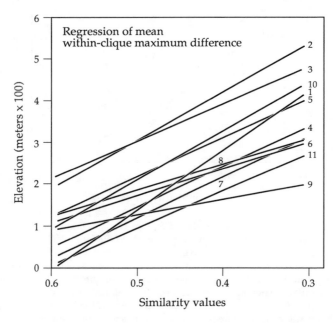

Figure 4.12. As might be expected, wider cliques show more elevational difference within cliques. However the eleven forests plotted here with clique averages showed altitude as a constraint much tighter in some forests than others (after Roberts 1984).

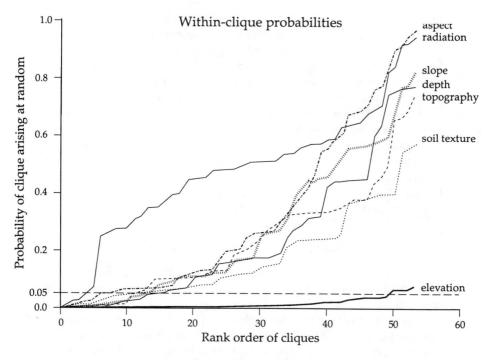

Figure 4.13. In Bryce Canyon the vegetationally defined clique width that allows 55 cliques in the data are considered here with regard to 7 environmental factors. For altitude, 50 of the 55 cliques have less than 0.05 probability of arising at random. The other factors forming the lines above refer to factors where the cliques stand a much greater chance of occurring at random with respect to the environmental factor in question. Clearly elevation is the overriding tightest constraint. Here individual clique values are plotted, whereas in figures 4.11 and 4.12 average clique values are considered (after Roberts 1984).

tional differences. The critical finding was that not all factors became useless at the same clique width. In the Little Rocky Mountains of Montana, parent material from which the soil was derived and altitude both persisted as predictors of clique environment for wider cliques than did site aspect. This means that relatively similar sites can have very different aspect with some facing north and some south, but will have a relatively narrow range of altitude across clique members. Although aspect makes a difference, the more important constraints that will generate firm predictions in the forested mountains of the Little Rockies are altitude and parent material. Roberts applied clique analysis in the mountains of Montana and in Bryce Canyon. In Montana he found that altitude constrained to different degrees depending on the particular mountain range (figure 4.12). In Bryce Canyon he found that altitude was far and away the most significant constraint (figure 4.13).

With that knowledge of the ordering of constraints, predictions can be made. Knowing which are the tightest constraints, we can devise experiments to test on limiting factors. With such well-studied experiments we can at least hope to find the critical mechanisms underlying community pattern and structure. Knowing the constraints allows the vegetation ecologist to ask questions that avoid implying a middle number specification of the vegetation. We are now in a position to ask questions that can generate predictions rather than mere description of just another special case. In Montana forests the prescription is: test for the influence of aspect only after normalizing for altitude. Understand altitude first, and avoid being confused about the effect of aspect.

Habitats and Niches: Plants and Animals

We have already introduced the notion of "niche" as an example of a word with rich meaning in ecology. As we suggested, it is a helpful but contentious notion. "Niche" refers to the position of a species in a community in terms of the role that it plays and the physical places in which one might find it. The contention comes from different emphases on role as opposed to physical place. G. E. Hutchinson described niche as a hypervolume where the axes of the space are resource variables such as prey size and abundance. He also included more strictly habitat variables such as temperature. Any given species occurs within ranges on these variables, and in this way it occupies a portion of the hypervolume. In some ways an animal niche, so described, resembles the environmental space used in plant community analysis. Nevertheless, there are striking differences between the way plants and animals are considered as community parts. In the BIOSIS search of the literature that Flather and we performed, we found that plants are associated with habitat while animals are given niches. The critical difference is the movement of animals. Habitat is a place in which the plant passively sits; this is incorrect, but it captures the common attitude that both the lay public and even ecologists have toward plants. A niche involves playing a role in the community, and the observable activity of animals sits easily with role playing.

The meaning of habitat applied to plants as opposed to animals is different because of their respective relative scaling in time and space. The habitat of even a small single mammal involves an area that is so large as to have little equivalence to that of a single tree. An area the size of an individual animal's range could be the size of the habitat of a plant population or an entire species of plant. It would almost certainly be much larger than the habitat of a single plant. Nevertheless, animal community ecologists have often deferred to plant community ecologists to give them the habitat for their fauna.

Using plant communities as the setting for animal communities is probably a mistake because animals regularly cross plant community boundaries in their daily routine. Working in the New Jersey pine barrens, Emile DeVito found exactly such a problem. His study site was a mosaic of pine community patches and oak-dominated, deciduous communities. Careful observation of resource use allowed him to assign some birds to either a conifer or a deciduous resource base. The problem was that despite clear patterns of separation on resource niches, conifer birds could be found unpredictably in oak forests, and vice versa. More puzzling, some patches of oak forest contained very few birds that used a deciduous resource base.

As the research continued the pattern became clear; birds would occupy an alien, inhospitable habitat so long as that patch of alien habitat was set in a matrix of another plant community that did indeed offer resources. On the other hand, birds would not go far down a peninsula of resource-offering habitat if it projected into an alien habitat type that offered few resources. This explained the absence of deciduous-dependent birds in some oak patches. Although deciduous peninsulas offered some resources to oak-dependent birds, those resources were insufficient in extent to warrant living in a matrix of pine forest.

DeVito discovered another interesting pattern of bird occupancy of plant communities that also pertained to problems of scale. Humans can observe birds using resources in a very local area. The model which is implied in such observations is one where birds are tied to habitats where resources are abundant. However, this was not what he observed. The explanation is that the bird must enter a site in order to use it. This decision is not made on data collected at the scale of a human standing within the forest looking at particular tree trunks. The birds may decide to come in through the canopy based on data collected while flying over the forest at over thirty miles an hour. Only after making that decision is there a possibility of making a second decision to stay, based on the sort of events that the ornithologist can see from a standing position.

Most animals experience their environment at scales different from the field biologist. Even reasonable attention to scaling can be misplaced because the animals lie in a space defined by criteria independent of human perceptual categories. It is reasonable to expect larger birds to relate to larger resource bases because of greater demands of energy for individual sustenance of a big body. The logical extension of this is that larger birds would be expected to occupy larger areas of forest. The conventional wisdom is that larger birds use cues that pertain to larger areas of forest when making decisions to stay in a patch. However, birds do not see the world through the eyes of ornithologists, and so the patterns in nature do not follow the model that relies principally on bird size. DeVito found that some very small birds, like the blue-gray gnatcatcher, relate to landscape-level vegetation of 50–150 hectares, a very large area indeed. Small birds will display this large-scale pattern of occupancy of the landscape when the resource base of the bird is narrow in the vertical dimension. A thin layer of resource needs to be spread over a larger area if it is to be sufficient for even a tiny bird. The critical information that such birds use to decide to stay in a patch appears to involve knowledge of a vast area. Conversely, a bird, large or small, using a vertically dispersed resource base will use local cues in choosing a site.

Plant community data often consist of tallies of identified organisms, and data from different species are collected in parallel. Plant community data are not usually collected to focus on populations by species separately. When plant data are collected one species at a time, the focus is on the population rather than the community. The movement of animals means that the individual is less tangible than in plant data collection. Animal data often take the form of encounters by sight or sound at sample places. Even though these are encounters with individuals, the experience is much more one of watching the flux of populations as their parts move past. Population units like the herd or the shoal stand out so that animal community data reflect populations much more than do plant community data.

Our analysis of the literature with Curt Flather allowed us to see if the same organisms are used for population studies as are used for community analysis. The answer is that they are not the same taxa as far as plants are concerned. Plant community work appeared to be typically of forest systems, and the plant taxa employed there were consistently used below expectations for plant population work. The attributes of plant communities seem not to be populations, but rather individuals, in keeping with the common mode of plant data collection. It is possible to abstract a population of trees belonging to a species in a forest, but it is very much an abstraction. Trees spend most of their time growing vegetatively side by side as individuals, while the reproductive events that bind them together as the parts of plant populations are relatively infrequent in the forest. Members of populations are bound together by connections through lineage, and sometimes by physical connections as in the case of clones. Herbs are plants stripped for reproductive action, both sexual and through tillers and rhizomes. Accordingly, herbs, not forest flora, were the plants we found that ecologists choose most often for plant population studies.

We found that animals were the opposite of trees, for all animal taxa were used below expectation for community studies. On the other hand, animals were consistently used more often than expected for population work. Animal community data are collected as samples of populations and so, unlike plant communities, animal communities have populations rather than individuals as their attributes. The constant flux of animals through a place give the animal community a texture very different from that of the plant community. The resource base of plant communities consists primarily of light, water, and mineral nutrients. That resource base is remarkably homogeneous; when you have seen one photon you have seen them all, and water is always water for all species of plant. The same is not true for the resource base of animals. They all need food, but food is a very heterogeneous commodity, very much beheld in the mouth of the eater. As a result, Richard Root, who studies insects on plant, prefers to scale down his communities to those animals which share a common resource base. All sap-sucking insects will see a plant in much the same way and will make a meaningful assemblage if pulled together into a community of sapsuckers.

The movement of animals makes them awkward objects of study. If they can be physically contained, they become much more manageable. Accordingly, resource islands have considerable appeal for animal community ecologists. Island biogeography started as an animal-based study and has remained such for the most part. Birds are particular favorites of island biogeographers. The firm boundaries of islands give the community a tangibility which is distinctly absent in animal communities set in a continuous matrix. Island biogeography is one of the most predictive of all ecological endeavors. The inputs to the system are well defined by invasion, while the removals are extinctions. Over the long run, the system arrives at an equilibrium species richness. The period of equilibration is long enough to involve astronomical numbers of individuals that succeed or fail to arrive, establish, and survive. That number is sufficiently large so that the individuality of species can be reliably subsumed inside averages. The constraints on invasion imposed by distance from the mainland are asserted with a reliability that allows prediction of species number from distance. Also, the constraints of island size emerge as reliable monitors of extinction, so allowing island size to be another predictor of species number.

Most animal communities are much less well behaved than island communities because open communities are not subject to readily identifiable constraints. The animal community is often a middle number specification. Root's suggestion of scaling down to only members who share a common resource base is therefore particularly helpful. Moving past the sample space is what allows a species to be included in a mainland community. Nevertheless, continental community members usually live in physical spaces that overlap only to a degree with those of community neighbors. Each species, in it own way, changes location depending on season and life stage. While a pond represents a well-defined faunal assemblage, members such as amphibia regularly leave the aquatic system and become part of the terrestrial community. Birds that occupy temperate habitats in summer overwinter in different parts of the entire globe. Time destroys the integrity of the sample taken at a place. In a sense, the bird community that we define in summer in northern climes can belong to different continents if we consider the full annual cycle.

Conclusion

The community is an important criterion for organizing ecological ideas. However, it is itself a complex notion, meaning different things for each major taxonomic and resource-sharing group. As the scope of a study is expanded in time, space, and heterogeneity, the nature of the community changes radically. The only universal that applies to all communities is the definition that we used at the outset. A community is a complex whose parts are organisms assigned to some taxon. The types of organism are the key to a given community. To be worth studying, the community parts must be accommodated to each other in some way, otherwise the community is only an arbitrary collection. However, different types of communities are variously integrated by spatial or temporal contiguity. Animals can come and go over time, and succession can alter the flora.

The community is something other than a place in space, for that is a landscape consideration. It is also other than a set of taxonomically undefined producers and consumers in mass balance, for those are the attributes of ecosystems. It is taxonomic identity that makes biota members of the community. Since the distinctions between taxa arise through evolution, evolutionary principles are the underpinning of the accommodations between community parts; evolution is the glue that bonds the community together.

5. THE ORGANISM CRITERION

———◆◆———

Of all the criteria, the organism is the most tangible. That tangibility gives powerful insights but is a double-edged sword. This chapter observes the blade cutting both ways. The conventional view of organisms surrenders itself to organismal tangibility, emphasizing the most readily apparent aspects of organisms. Behind this emphasis is a veiled anthropocentrism. There are good biological reasons why humans can recognize each other; members of our own species are amongst the most tangible parts of our experience. In emphasizing tangibility of organisms, we take full advantage of our facility to identify our own kind.

This chapter will follow the same plan that has been used in describing other ecological criteria. The organism needs to be considered as just one criterion amongst many for looking at ecological systems. First we will lay out the primary characteristics that make the organism distinctive. Then we will put the organism in its environment. There follows a strictly structural account of organisms as mechanical systems. Plants and animals represent two different ways of being an organism, and so there follows a section contrasting them in terms of scale. The genetical basis of organisms makes them a most distinctive type of entity. The organism concept links ecology to genetics; accordingly, this chapter ends with genetics and evolution.

Let us now flesh out that plan. At the beginning of the chapter we will employ the same process-oriented approach that was used to deal with the ecosystem criterion. In terms of the layer cake of the first chapter, we will cut across horizontally to see the extent to which an organism is just another sort of ecosystem. The importance of the system boundary arises again here. The organism can be easily seen as an integration of a closed set of processes. Unlike the ecosystem and community, the boundary of the organism is usually tangible.

Having defined the organism in terms of a collection of internal processes, we can look at the organism as a part of larger systems in which it is set. Other chapters have already put the organism in the context of communities, ecosystems, and landscapes as distinctive contexts, so this chapter does not repeat that exercise. However, there is a more general issue as to how organisms fit into their physical and biotic environments, independent of the particular conceptual device (e.g., community, ecosystem, or landscape) used to approach that environment. This chapter puts the organism in its setting from that more general point of view.

We take a process-oriented view that defines the organism as a set of parts responding to external stimuli. An interesting point of tension arises here, for the interplay between the organism and its context is an excellent way of defining the organism, quite apart from the role it plays in the larger system. In a sense, the essence of the organism is not captured when we see it as a collection of parts or internal processes, but is found when it becomes a whole that is the interface between the parts and the context. The way the organism acts upon its environment in response to stimuli is the sum of its internal functioning. Arthur Koestler coined the phrase "Janus-faced holon" to emphasize the part/whole duality raised here. We take care to avoid a naive stimulus-response of the behaviorists in psychology.

As a counterpoint to a process-oriented description of the organism, we go on to look at the organism in strictly physical structural terms. This leads very comfortably to a consideration of the organism criterion at different scales. Many researchers have looked up and down the organism column in our grand conceptual scheme, and we end with reports of their findings.

Reifying the Organism

The archetypal organism is human, and other organisms variously represent departures from ourselves, roughly in the order: cuddly and childlike; warm but big; scaly and cold; immobile; microscopic. The further away from being human is the organism in question, the less do the more formal attributes of organisms apply. These attributes are 1) a genetic integrity which reliably coincides with 2) a physical discreteness in space. Plants reproducing by vegetative means often display strong interconnections between parent and offspring, so the discreteness characteristic breaks down in plants. These problems are not restricted to plants, for the branching colonies of the most simple multicellular animals, bryozoans and coelenterates (figure 5.1a, b), present a similar ambiguous condition. Some sponges can be put through a sieve and still reaggregate to form a whole sponge again. Is one dealing with an organism throughout that process? Clearly there is cause for some equivocation no matter what the answer.

If we are correct in our surmise that the organism is an anthropomorphic construct, it should be no surprise that conscious goal-seeking is very much part of the organism concept. This is not intrinsically an unscientific view for, as Erwin Schrodinger suggested, biology without purposiveness is meaningless. For example, the notion of pathology depends upon a role and purpose which is not met in the diseased subsystem. Evolution by natural selection has amplified an explicit functional directedness in organisms. Organisms do things for a reason. However, focus on purpose and consciousness unduly segregates the organism from the other critical ecological entities. Landscapes, ecosystems, communities, and populations do not seem to manifest centers within them for active planning and preemption, but that should not preclude us from using analogy to compare the organism with other ecological entities.

There is something to be said for considering organisms in terms equivalent to those for ecosystems or communities, as just open systems that fit into a context. Certainly the active decision making that is part of being an organism is worthy of consideration, but the balanced view we have in mind will not overemphasize the central processing units to set organisms unduly apart. As much as possible, we will treat organisms much like ecological intangibles such as ecosystems or communities, as a set of interacting processes that show a certain degree of closure.

One of the costs of tangibility in organisms is the invitation to reify them. Let us use the ecosystem in contrast to the organism as an antidote to naive ontological assertions. Organisms have many of the matter-energy processing and cycling properties of ecosystems, so ecosystems make a helpful point of comparison and contrast. Organisms and ecosystems present the obverse problems, reification versus complaints of arbitrariness. Just because one can see organisms, there is no scientific reason to assert their ontology. Ecosystems may or may not be real; however, there is no reason beyond an act of faith to say that organisms are more real than ecosystems or any other ecological intangible. Both ecosystems and organisms show strong internal connections and

A

B

Figure 5.1. Primitive animals bud and branch in a manner similar to plants; the branches can break off and become unambiguous separate organisms. The branching colonies are altogether more ambiguous, straddling the line between organism and population. A. The parts and B. the whole colony of *Membranipora membranacea,* a bryozoon (photos D. Padilla).

weak external connections, which indicates that organisms and eco-systems are equally worthy objects of scientific investigation. We are tempted to reify the organism because it is tangible, and go on to assert that such and such is true because the organism is real. We take the op-posite point of view, and say that there advantages of intellectual flex-ibility to explicitly recognizing the organism as something arbitrary.

All sorts of modern findings challenge the status of the organism as an absolute. Certainly there are abundant data that indicate the eu-caryotic organism is an arbitrary assemblage, a collection of other organ-isms of the procaryotic type. Thus, even the archetypal organism, the human creature, is an aggregate of other organisms at the cellular level. The mitochondrion and chloroplast each has its own genome distinct from that of the nucleus. This puts a wrinkle in one of the key charac-teristics of organisms, the eucaryotic organism's genetic integrity. Li-chens present a similar problem, but since they are clear symbionts, one could assert that they do not count as organisms proper. However, the discovery of promiscuous DNA calls into question the genetic integrity of many living things that had been good organisms heretofore. It has even been recently suggested that land plants are better seen as reverse-phase lichens, with the algal matrix supporting a secondary fungal part-ner. The nucleus of the vascular plant cell is suggested, in this modern speculation, to be an amalgam of algal and fungal genetic material, with the fungal cellular morphology emerging in certain specialized cells, like the germinating pollen grain. The genetic integrity of the organism slips away from us, and organisms begin to look more like happen-stance collections.

To avoid confusion, we treat the organism as a heuristic. Certainly the organism is a special entity, but the manner in which it is special does not stop us from asserting that the organism is a convenience. For this argument, consider another special but arbitrary entity, the species. As in the organism, there is also a veiled anthropocentrism in the spe-cies concept. The perfect species breeds freely amongst species mem-bers but suffers significant infertility in the parents or the offspring when mating between species. Furthermore, there is significant mor-phological homogeneity inside species and consistent morphological differences between species. Zoologists, and vertebrate zoologists at that, feel comfortable identifying the species as being the only taxon which is other than arbitrary. Botanists know better; hybrid swarms, clones, and species with inconsistent chromosome numbers all make the species as arbitrary as the genus, family, or order. Nevertheless, the species is the foundation of the whole taxonomic system in that the species name is overwhelmingly the one given to most organisms in most discussions. The name of an organism is its Latin binomial.

The species is special just like the organism, but it is not special because it deserves to be reified. Rather, it represents the level of variability that humans can readily tolerate in a collection. We feel comfortable looking at a member of a species and saying that when you have seen one, you have seen them all; that is exactly the meaning of a species name. The distinctiveness of species is in ourselves, not in nature; the same goes for organisms. We have a large vested interest in organisms because we fit the criteria for the most typical of organisms ourselves. If the organism is based in nature, it resides in human nature.

Emergent Properties in Organisms

Scientists pursue something more than a reflection of their observation protocol; it is much better to look for entities with emergent properties. There is nothing mystical in what we mean by emergent properties. They are found by identifying a relatively discrete entity using one observational criterion and then changing the protocol. Anything new implied by the new observational procedure is an emergent property. Without emergent properties in the entities it studies, science would have no generality, it could not be predictive, and it would only reflect the happenstance of how observations are made. What science seeks is entities that have a useful generality stemming from the way they can be observed meaningfully in many ways. The presence of some properties can be used to predict others.

Although the organism concept is not helped by reification, organisms are robust observables with emergent properties. The coincident mapping of distinct characteristics onto a single structure makes those characteristics emergent properties. Accordingly, emergent properties of organisms are: 1) a certain physical discreteness; 2) genetic homogeneity; 3) recognizable physiological subsystems that perform various service functions, like circulating resources; 4) coordination of parts, even in the most simple organisms; 5) irritability or response to outside stimuli; and 6) reproduction with a certain genetic consistency.

The last in the list is what make individuals the favorites of evolutionists as the default level of selection. The first two characteristics map onto most organisms, particularly mammals. In many species, an individual is apparently very capable of recognizing self from not-self. This is particularly true for placentals, like ourselves, where the act of reproduction nibbles very close to overt, deleterious parasitism. The control systems of internal housekeeping have to be unambiguously separated from those of defense against invasion in placental reproduction. Even so, in the case of rhesus-negative blood groups, confusion arises.

Not all the things we call organisms have all the properties of the ideal organism; for example, some creatures lack a convincing structural separateness from their fellows. Nevertheless, things that meet more than one of the standards for being considered an organism can be helpful if not perfect examples. Not all the entities we consider organisms will be good by all standards. We are concerned with the ecological significance of the organism criterion in general terms, rather than a pursuit of the perfect definition of an organism or the perfect entity that fits such a definition.

The Critical Subsystems

To provide some redress for an overemphasis on the structural aspects of the tangible organism, let us proceed with a process-oriented approach. James Miller has suggested, in his "Living Systems Theory," nineteen subsystems that perform various analogous roles across a range of highly organized, scaled entities from the cell to the international global political system (table 5.1). In Miller's scheme, organisms seem to be the archetypal biotic system, and the nineteen analogies appear to be mostly based on recognizable things that organisms do. Its appeal is that it is a distinctly process-organized approach which identifies the critical things that need to be done. It posits subsystems that perform those tasks. The tasks are of a very general sort, pertaining to general system properties like handling undefined information or spatial relationships, rather than referring to actions directed toward named, scale-specific structures like brains or limbs. Being role-oriented rather than entity-focused, the set of subsystems avoids being locked into a particular type of system defined by physical structures. It is a suitable device for beginning to look at organisms with a wider view.

Despite the generality of the subsystems, there is a heavy anthropocentrism in the scheme. There is a significant emphasis on active, even conscious decision making in the entity. Miller's scheme does appear helpful in recognizing system equivalences in biotic aggregations up to the organism. However, his larger aggregations proceed not up to ecological entities, but up to human social systems, eventually at a global scale. At least the anthropocentrism is explicit.

When one sees Miller's larger social aggregations, above the level of the organism, the reason the subsystems take the form they do becomes clearer. His perspective on organisms and cells is one where they function as analogs of human organizations. This is not wrong, but it does

Table 5.1
Miller's Critical Subsystems

These subsystems are of a general type, although they are particularly applicable to organisms. Miller applies them to Cells, organs, organisms and four levels of societal systems up to international politics.

1) *Reproducer* creates new versions of the system to which it belongs.
2) *Boundary* contains the system and protects it from its environment, permitting only some inputs.

Matter-energy processors

3) *Ingestor* takes in material: mouth, sea port.
4) *Distributor* moves material around inside the system: blood moving sugar, road system.
5) *Converter* changes inputs of matter-energy into more useful forms: teeth, sawmill
6) *Producer* creates associations of materials used for repair and growth: leaf, industry
7) *Matter-energy storage:* fat, warehouses.
8) *Extruder* removes waste: urinary tract, sewage works and landfills
9) *Motor* moves the whole or the parts: muscles, trucks
10) *Supporter* maintains spatial relationships: skeleton, buildings or the landscape

Information processing subsystems

11) *Input transducer* senses the environment: skin nerve endings, satellite dishes.
12) *Internal transducer* transfers signals inside the system, often to a different medium: synapses, glands, telephone exchanges.
13) *Channel and net* moves information around inside the system: blood transporting hormones, nerves, telephone lines.
14) *Decoder* converts inputs into a private internal code: ganglia, foreign news analysts, newspapers
15) *Associator* performs first stage learning: short term memory, CIA
16) *Memory* performs long term learning: reflexes, habit forming, scientific research institutions, libraries.
17) *Decider* is the executive information system: brain, political administrations.
18) *Encoder* translates internal private code into a public code used in the environment: speech centers, translators, speech writers.
19) *Output transducer* sends out information: voice, policy statements, embassies, radio transmitters

explain why something as simple as irritability is such an involved process in the scheme. The analogy to social systems also explains why something as integrated as the bloodstream is split into two subsystems. One bloodstream subsystem is for delivering energy in the form of sugar, while the other role is in a separate subsystem that delivers hormones as part of information transfer. The separation of railway lines for delivering coal from the telephone system for moving information is the societal analog of the bloodstream.

Management systems appear particularly well suited to Miller's scheme, by virtue of the human component which decides management action. This is unusual given the fact that most ecological criteria are ill suited to Miller's scheme. Management entities form an ecological hierarchy of their own, independent of the academic ecosystem and community hierarchies. The management hierarchy goes as high as the problems of international treaties addressing acid rain, as well as Amazonian deforestation that sit at the door of the World Bank and its policies. It is less clear where we might find the decision-making centers in communities, ecosystems, landscapes, and biomes, assuming that decision making is a reasonable attribute to ascribe to such systems in the first place. The organism analogy can be pressed beyond advantage in ecological systems at large.

We are a little cautious of buying into Miller's scheme whole hog for fear of leaving the impression that our reader is required to believe in the necessity of all nineteen analogies (actually, yet two others have been proposed, the timer and the accountant). It is clear that the number and type of generic biotic subsystems will depend on the perspective which is taken for the purpose at hand. In the following paragraphs we give examples of how the nineteen subsystems relate to more conventional accounts of organisms.

There is not always a one-to-one mapping of the subsystems to many generally recognized properties of organisms. Some of the subsystems do, however, give to the whole one of the properties of organisms cited in introductory texts. The "reproducer" is one such subsystem; Miller says it is "the subsystem which is capable of giving rise to other systems similar to the one it is in." Also, movement is achieved with a one-to-one mapping onto the "motor" which corresponds to the musculature. Structural integrity comes from the "supporter," the skeleton in vertebrates.

Other properties conventionally ascribed to life, like the necessity of external energy resources for living systems to survive, are manifested in a series of Miller's subsystems. This set takes in matter-energy, prepares it for use, uses it in metabolism, and finally expels waste material. The names of these subsystems speak for themselves: "ingestor," "converter," "distributor," "producer," and "extruder."

Similarly, the responsiveness of organisms to environmental stimuli, the property of irritability, corresponds to the aggregate action of several of Miller's subsystems. Reception of the signal from outside is not just a matter of simple detection of the primary force, but also includes moving the signal to appropriate input centers, and translations to make the input meaningful. There are centers for coordinating and executing response to the input, as well as the devices for exhibiting actual behavior measurable from outside. Furthermore, there are different echelons within the various subsystems, each performing the subsystem's functions but at its own level of sophistication. The lower echelons do whatever it is that the subsystem does in a rudimentary fashion, while other echelons are highly integrative and intricate in their operation.

As an example of how complex Miller's system can become, consider irritability. Irritability involves first the organism internalizing a stimulus through the "input transducer." This subsystem converts the stimulus into a matter-energy flow suitable for transmission within the organism; you stub your toe and the nerve endings detect pressure. Then the "channel and net," involving nerves or vessels, transmit signal around the organism. The "internal transducer" converts information already inside the system into suitable matter-energy forms for further transmission within the organism (e.g., chemical transmitters at synapses). The "decoder" changes the code of the input signal to an internal private code meaningful only inside the organism. The pain decoder for the stubbed toe involves the central nervous system, although it is not known where that function is located, nor even the level in the neurological hierarchy that coding for pain occurs. All these subsystems are involved in perception of the insult. Once it is perceived, the signal then leaves a mark in the system. Closely related aspects of brain function are divided into first- and second-stage learning subsystems in Miller's scheme, to wit the "associator" and the "memory."

The response of the organism involves a different collection of subsystems. The "decider" is the executive subsystem that coordinates control of the entire system. In stubbing your toe, the decider occupies at least two sites, at two echelons of decision making; one is the reflex arc in the spine, while the others are found in the centers of the brain involved in cognition. The reflex arc decides if the signal is strong enough to warrant an output to the muscles which pull the foot back on reflex. The cognitive center orchestrates more complex responses, like the decision to pick up the thing that hurt your foot and vent your anger. Even after the decider has deemed the appropriate response, the organism still does not manifest irritability. The "encoder" takes the private code of the various information-processing subsystems and translates it into a public code which can be read by other systems outside; the speech centers organize the expletive oath. The "output transducer" then transmits in the public code. At this point the organism gives the response which is an irritation in the face of the original stimulus, the stubbing of the toe; the mouth opens and the vocal cords go "Ouch!"

In organisms the "boundary" fits easily and comfortably, giving one of the critical characteristics of organisms, namely physical discreteness. The skin is the boundary in vertebrates. Medicine recognizes it as an organ rather than a place. This would meet with Miller's approval, it being a process-oriented appellation.

In the conventional ecological hierarchy, the "boundary" is probably the most troublesome of the subsystems. The early community ecologists recognized integrity in the community, and used an organism metaphor to express the fact. Clements' knowledge of the critical process that held the system together led him to refer to the vegetational species association as a "superorganism." That analogy was overstated and unfortunately came to be used more as a homology. Eventually, critics found that all the obvious characteristics of organisms were not obvious in Clements' community conception. One of the foci of the attack was on the absence of a discrete boundary to the system. As we have seen, the physical boundary at the edge of a woodlot is not a community boundary; it is rather a landscape boundary, like a fence. The boundary of the community itself is not only intangible, but it is difficult to see even after formal analysis of the vegetation. Vegetational composition found in ordination is a continuum, with obvious boundaries in composition being the exception and not the rule.

Organisms as Perceiving and Responding Wholes

For all its detailed process- and information-based emphasis, Miller's scheme only tells a part of the story of organisms. It is distinctly part-oriented and barely addresses the way the organism relates to its environment as a cohesive whole. A much older work was originally published in German by Jacob von Uexkull and contained the term *"Umwelt,"* a word not easily translated into English. Von Uexkull's translator used the word "self-world." He was trying to convey the notion of the private world in which each type of organism lives, a world determined by what the organism can detect and how it responds to stimulus. This is a very different conception from that of Miller, for it focuses on the organism, but still includes the environment as part of the system. The functional environment is defined by the organism itself. Anything that the organism does not perceive is not part of the environment in the *Umwelt*. We will use "self-world" and *"Umwelt"* interchangeably.

Von Uexkull makes an interesting contrast between nonliving physical objects and organisms. He points out that living material only responds in a particular way to even a large range of stimuli. For example, muscle responds with contraction to a wide range of causal agents. By contrast, a nonliving entity like a bell clapper only responds in its functional role if it is swung back and forth in a particular manner. To all other stimuli like acid, alkali, or electricity, it responds like any other piece of metal. The response of the organism is canalized. If it responds, then it always makes its own specific response. All external agents to which it does not respond may as well not exist.

In von Uexkull's scheme, all the complicated input pathways and special subsystems of Miller disappear. In fact, von Uexkull's organism becomes an input/output device on a par with the entities in the environment to which it responds. While one is cautious to embrace anything resembling the behaviorists' rude stimulus-response models, there is a certain appeal to von Uexkull's perspective, as a rescaling of the organism so that it functions as a unified whole. The example he gives, the wood tick, is enlightening (figure 5.2). It indicates a much greater sophistication in scaling than psychological stimulus-response models, for which it might be mistaken.

The female wood tick is modeled as being part of a perceptual arc and

Figure 5.2. Wood ticks (photo U.S. Forest Service).

part of a motor arc. Joining these two arcs on the opposite side, and so making a circle of response, is the object in the environment upon which the tick operates (figure 5.3). This is a cleverly simple scheme, for a track once around this circle is the phenomenon of irritability, the same phenomenon that required at least seven of Miller's subsystems. The tick hangs in a tree just above the height of its prey. The first stimulus is butyric acid emanating from the skin glands of the mammal. The tick responds by dropping from the tree. The second stimulus is the shock of hitting the hairs of the mammal. This extinguishes the olfactory cue and stimulates a response of running around. If the tick lands on something cold, she has missed her prey and must climb back up the tree. If she runs around and finds a warm membrane, the heat stimulus causes her to start her boring response. The tick has no sense of taste and will puncture any soft membrane and consume any fluid of the right temperature.

What fills von Uexkull with wonderment is the fact that of the many signals that could be taken to indicate the presence of a mammal prey item, only three are read, and then in a particular sequence. Only three signals stand out like beacons to guide her to her prey. It is the very paucity of signals in the tick's "self-world" that allows her to make the unerring response. True, there is some luck involved in finding the prey, but another critical factor comes to the tick's aid, time. Von Uexkull reports that ticks have been starved for eighteen years and still survive.

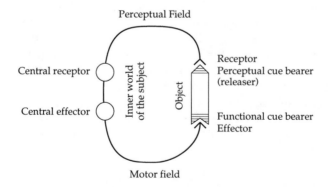

Functional Cycle

Figure 5.3. The organism and the environment upon which it operates form a cyclical loop of physical action and input to the animal from the world acted upon (after von Uexkull 1957).

The metabolism of the tick has been scaled almost beyond comprehension, so that its "self-world" works.

In a charming walk through the worlds of animals from insects to humans, von Uexkull points out repeatedly the striking differences in the perceptions important to the organism. He shows how visual cues readily noted by us, such as a fertile female insect under a soundproof cover, pass unheeded by males who one might have thought would have an interest. Instead they flock around a loudspeaker that is carrying the appropriate sexual signal in their *Umwelt*. He reports mother hens pecking at distressed chicks instead of coming to their aid, only because the wrong cue was given. He shows how slow organisms ignore fast cues.

The important point in all this is that organisms are deeply context-dependent. Von Uexkull's organisms extend beyond their tangible bodies to become one with their context. By looking at organisms this way, we see beyond the physical, tangible entity and observe a new entity, the scaled "self-world" that is often only fleetingly detectable by us. We cannot see the organism's *Umwelt*, although it is probably more important for our understanding than either the tangible creature itself or a full accounting of Miller's subsystems inside it. Von Uexkull presses upon us a rescaling of organisms that gives a glimpse of the relationship of the organism criterion to other ecological criteria.

Ecosystem, community, population, and landscape each require a different reading of the organism's "self-world." The contributions of organisms through their "self-worlds" to communities cannot easily be generalized. As contributors to ecosystems, animals work through the resource objects in their world. They pick up cues about food, be it a sac of blood in the coating of fur for a parasite, a hunk of protein on legs for a carnivore, or green fluids in cellulose cases for a herbivore. They go through a series of cycles where they handle the resource through clinging and burrowing for the parasite, stalking or chasing a carnivore, or searching for the herbivore. There are two critical actions. The first involves the motor action of ingestion, while the second is defecation. The food item thus changes ecosystem compartments in its passage through the animal's *Umwelt*.

In general, small organisms have a small, spatially defined "self-world," although a bird views a wider landscape than does a large mammal. It is the size of the *Umwelt* of the organism, relative to the size of our own, that determines which organisms are used for which type of ecological investigation. That is why birds are used for island biogeographic studies. Their "self-world" is larger than our own, so we use them disproportionately often for studying the ecology of going to and from places. The organism's occupancy of the landscape is determined by the scale and texture of its *Umwelt*.

The difference between communities and populations corresponds to an accommodation between "self-worlds" that are different in communities but the same in populations. Inside populations, organisms use a perceptual-motor handling of objects belonging to its own species. There is a resonance with a genetic basis between *Umwelts* of members of populations which does not apply between community members. One might expect similar *Umwelts* to be able to read and respond to each other over a larger spatial scale than *Umwelts* which are different. Accordingly, one might expect populations to have a larger spatial scale than communities. Note how competition, a principal process inside communities, is a very local affair. A plant competes with others over only a very short distance. On the other hand, pollen that is a minister of exchange between *Umwelts* within a species can come from great distances. Also, insects may have a small food resource "self-world" in which they compete for food, but can manipulate each other with pheromones over several miles. Populations are held together by processes that operate across a wider spatial scale. In that sense, we would expect communities to be generally smaller scaled than the populations of the species whose fragments are represented in the community. The conventional ordering of communities above populations appears to break down.

On the other hand, communities may well be scaled larger than populations in a temporal sense. A signal between different "self-worlds" may take longer to be exchanged than between similar "self-worlds." Accordingly, interactions inside communities are generally slower than inside populations. We might expect evolution to take longer to build community structure than for it to regulate and modify populations. This would put communities above populations in the ecological hierarchy in accordance with the orthodox view.

The above implies a weaker bond strength within communities as opposed to populations, again corresponding to the conventional wisdom that puts populations inside communities. Rather than argue that one is above the other, populations above communities in the spatial hierarchy, or communities above populations in the temporal hierarchy, we prefer to stay with our original position that they are different criteria rather than alternative hierarchical levels. Accordingly, there is no need for them to be ordered one above the other in a hierarchy. The difference can be translated to the different *Umwelts* within as opposed to between species. The interference patterns, to which we alluded in our consideration of communities, is an interference pattern between differently scaled "self-worlds." The notion of *Umwelt* and "self-world" are the means of putting organisms into the larger ecological framework, linking them to other criteria.

Basic Structural Units in Organisms

We have worked through a process-oriented approach to organisms, both internally with Miller and externally with von Uexkull. There is, however, another side to organisms concerning their physical structure. Everday experience influences the choices scientists make in constructing their models. Our models for physical form in biology are also influenced by commonplace structures in modern society: levers and girders. There is something to be said for biologists turning to architects for their insights into biological constructions. Earlier in this text we turned to Peter Stevens, an architect, when we were seeking a unification of landscape pattern. The structural model employed here goes beyond the familiar and commonsensical girder and lever models. We turn instead to counterintuitive, tension-compression structures of architects like Buckminster Fuller. This class of structures has stimulated Stephen Levin, an orthopedic surgeon, to see organisms in a different light. We present Levin's model below.

Feeling uneasy about the conventional model of his colleagues for neck and back problems, Dr. Levin went to the Natural History Museum in Washington, D.C. to see if looking at dinosaur skeletons could help. He left the exhibit troubled by the unlikeliness of a lever-girder model for a dinosaur's neck and pensively strolled across the mall. Then, minutes after looking in puzzlement at the neck of a great extinct monster, he was in front of the Hirshhorn Museum, where he encountered the breathtaking tension-compression sculpture of Kenneth Snelson. Like Newton's apple, the statue changed Levin's view so he would never see organisms and skeletons the same again. The statue, *The Needle Tower,* is just twenty by seventeen feet at its base, but is as high as a four- or five-story building. It is essentially rigid and consists of cables and light girders. The remarkable feature of the design is that the girders do not touch each other, being held in a helical tower by a web of cables. The statue is neither suspended from anything rigid above, nor is it a stack of compression members sitting on top of each other.

From this experience Dr. Levin saw the underlying model for not only backs and necks, but other biological structures including knees, cell packing, and bacteriophage viruses. Most of us are awestruck by Olympic gymnasts, but we are not as impressed as we should be. With a girder-lever construction to the body, such activities are impossible.

Levin's model also accounts for unlikely behaviors like gymnastics, because his models are gravity-independent.

Levin suggests the icosahedron as the basic unit of construction in organisms. When he says basic, he means the way that small icosahedra pack to give a larger icosahedron ad infinitum. This nesting allows the form to be basic at many scales and levels of construction. An icosahedron is a regular solid with twenty triangular faces. Triangles are important because they are the only two-dimensional forms where pressure will not deform them, so long as the sides remain connected, straight, and the same length; a rectangle, for example, deforms to a parallelogram (figure 5.4a, b).

The icosahedron has thirty edges and twelve corners (figure 5.5a–c). If the edges are rigid, then pressure at any point transmits around the thirty rigid surface members, placing some under pressure and others under tension, in a regular fashion. The twelve corners each have three edges that come together at a point, which is the corner. Some of these edges are under compression while others are under tension, depending on the position of the corner relative to the loading of the whole icosahedron. It is possible to transfer all compression away from the outside of the structure.

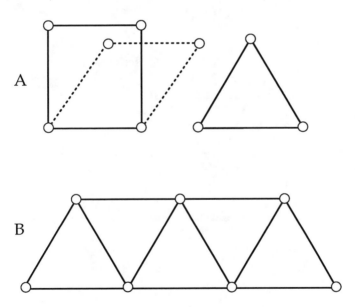

Figure 5.4. A. Triangles are the only two-dimensional structures which do not deform under pressure. Rectangles, for example, deform to parallelograms. B. Plane trusses may be readily made from triangles.

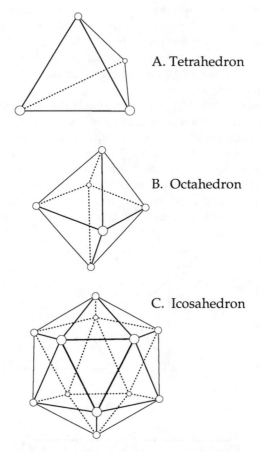

A. Tetrahedron

B. Octahedron

C. Icosahedron

Figure 5.5. A.–C. The relationship between the icosahedron and the smaller triangular faced structures (after S. Levin 1986).

The trick is to connect each of the twelve corners to its opposite number with one of six new compression members spanning the middle of the icosahedron. Opposite corners push each other away from the center in opposite directions, working through the new rods in the middle. The "opposite" corners are not strictly opposite, so the connecting rods are slightly eccentric and slip tangentially past each other without touching. These compression members push the corners away from each other. In this way, all edges joining the corners become tension members, which does not sound astonishing in itself. What does defy intuition is the fact that with the edges universally under tension, all thirty can be replaced with cord or cable. Together these tension edges hold apart the six compression members spanning the middle. The tension icosahedron, as it is now called, is a rigid structure but none of the compression units touch; rather they float, suspended by the skin of edges, all under tension. Pressure applied to this structure causes an increase in tension around all the edges almost uniformly, and this distributes the compression load around all the six suspended compression members evenly (figure 5.6).

The tension icosahedron has counterintuitive deformation characteristics. Remember that pressure applies tension all around the surface. As one presses down on the form against a surface, instead of flattening like a pancake, it grows smaller and becomes more compact. As one tries to pull the tension icosahedron apart from opposite sides, instead of stretching into a sausage, the whole thing just becomes bigger; that is to say, not only does it become longer between the poles under tension, but it also becomes proportionally fatter around the girth and so keeps its shape.

The relationship between stress (say, pressure) and strain (deformation) in the tension icosahedron can be graphed as a curve. That curve is radically different from the same relationship in Hookean girder-lever structures (constructions where compression members press against each other, as in office blocks or toy model building kits). Hookean material deforms at a constant rate with increases in stress until there is a short period of slightly faster deformation, immediately preceding system failure (figure 5.7a–d). In contrast, the tension icosahedron resists deformation with the first application of stress, but soon deforms considerably. This rapid deformation is not, however, an indication of incipient system failure, for after a short period of rapid deformation with increasing stress, deformation essentially stops. A lot more stress causes very little further strain. The tension icosahedron is very tough at high stress because it distributes the load around the whole system evenly. Rubber and animal tissue have stress-to-strain relationships like

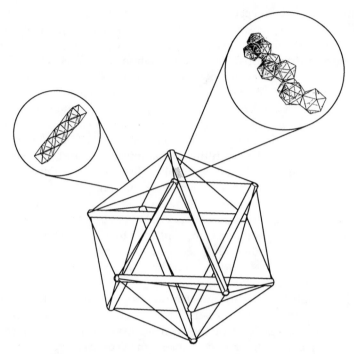

Figure 5.6. Tension icosahedron. A continuous tension "skeleton" with the compression elements suspended within the tension network. The compression elements do not touch one another—discontinuous compression. Lower level stacks of icosahedra comprise the compression and tension members (after S. Levin 1986).

that of an icosahedron, not the linear stress-to-strain relationship of a Hookean girder system.

The tension icosahedron can form hierarchically nested structures. First, consider a three-dimensional increase in size. An icosahedron is made from packing spheres round an empty space, and connecting their centers with straight lines (figure 5.8). The icosahedron approximates a sphere, and so icosahedra formed from twelve subunits can themselves be packed in dozens to form yet a higher order icosahedra ad infinitum. Plant parenchyma cells pack together in this manner to make tissues for larger units like organs.

Now consider a one-dimensional aggregation to form a rod. Beyond packing to make higher and higher order solids, it is possible for icosahedra to sit on top of each other to form a column with a spiral twist to

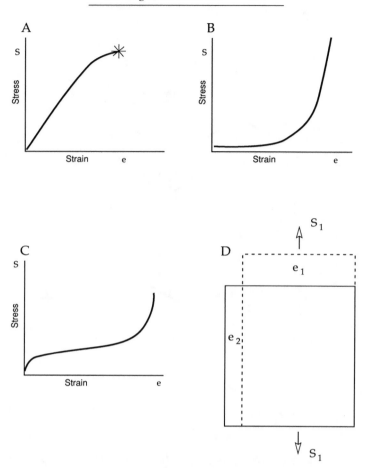

Figure 5.7. A. Typical stress-strain curve of a girder/lever Hookean material with compression members touching (after S. Levin 1986). B. Stress-strain for typical animal tissue and the icosahedron (after S. Levin 1986). C. Stress-strain curve for typical rubber (after S. Levin 1986). D. When a solid is stretched by a tensile stress s_1, it extends in a direction of s_1 by primary strain e_1, but also contracts laterally by a secondary strain e_2 (after S. Levin 1986).

it (figure 5.9). An icosahedron connected into the column fits by triangular faces that are not quite opposite. Think of the separate icosahedra as holding these two triangles apart by tension rather than compression. A column of icosahedra is a column of these triangular interfaces held apart by tension. Therefore, the column is not a conventional rod, but

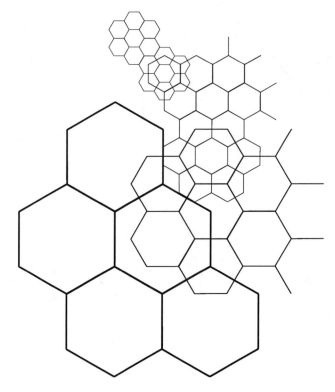

Figure 5.8. Icosahedra can pack to ever higher levels (after S. Levin 1986).

a tension-compression system that transmits stress throughout the "beam" evenly through the tension web.

The remarkable effect of building higher-order structures from tension units is to make the strain within the aggregate units independent of gravity. Stand icosahedron tower on one end and it suffers the same stresses and shows the same strain as if one stood the tower upside down. Snelson's *Needle Tower* could stand just as well on its top. This is because the pressure on the base, whichever end it maybe, is transmitted throughout the whole tower. That explains how humans can stand on their hands with no ill effect. It also explains how we can stand on our feet, which is by no means a trivial exercise. The living bone units in the foot apparently bearing the load are far too small and soft to do the job were the skeleton of Hookean girder construction.

It seems that joints in vertebrate skeletons are not simple levers, de-

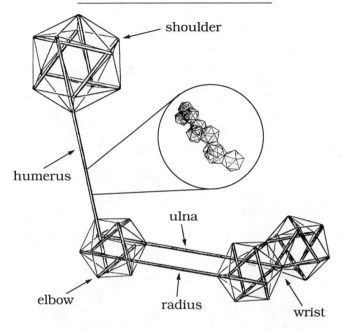

Figure 5.9. Icosahedra can be fitted into spiral twisting columns that become compression tension beams. Strain on any part is transferred evenly to all parts and so it is not limited by the normal constraints of beam construction. Two levels of icosahedra are used to model the human arm (courtesy S. Levin).

spite the conventional wisdom. It is well known among specialists in the knee joint that it functions like those children's toys where a string of tiles is held together by three strands. The tiles apparently form a single ribbon except that, counterintuitively, simply folding the ribbon allows tiles at the fold to flip over and make a branch off the main axis. The analogy is valid, and the knee does flip back and forth in that fashion (figure 5.10). The problem for the orthodox model of the joint as a simple lever with compression across the knee joint is that it is at odds with the other conventional wisdom of the knee as a flip joint. For the knee to work as a flip joint, which it does, it must be under tension, not compression. The knee and the other major joints are tension icosahedra, not simple articulation points between levers. The membranes around the fluid in the knee are not strong enough to act like a cushion air bag in an automobile crash. The sinovial fluid cannot be compressed to give support between the shin and the thigh, and yet there is always space between the two bones, even when the knee is loaded. With an ico-

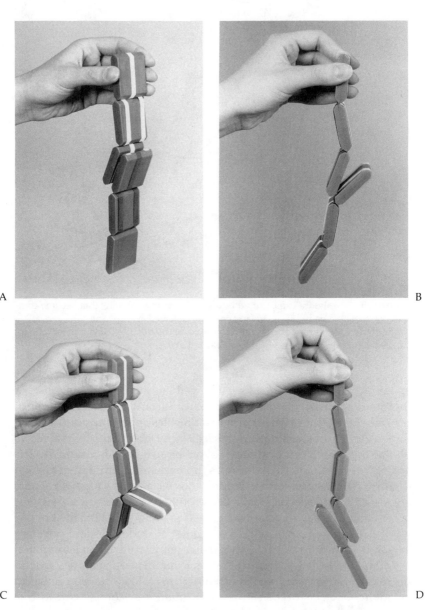

Figure 5.10. The children's toy of flipping tiles operates under the same principles of tension as does the knee joint. The sequence A.–D. shows the stages of the count-erintuitive flipping of the tiles. E. shows a tile held under tension ready to flip either way (photos C. Lipke).

E

Figure 5.10. (*Continued*)

sahedral model, the load is taken by the ligaments. In knee surgery, Dr. Levin would tighten the ligaments, not to pull the joint together, but to pull the bone compression members apart.

Movement across joints is achieved by manipulating some of the tension members so that the icosahedron column changes shape. Pathology comes from damage to tension members such that the compression members are not held apart in the correct fashion. Release all tension from a tension member and the icosahedron will collapse. When orthopedic surgeons have to fuse an ankle damaged beyond repair, they have an unreasonable amount of difficulty getting the bones to knit. The ankle bones are soft and removal of the cartilage should occasion rapid fusion, but it does not. Often fusion requires pins to hold the bones together so that a bond can be formed. If the ankle joint is icosahedral, then the bones being pulled apart by the ligaments would explain the problem. Dr. Levin found that cutting the ligaments, which are superfluous after fusion, would collapse the system and cause the desired freezing of the joint.

The neck of a dinosaur or a giraffe makes no sense as a multiple-lever system, but as a stack of icosahedra it becomes reasonable. Although the human back is a less tenuous structure than the neck of an ostrich, it too requires an icosahedral tower to explain its loading capacity. Back pathology is a matter of weakened tension members in an icosahedral stack. The soft discs between the vertebrae are not compression members, but are shock absorbers for when the icosahedra are exhibiting high frequency pulses of strain. Weakened tension members slacken the stack, putting the disc under chronic low-grade compression, for which it was not designed. Often discs rupture when the victim is doing nothing that might lead one to expect strain. The reason is that the primary cause is unfelt chronic compression. Dr. Levin now knows that the rationale behind even successful surgical procedures is completely wrong, and finds himself having to think through a new tension-compression model for each case. This distracts him from the parts of the surgery requiring focused manual dexterity, and so he has given up surgery, now treating backs only with less invasive means. Some of these therapies involve stretching exercises which are focused with precision on the particular tension member responsible for the pathology. The sort of generic stretching prescribed by conventional practice is much more hit or more, because it has no rational predictive model.

The structural model for organisms presented here should be seen as the three-dimensional equivalent of the explosion, meander, and branching patterns we used to explore landscapes. Certainly icosahedra occur widely across many scales, giving the organism criteria a robust unity over a range of grains and extents. Icosahedra seem to apply all the way down to the scale of bacteriophage viruses (figure 5.11) which are clearly icosahedral, and all the way up to the packing material that makes up giant redwood trees. The scale independence of the icosahedron model gives it enormous appeal for our purposes, as we move up- and downscale in the context of the organism criterion.

Figure 5.11. A bacteriophage virus with a characteristic icosahedron form.

The Form of Plants and Animals

There is much more to the form that organisms take than just the abstract generalized unit of construction, the icosahedron. Levin's model gives a wonderful unity to organism construction, but it does not address the splendid diversity of forms that makes biology at the organismal level such a carnival. Let us turn to some of the principal patterns of difference across the organism criterion.

One of the recurrent ideas in this book has been a contrasting of plants and animals. We have done this not to maintain traditional academic demarcation lines between botanists and zoologists, but to allow a proper unified treatment that recognizes the concrete differences in scaling when they occur. Sessile animals do appear plant-like because the environmental experience is similar. Plants possess radial symmetry even when they have a top different from the bottom; even with roots at one end and stems at the other, a view from above a tree yields a circular structure. This is in contrast to advanced mobile animals which are bilaterally symmetrical. However, to say that animals are bilaterally symmetrical is to look at the problem the wrong way around. This is because the concept of symmetry recognizes the similarity of the sides, rather than the differences between both the front and back and the top and bottom. Those differences are a much more insightful matter. Bilateral symmetry is more interestingly a bipolar departure from radial symmetry.

E. J. H. Corner has made a pretty argument that explains the patterns of symmetry we see. He says that radial symmetry is the default condition for organisms. In radially symmetrical organisms the same developmental model can be used for development in every direction. This is the form commonly found in phytoplankton that float in the brightly illuminated surface layer of the ocean. They are so small that the photoic zone is spatially enormous to them and functionally three-dimensional, even though it is only the skin of the ocean. Since the cells rotate as they float, their environment is not polar; light only incidentally comes from above. With no polarity in their environment, there is no universal stimulus for plankton to be anything other than radially symmetrical.

The center of the ocean is a nutrient desert, so productivity over most of the ocean is painfully low. However, close to the land the waters are not only nutrient-rich, they are saturated with gases. Since close to shore there is a solid, brightly lit bottom, there is no danger of sinking out of the photoic zone, and every reason for cells to settle on the bottom. Should that happen, the plants find themselves in a polar environment where light comes from above and the substrate is below. Bottom-colonizing plants now live in a two-dimensional environment and so space becomes a limiting factor. Growth up into the third dimension is the only solution to crowding, and so the organisms on a surface become polar, although simply so at first.

Small organisms interfere with their environment very little. When they grow bigger, their presence modifies the environment. As they grow bigger two things happen to make their experience of the environment more polar. First, as they grow further from the plane of the substrate, the top of the plant finds itself in an environment increasingly different from that close to the substrate at the base of the plant, simply by virtue of being increasingly distant from it. Second, the large upper part of the plant modifies the environment on the plane by shading it. With this increased polar experience, the plants have every reason to differentiate between top and bottom. Increasing size also changes the organisms' surface-to-volume ratio. The way to correct it back again, to one where there is enough surface to support the increasing mass of the plant, is to change shape by becoming flat. Space being in short supply, the large plant faces three problems which it solves by morphological differentiation. The leaf or frond is the flat structure that increases photosynthetic surface. The stem or stipe lifts the photosynthesizer into the canopy. The root or holdfast keeps the plant in possession of its site.

There are remarkable parallels between algae and land plants across unrelated groups for reasons of the physical structure of the environment. Plants, particularly big ones, are polar because their environment is polar. Remember, however, that a tree viewed from above is as radially symmetrical as a plankton cell. That is because there is no directional difference across the plane in environmental quality. Animals also live on the plane and so they too have a top and bottom. The critical difference between plants living on a surface and animals is the strongly directional movement in pelagic animals. Because the front end arrives in the new place first, there is a temporal polarity to the experience of the moving animal. The front is the best place to put sense organs, and so into that position they have evolved. The bipolar environment dictates the bipolar animal.

Size produces another architectural difference between plants and animals. This time the argument turns not on motility but on energy sources. Because animals depend on a high-energy food source, their cells have to remain soft and naked so that they can take particles into the body. Plants, on the other hand, only require small molecules, and so they can surround themselves with dead exudate that is the fibrous cell wall. Much of the mass of tissue of the land plant is dead, the living part being encased in a dead cellulose fortress (figure 5.12).

As an organism grows, its mass increases by a cubed function of its

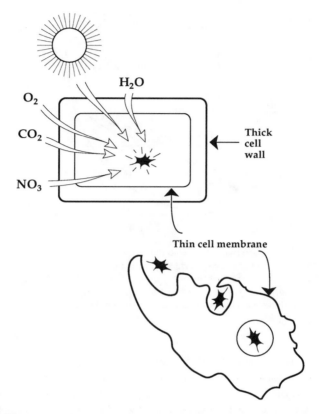

Figure 5.12. Because the plant cell uses only small molecules, it can exist inside a tough supporting cell wall. Animal cells, however, need to be capable of ingesting large high energy food particles.

length, while its surface only increased by a squared function. The larger the surface, the more mass can be spread to give a relatively low pressure on the surface. Therefore, as the surface-to-volume ratio becomes smaller because of increased size, the organism runs into structural problems. Being made of tough woody material, the plant just keeps growing. The animal, being made of soft flexible material, quickly presses against structural limits (figure 5.13).

If the animal toughens up its outer surface to keep from falling apart, it will be unable to absorb food particles. The outside of an animal has two jobs which interfere with each other. First, the skin keeps the animal's insides within. Second, the outer surface must be an interface for food and waste products. The solution to the dilemma is to divide the

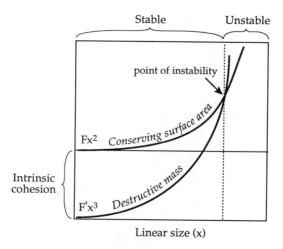

Figure 5.13. There are two facets to stability of a structure: intrinsic cohesion which is constant for a given material; surface area that increases as a squared function of length. The destructive force is weight coming from mass which increases as a cubed function of length. Despite the headstart of intrinsic cohesion, eventually the cubed function of weight overtakes the squared function of surface and the structure falls apart.

outer surface into two parts, one structural and the other nutritional. It is a radical resolution; the animal turns itself inside out, or rather outside in. Early in development of large multicellular animals, the process of gastrulation tucks part of the outer surface of the embryo inside to make the gut (figure 5.14). As a consequence, the animal has an inward-looking circulation system. Being a bag, the animal fills up and so manifests determinate growth as a rule. Plants have an open-ended circulation system and indeterminate growth because they are not forced into the animal's inward-looking morphology and physiology.

The Worlds of Small and Large Organisms

The question of size of organisms emerges again, but this time not so as to separate plant and animal experience. It pertains to the difference in the critical physical forces that press upon big as opposed to small organisms in general. Water and gravity have different meanings depending on the size of the creature involved. First, let us consider gravity.

The bigger the organism, the more marginal its existence in the face of gravity. Mastodons used to break limbs regularly, while beached

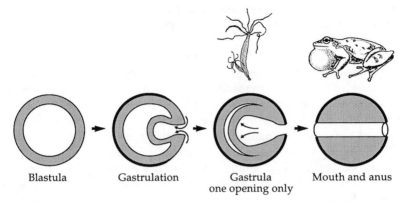

| Blastula | Gastrulation | Gastrula one opening only | Mouth and anus |

Figure 5.14. In the process of gastrulation the hollow ball of cells tucks itself to form a cup. The cup becomes the gut in the mature animal.

whales suffocate under their own body weight. By contrast, small creatures fly in air without wings. As J. B. S. Haldane observed, in a fall to the bottom of a mine shaft an ant flies, a mouse is stunned, a man is broken, and horse would splash. Even under a static load, larger organisms need to be formed in a distinctive way that accommodates the effect of gravity on their bulk. Both elephants and large plants are supported by very solid columns. Sequoia trees and hippopotami both look stocky because both have heavy loads to bear. Should the break occur, then it involves a separation across a cross-sectional area. With a linear increase in height, the thickness of the support structures must increase on a squared function if the strength under static loading is to remain constant.

While trees are stockier if they are large, actually they do not increase the base of the trunk on a squared function. A big tree does not have the same proportions as a sapling, but it also is not thicker at its base by as much as a squared proportion to its height. In terms of static loading, bigger trees are not as strong as small trees. They increase in thickness not on a squared function, but with an exponent rather less than 2.0. The exponent for thickening is only 1.5, indicating that smaller trees are overbuilt for the stresses that a structure of their size would experience. This overbuilding of trees emerges as one of the strategies for staying upright. The overbuilt bulk of a tree sits as ballast at the bottom of the tree; it remains upright on the same principle as lead-loaded chess pieces or a plastic, round-bottomed clown doll that pops back up again when a child hits it. The exponent of 1.5 keeps constant not strength under static loading, but rather a flexibility factor as trees increase in size.

This constancy of flexibility indicates two problems for trees that are more important than strength under static loading. The first is concerned with form, while the second a matter of dynamic loading. First, if a tree keeps the same flexibility as it increases in size, then the limbs will bend to the same degree no matter what the size. This means that the form of the canopy remains constant as the tree grows bigger. Apparently, keeping a constant outline to the canopy is important.

One explanation for the singular constancy in canopy form might be the modular construction of leaves. That modularity requires a conservative plan for growing a particular shape of leaf, as well as one for placing leaves to account for self-shading effects. Neither blueprint can be continuously reworked for all sizes of the tree. Only exceptionally do plants change leaf form with age, and then they are usually climbers like ivy. The life form is climbers focused on getting to the light. Of course trees struggle for light too, but there are more facets to their life histories, more factors that matter. Thus fixated on light as the one issue, climbers appear prepared to indulge in the extreme adaptation of change in leaf form. So are roset herbs where a few inches make all the difference. In general it is easier for the plant to keep the context of the leaves constant by holding canopy architecture constant, something which can only be achieved in a growing system normalized on flexibility rather than strength under static loading.

The dynamic loading consequences of flexibility pertain to air movement. The sheer weight of a tree is not what brings it down; rather it is the wind. Clearly, absorbing some of the force of a gust of wind in bending is helpful, and constant flexibility simplifies these considerations. Also the ability to deform means that the tree can change its drag coefficient at the critical time when the effects of the high exponents for drag itself appear likely to have disastrous effects. Drag changes with wind speed nonlinearly. That is, a few more miles an hour of wind at high velocities has much more effect than the same increase in wind speed from calm conditions.

There is a dimensionless number that corrects for increasing effects of wind. That dimensionless number normalizes around the exponents for drag itself. This is constant for a given shape. The only thing that can be done to circumvent the worst effects of very high winds is to change shape, that is, change the dimensionless drag coefficient. At low wind speeds trees flutter their leaves, and this actually increases the drag coefficient. At low winds the change in shape actually makes things worse. However, at higher wind speeds, the drag coefficient plummets as the plant streamlines and reduces the relative impact of the wind. With increase in wind speed, the drag itself will increase, but not as much as it would have increased had the tree kept its original shape. The change in form that drops the drag coefficient alleviates the danger, but only at its worst. By normalizing on flexibility rather than on static load, the effect of the form of the tree in high wind is predictable. Predictability invokes an upper-level constraint that must pertain over the period of the prediction. Only factors that remain constant for a long time can be coherently selected in evolution. Thus, tree flexibility engenders a predictable framework in which the various strategies of different tree forms can be selected.

The fluid dynamics of air and organism size is only one place where fluid dynamics comes to bear on scale problems of organisms. The Reynolds number takes into account the size and speed of the object and the viscosity of the fluid. All those different units are canceled out to give a dimensionless number, the Reynolds number, that predicts the turbulence of the system. For very small organisms, water is functionally very viscous. That is to say, the Reynolds number is small. Accordingly, for small organisms, swimming actions like a breast stroke kick do not work. Kicking fast on the stroke does no good because the return stroke pulls the organism back to where it began, even if the return stroke is much slower than the original kick. Animals of our size sink struggling in quicksands or mud because wet sand and mud are viscous. Microscopic animals in water are like us in quicksand. Accordingly, they swim by other means than us or fish; they often screw themselves through the water like a propeller. The world of small organisms is not just a small piece of our own, it is different.

Let us turn to other physical properties of water as it affects organisms of different sizes. One of the most distinctive things about our world is the ubiquitous presence of water. Water is a very peculiar substance. Because of its particular molecular and atomic characteristics, water allows visible light to pass. That is why visible light, as opposed to any other wavelength, is important to biological systems for photosynthesis and sensory perception. The density of the commonest form of solid water is less than liquid water. Therefore, ice floats and saves water bodies from freezing solid, thus enabling aquatic species to survive cold winters. For the most part, water is macromolecular, with bonds breaking and forming readily. That is why water, despite the small size of the single water molecule, is liquid at temperatures that prevail on most of the surface of the earth. The critical property of water for the present discussion also follows from its macromolecular form that imparts its spectacularly high surface tension.

If water had a more normal surface tension, then the ocean would not have significant waves (cf. pouring oil on the water) and erosion would be minimal. Furthermore, raindrops would not form, thereby breaking the water cycle. The ecosystems of the earth and its landscapes would be radically different. The critical danger for an ant is from the effects of surface tension, a force hardly detectable by large organisms. Although an ant can lift many times its own weight, it is almost completely incapable of escaping from a drop of water, and would almost die once it was submerged in any body of water as large as a half-full petri dish. A needle of solid steel can be readily floated on the surface tension of water. This is the principle on which some insects depend when skating across the surface of water as if it were ice.

Quite unlike the ant, some of the very largest organisms may not be threatened by surface tension, but rather they depend on it for their daily functioning. Trees are commonly higher than 32 feet, the height of a column of water that can be sucked up a tube of macroscopic dimensions. Above that height sucking harder only creates a near-perfect vacuum above the water column. The tallest trees can pull water up hundreds of feet only because the woody tissue of the trunk consists of microscopically fine columnar spaces. Because they are so fine and exhibit such large surfaces inside these tubes relative to the volume, the water in these columns is held together by surface tension. Under as much as 15 atmospheres of suction, water is hauled to the tops of trees as if it were piano wire, not a liquid.

Comparing the size of organisms amounts to moving up and down the organism column in the layer cake figure that we used in chapter 1 to set up the ecological criteria. Knute Schmidt-Nielsen finds physiological and behavioral patterns that are remarkably simple across a wide range of organism sizes. If one plots the log of organism size against the log of a series of metabolic or behavioral traits, then the line is often straight. These unexpectedly simple effects of size, measured sometimes by mass, other times by linear dimensions, include: speed of running or swimming or flying; the energetic cost of the same; oxygen demand; number of heartbeats per minute or per lifetime; longevity; and many more (figure 5.15). Sometimes the line is only straight for a given class of organisms, for the fundamental differences in strategy for dealing with a given problem may change the slope of the line. Sometimes the equations for the various powers in the relationships make intuitive sense

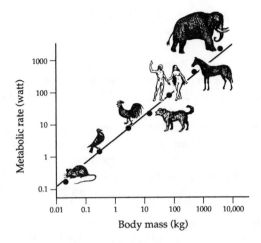

Figure 5.15. The size of an organism relates to the speed of many aspects of the organisms' functioning (data from Schmidt-Nielsen 1984).

because of squared surface areas and cubed volumes. However, the form of some of the simple relationships is less than intuitive. The tangibility of the organism criterion allows us to see some remarkably deep patterns.

Life History, Scaling, and Speciation

Corals represent a most distinctive relative scaling of movement, longevity, reproductive strategy, and environmental catastrophe. So peculiar are these conditions that coral has a very unusual pattern of genetic heterogeneity and even speciation. Corals are interesting to the ecological hierarchist because they are animals that spend most of their life history with a spatiotemporal scaling very like plants. Corals live as colonial creatures, vegetatively reproducing to make more coral mass that is genetically uniform. Great ecologically significant structures come from a few genomes. Given the long life of this clonal activity, it takes a very long time for any sort of competitive equilibrium to occur; in fact, in the present macro-climatic variability it does not occur. Corals in a given place are one great big founder effect.

In many ways, coral resembles redwood trees. Redwoods live a long time as individuals, but more than that they reproduce vegetatively when they are dying. Often, coast redwoods occur as massive fairy rings, where the old parent stem may have been dead and rotted away a half-millennium or more. They can be viewed as hedging bets against ice ages.

According to Donald Potts of the University of California, Santa Cruz, there is a major change in sea level about every 3,500 years. Coral are very sensitive to depth of the water in which they live and this is reflected in distinctive patterns of growth, like the circular attoles in reef formation. When there is a major change in the polar ice that affects sea level significantly, coral can often not accommodate through normal vegetative growth. Therefore, new sites must become established by means of sexual reproduction involving the medusa or "jellyfish" stage of the life cycle. These sexual stages are produced most of the time that the vegetative growth of old genomes persist, but their success is minimal in the face of the old clones holding the favored sites. However, when the sea level changes radically, old sites become uninhabitable and new sites must be colonized by sexual means. Presumably there is some selection between genomes at those times when the new corals are small. If selection at that rate were to continue, then a normal process of differentiation between successful lines could ensue. However, almost immediately the K strategy of cautious coral colony growth takes over and the system is once again denied competitive exclusion.

As a result of the extended life of the first colonists after a change in sea level, normal processes of speciation do not occur. Consequently, Potts can report enormous amounts of genetic heterogeneity in corals, without the discontinuities one might expect as speciation proceeds. Their genetics are so rich that corals appear to be veritably bursting to speciate. The peculiar relative scaling of the phases of the life cycle of corals, coupled with synchronous global disturbance with changes in sea level, appear to make coral biology operate at a small biological scale, even when one considers their global speciation. In corals their whole world is perturbed synchronously, so the global-scale biology is dominated by happenstance at those local moments in time. Presumably corals show more usual patterns of speciation during eras when changes in polar ice are not a normal perturbation three times in 10,000 years.

This situation parallels the peculiarly small-scaled ecology in the community structure in the isolated valleys of the Bitterroot Mountains. Recall McCune's study that revealed a system that is too small-scale to encompass normal organizing forces to produce an orderly set of communities. Usually plants at a site are what is left after selection through ecesis of a constant stream of propagules in that environment. However, the limited input of propagules into the isolated valleys of the Bitterroots produces a vegetation that can only be described in terms of historical accident rather than environment site factors. Dominance by happenstance means that the system is functioning at a smaller biological scale than the spatiotemporal scale would indicate.

Genetic Versus Structural Definitions

While the organism is usually genetically homogeneous, this does not mean that the genetic aspects of organisms relate simply to their form or life history. We have made reference before to our study of BIOSIS with Flather as to the choice of organisms appearing to be related to the sorts of ecological study. Anne Fausto-Sterling and Gregg Mitman show how the course of the study of genetics in relation to organismal form has also been enormously influenced by the choice of organism. Their historical analysis shows that at the turn of the century, the flatworm *Planaria* was the organism of choice for investigating heredity and development.

The critical feature of the flatworm is that it does not sequester its germline, the cells whose descendant will become gametes, from the other lines of cells that will become the main body of the organism, the somatic cells. This distinctive but by no means unique genetic arrangement greatly complicates the animal as an experimental model. For example, flatworms have remarkable powers of regeneration; slice the head many times and each slice becomes a new head. It can reconstitute a new organism from a very small piece of its body and still develop in roughly the same orientation. The model for development is forced to be more outward in its posture; the system demands holism. The genetic expression of the flatworm is much more environmentally oriented, much harder to place as unequivocally Darwinian as opposed to Lamarkian.

By contrast, the fruit fly does sequester its germline, and from that comes the simple genetic models in favor today. Had not the fruit fly become the experimental organism of choice, the central dogma of the primacy of genetic influence would have emerged in a subtler form than the strident version that is now the conventional wisdom. The fruit fly is much better behaved as a model and leads to reductionist genetics of inheritance and development. Because it allowed answer to local, focused questions, the fly gained the ascendancy in an acrimonious battle that was basically won for a reductionist genetics early in the century.

Let us assert that we do not view a reductionist genetics as wrong, but it is most distinctive and greatly limits the scope of questions that are seen as worthy. We would emphasize that reductionist genetics is only a model and with our firm stance against naive realism, power of explanation overrides any consideration of ontological truth. Had *Planaria* been the winner, we would be asking broader questions about genetic control and development. It would be a much more ecological environmentally ordered scheme than the one that prevails today. Furthermore, the issues raised by the flatworm model have not gone away, they have merely been neglected by a paradigm that is powerless to address them.

Much as flatworms offer a very different window on ecological genetics, plants too offer a most distinctive model for discussing the relationship between genetic and morphological integrity. That is one of the reasons why plant population genetics has such different protocols and questions from those occurring in animal studies. We have indicated before that plants are more variable than most animals, particularly vertebrates, when it comes to the discreteness of the individual. This plasticity in plants leads to an important distinction. The term "organism" is inadequate in plant demographic studies because it is unclear whether one means the genetically unique plant derived from a single seed, or the physiologically autonomous stem systems that come from the shared root system. Grasses are particularly difficult in this regard, for it is so common for separate shoots to emerge that they have a special name, the tiller. It is also a verb, to tiller, which is growing by creating tillers.

In plant demography one needs to be able to count units. Organisms are too variable to be those units for many species, so two new terms have been coined. The "genet" is the unit of genetic integrity, the entity that comes from a single seed. The "ramet" is the structural unit, which for the most part does not correspond to the genet. Each tiller in a grass is a ramet. It is the discrete photosynthetic unit, the shoot, often with its own root system. It is a lower-level entity compared with the genet because several shoots usually emerge from the plant that derives from a single seed.

It seems that stems are remarkably autonomous and need to be self-reliant, for they offer relatively little assistance to each other. John Harper and his students have investigated competition using genets and ramets as the units of selection. If one grows plants together at a sufficient density, then there will be mortality. In a series of experiments Harper found that mortality in genets was independent of that in ramets. In an experiment where genets were continuously lost, the loss rate converged for several different initial densities. Ramets could be produced by stronger root systems and so at low densities strong root systems increase ramet number. At high sowing density, the ramet number declined over time. Instead of converging on a constant loss rate of ramets, all initial densities converged on a constant ramet number. In many plants, ramets are lost before genets in a self-pruning of the clonal growth. There is a hierarchy of competition, with ramets feeling the constraints first.

Under competition stress, neighboring plants do not bear the brunt of deprivation evenly. Some whole plants remain large while an increasing proportion are stunted. The distribution of plant size is the first thing to show that there is stress. The distribution skews to show a few very large individuals, even before there is mortality (figure 5.16). As time passes, the plants grow bigger and increasingly interfere with each other. At first there is no genet mortality. When eventually genet mortality starts to occur, it is related to plant size in a very conservative fashion. In a log-log graph of plants per square meter on the abscissa, and mean genet weight on the ordinate, plant growth gives a trajectory straight upward (figure 5.17). Eventually the plants reach a size when someone is going to die. As the plants grow bigger the trajectory angles to the left. With further growth up the ordinate, more plants die, taking the graph down the abscissa. This happens with remarkable regularity, and the population proceeds up and to the left on a slope of 3/2. Although not a universal, this slope is manifested by many plant species. Some slopes are found to be more shallow, but 3/2 seems to be a limit. The universal log-log straight line is reminiscent of Schmidt-Nielsen's scaling for animal size and activity.

There is hierarchy in plant reproductive units within individual plants when resources are limited. As a rule, the last plant parts to suffer constraint are the individual seeds. As plants are forced to economize on reproduction, the first economy is on flowering shoots. Only under more severe deprivation is there a reduction in the number of flowers on surviving shoots. Only under even more difficult conditions is there an economizing on seed number per flower. In some species there is little evidence that there is any reduction in the size of seeds. Rather than make a small seed, the plant makes no seed. The presumption is that the plant is gambling on things getting better, rather than placing a losing bet on a small seed that can never make it to maturity.

Plants are very conservative when it comes to form and reproductive effort. If one identifies the proportion of an annual plant that is put into stems, roots, leaves, and flowers, the relative proportion remains remarkably constant under stress. There is an immediate shift to smaller plants under competition, but the proportions of biomass invested in each organ is more or less constant. Eventually there is a shift to more root production, but the amount of effort put into reproduction remains constant. Eventually under desperate conditions the average plant ap-

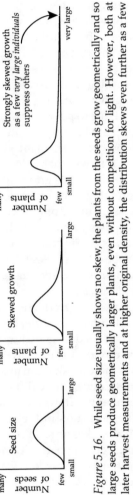

Figure 5.16. While seed size usually shows no skew, the plants from the seeds grow geometrically and so large seeds produce geometrically larger plants, even without competition for light. However, both at later harvest measurements and at higher original density, the distribution skews even further as a few plants suppress all others.

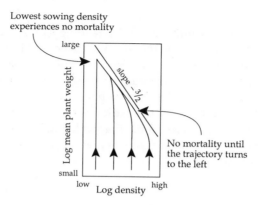

Figure 5.17. A typical result of sowing seeds at various densities. As plants grow each density of sowing follows a verticle trajectory until the various densities reach the thinning line of slope 3/2 or less. The trajectories turn left as plants die and follow the thinning line while the average plant becomes larger.

pears to put less into flowers and fruits, but this is deceiving. The average plant is not representative because most plants under the worst conditions fail to reproduce at all. Those that do make a run for it, and produce flowers and fruits, do so close to the proportion of effort employed by full-sized plants.

Plants are much more malleable than animals, both in form and in means of reproduction. They make ideal subjects for investigating growth and death in the tussle between survival and reproduction, and between parts of an organism and the whole. The critical feature of plants is their modular construction: leaves, branches, tillers, and vegetative propagules like strawberry runners. It emerges that the modules inside a plant go through much the same sort of competition and selection as the whole plant. With parts behaving like the whole, plants bring home the message that the organism is an arbitrary conceptual construct, albeit a useful one.

Conclusion

The organism being a tangible entity, it offers a wealth of signals to human observers. As with other easily read objects of study, there is a problem in working out what in the organism is generalizable and what comes from our own humanity. We are likely to be overpowered by its obvious signals so that we miss more interesting and subtle aspects of organismal ecology. A whole new world emerges if we try to look at the organism in the bubble of its *Umwelt*. The physical organism itself is isolated and meaningless until we watch it reading its "self-world." There is much to be said for allowing the organism to be seen not as a tangible, but by means of the closed loops and processes behind the structure. Let it turn into a set of fluxes. By struggling against the tangibility of the organism throughout this chapter, we have endeavored to enrich the concept of organism. We tried to make it more than a reification of what the naturalist finds on a walk. If this chapter has achieved anything, we think it has laid the ground work for integrating organismal biology into a wider view of ecology.

6. THE POPULATION CRITERION

————◆◆————

Of all the criteria for ecological observation, the population criterion appears to fit most neatly between other criteria. It is a collection of individuals, and because it usually involves only one species, the population may be considered as contained within the community. Nevertheless, like the other criteria that we have discussed so far, the population is a depauperate notion if it has to be couched in terms of a simple, intermediate degree of aggregation between two other criteria. The population criterion is richer when it is viewed as a distinctive way of observing ecological phenomena, rather than a staging post between other criteria such as the individual and community. Populations share with communities the fact that both contain individuals.

As with several of the criteria considered so far, defining "population" can be done by contrasting it with other criteria. From chapter 1 recall the layer cake which separates ecological criteria as horizontal sections from scale represented by height (figure 1.13). We start by cutting across the cake so as to hold the scale constant while making a comparison to other criteria operating at the same scale. The first comparison is to the community. Relative to the community, the scaling of the population in time and space is distinctive. The single-species characterization of populations as opposed to the multispecies characterization of communities leads to different types of occupancy of landscapes. The second contrast is between populations and landscapes.

The relationship of communities to populations is recurrent through this chapter because they are closely related concepts that must be teased apart with care. The recurrence of communities here is also because the relationship has two distinct parts requiring separate treatments. The two relationships are: 1) the community as just an alternate way of looking at the same sector of the universe at a given scale (population and community on the same layer of the cake); and 2) the community as a context for population considerations (the community on the next layer up). Sandwiched in this chapter between those two treatments of the community-population relationship, we turn to the population alone at different scales. This is, of course, our third way of considering an ecological criterion: hold the criterion constant and change the scale. This scattering of community considerations across the present chapter indicates how our organizing scheme differs from conventional treatments of multiple criteria in ecology.

Populations and Communities at the Same Scale

Processes that define individuals as community members involve, to a large degree, interactions between members of different species. By contrast, populations have members that are, for the most part, bound together by interactions that apply strictly within single species. Involving more than one species does not necessarily put the community above populations in the general ecological hierarchy. Many processes pertaining to populations as opposed to communities are not particularly narrow spatially, nor do they necessarily arise at a smaller temporal scale. For example, the fact that breeding is generally restricted within species is not a matter of scale. Breeding is an interaction not available to community members of different species. The irrelevance of breeding as an interaction between members of different species says little about temporal or spatial scale, and so it need not define populations as being below communities in the ecological hierarchy.

Competition as it applies to individuals in populations as opposed to communities is only a matter of scale in a very restricted sense. Consider the following truisms: interspecific competition cannot occur inside single-species populations; interspecific competition does impinge on individual population members, but as an interaction with something from outside the group; any competition that is not interspecific is intraspecific. There is not a necessary hierarchy of competitions, because interspecific competition need not occur at a higher level than intraspecific competition. Interspecific competition does not necessarily take longer to happen than intraspecific competition, nor must it occur over a wider spatially defined universe. They are merely different types of competition with approximately the same spatiotemporal qualities. We need not construe the within-community character as belonging to a higher level than the within-species, within-population character.

Furthermore, the continental population of a tree species is often wider in space and more persistent in time than the community it dominates. For example, the balsam fir population defined by species range is scaled larger than its coniferous communities. Therefore, we will not assert the community as the necessary context of populations, except where that point of view is enlightening. Accordingly, the population has its own independent properties. Interactions binding together population members are not so much more local than those defining communities, as they are interactions of a different type. This chapter will consider the types of processes making organisms and demes attributes of populations, as opposed to any other type of ecological assemblage.

The Taxonomic Requirement

With a few exceptions, the minimum requirement for members to be from one ecologically defined population is belonging to the same species. This is not a necessary restriction, for the population concept could be useful with members belonging to different biological species. The use of the term "population" in statistics captures this more general notion of population. Given the proviso of taxonomic homogeneity, ecological populations then revolve around two major organizing principles giving two types of populations. The first consideration is spatial contiguity: members are aggregated. The second consideration involves a shared history of some sort. Often this echo from the past amounts to a level of genetic relatedness; members share ancestors. However, the historical connection could be some other bond which is not genetic, such as being in an infected population; the disease agent may be of one genetic strain, but the population of the infected need be neither closely related nor genetically homogeneous. Most populations amount to a mixture of these two types of prescriptions, there being a continuum between historically and spatially defined populations.

Note that the genetic relatedness prescription is different from the original requirement that members be from one species, although clearly membership in a species does involve a degree of genetic correspondence. The original within-species requirement need only incidentally involve shared ancestry. When a population is of the special type that emphasizes relatedness, the genetics are often very limited, perhaps so local as to involve only a single allele shared by all members. Clearly at that point, belonging to the same species is only peripherally a matter of genetics.

Populations, Communities, and the Relationship to Landscapes

Defining the population as a collection of individuals belonging to the same species says nothing about the time frame over which populations might exist, and nothing about the spatial coherence of the population on the landscape. It is therefore surprising that "population," a term grossly defined only by the taxonomic homogeneity involved, should have such particular connotations with regard to time and space. Let us explore the temporal and spatial specificity involved in the population criterion.

While it is possible to consider individuals scattered through time as members of a population, for the most part population refers to a temporal cross-section, an instant in time. This does not preclude populations existing over time, but at any moment along the entire study, the population is identifiable as being delimited in space. Even when we define the population temporally, such as an epidemic, we still find it helpful to map literally the progress of the disease (figure 6.1). History books

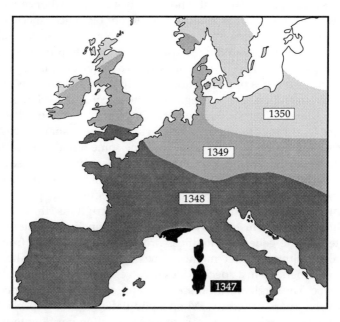

Figure 6.1. Even though populations can be dynamic, it is commonly helpful to circumscribe them on the ground. The population of infected individuals in the great plague of Europe is an example.

map where the Great Plague moved in waves across Europe during the fourteenth century.

Where spatial organization is inappropriate, perception or conception is simple only if temporal sequences anchor the framework. Time and space play something of a zero-sum game for control of our conceptions; tighter constraint in time appears to relax the constraint in space, and vice versa. In descriptions of populations, the temporal limits are not generally specified explicitly, even if the population is allowed to behave over time. Thus spatial contiguity is used frequently as a principal characteristic of population limits. For the most part, we at least think of a population as having a spatial limit, even if actually drawing the line around the edge would be problematical. On the other hand, even when the period for the end of a study is explicitly stated, it is the same population that sits unobserved in the next unit of time, and we are cognizant of that when the arbitrary time limit is set. The only time when a temporal limit to a population has simple intuitive meaning is in the rare event of extinction.

Populations can be mapped onto the landscape. Even so, it is probably a mistake to focus upon populations as things on a landscape. The factors causing change on the landscape over time are often not those that pertain to population dynamics and vice versa. The processes of exchange that hold together members of a population may work over periods of time long enough for the population to move over undefined tracts of land. The spatial criteria for populations must relax as time takes over. Accordingly, the population may not belong to any place on the landscape in particular. A spatially defined entity such as a migrating herd of caribou is only trivially defined by its place on the landscape at the instant of an aerial snapshot. Even tree populations may, for some purposes, be more interestingly considered as the camp of an arboreal army on the march since the last ice age.

We have already warned against considering communities as things on a landscape. For populations the caveats are different, although no less important. Both populations and communities involve spatial and temporal considerations together, but in different ways. Recall that in the community an interference pattern arises from different species moving across the landscape at different rates. Populations also involve interactions of time and space, but do not require an interference pattern. This is because of the relative scaling of the respective parts of populations and communities. Being all of the same species, members of a population occupy the landscape all at the same scale, or at least at a scale more similar than those belonging to community members. All members of a given population move at about the same rate. Therefore, space and time do not interfere with each other in populations.

Because space and time can be teased apart in populations, a purely spatial account of a population is possible. Certainly freezing the population on the landscape does not do justice to its temporal aspects, but the map is still valid, and it is of the population itself and not a surrogate. The same cannot be said for a spatial map of a community. This is because the community itself is an intangible interference pattern from which the temporal texture cannot be dissected. A woodlot, which can be mapped, is an exemplar of the community but is not the community itself. The exemplar and the thing itself belong to different logical types. A woodlot may contain many species like the community, but it is a landscape entity with spatial contiguity as the critical factor. The temporal texture coming from the different scales of the occupancy of the landscape in the community itself has been defined away by making the woodlot a spatially limited entity. The integrity of the woodlot comes from its spatial coherence at the spatial scale of its boundaries. The integrity of the community of which it is a landscape surrogate does not fit inside the woodlot's fences. Some species occupy the landscape at scales larger than the woodlot, while others occupy only a corner of it.

Because the temporal aspects of both populations and woodlots can be dissected, both can be mapped onto a purely spatially defined system. Communities and ecosystems cannot be so represented because the map cannot allow for the intrinsic temporal aspect of the community and ecosystem interference patterns. As a thought experiment, try to imagine a spatial map of a nitrogen cycle; it cannot be done because there is a temporally defined process which cannot be accommodated in space alone. It is no more valid to try and map a community or ecosystem on the ground than it is to try and find exactly where a wave of light exists at an instant in space: particles, yes; waves, impossible!

The Processes Behind the Pattern

The patterns raised above beg for an account of the underlying population dynamics, and we turn to that now. This account will involve looking at the population at different scales, moving up and down the population column. The relationship between populations and communities will reappear in a later section when we take diagonal slices of the cake to look at both communities as the context of populations and organisms set in the population context.

We have asserted two types of populations: spatially aggregated and historically explained populations. We have also noted a continuum of admixtures between the two types, namely populations that are variously both spatially aggregated and bound by history. In these populations often the spatial contagion has a historical explanation reflected in the processes that have generated the population over time. Concentration of individuals in a local place often indicates a reciprocation between individuals that makes them members of a meaningful population as opposed to just an arbitrarily defined collection. As suggested above, often these exchanges are processes associated with reproduction. In asexual reproduction of plants, the process may leave the members of the population literally connected to each other by roots, runners, stolons, or other such modified stems. Sexual reproduction involves exchange, often between neighbors, so that there are connections between members through ancestors that formerly occupied the same general vicinity. Thus, some spatially contagious populations can be characterized also by a certain genetic homogeneity.

Despite the genetic basis of some local populations, relatedness need not be the critical organizer of spatially contagious populations. Nevertheless, in these genetically undefined populations there may still be a historical explanation for the aggregation. Consider propagules invading an area from several genetically distinct parent populations. The resulting organisms can be bound into a single population by a common requirement for certain environmental conditions at the new site. Thus, the process of establishment might fill a patch of ground with a population independent of the lineage of the individuals. Note how the underlying process that generates a population is historical; the pattern is that which is left behind after historical events have passed. In this case, the pattern is the aftermath of the filtering through ecesis.

Some populations exhibit spatial contiguity with no real historical explanation, reproductive or otherwise. In these situations, the spatial aggregation is constantly updated. The aggregation reflects a continuing process. Cases in point would be herds, flocks, or shoals which are held together by behavioral cues between individuals.

Much in the way that we looked at the landscape criterion at many scales, it is possible to see populations nested inside bigger populations. When we do see such nesting there is no reason to expect the differently scaled populations to be of the same type. Consider the example of the pronghorn antelope. At the lowest level it is a genetically defined collection of the doe and her fawns. The next larger aggregation is not genetically, nor even historically defined. This is the herd, an entity with floating membership serving to avoid predation. It is held together by conspecific cuing. The members are not necessarily closely related.

At the next level is the antelope population, as defined by habitat.

There is a degree of genetic homogeneity to this grouping. Because favorable habitat is discrete, there is little mixing between populations at this level. For example, antelope cannot survive in deep snow and only very occasionally will an individual cross difficult, drifted terrain to join another isolated population. Most would not survive the passage and normal conditions would not lead an individual into the unfavorable habitat in the first place. Only individuals disoriented by a chase or an accident would even find themselves in a position to make the crossing. Note that although the herd is not defined on relatedness, the larger population can be so defined. Since one contains the other, one might infer a contradiction here, but there is none. Relatedness is a matter of degree relative to that which occurs in a larger aggregation. One looks for the largest aggregation showing a certain degree of relatedness and then ascribes "relatedness" as a criterion for that level only. Given the genetic heterogeneity in the population circumscribed by a habitat, the lower-level herd is not particularly less heterogeneous even though it is smaller. However, relative to the genetic heterogeneity across a collection of habitat-isolated populations, one habitat-isolated population is indeed more homogeneous than the collection of them. Therefore, the habitat-isolated population can be said to be defined by a certain level of genetic heterogeneity.

Finally, there is the population defined by the range of the species. The species range need not be defined by the limits of favorable habitat. It is quite common for a species at its range to be well represented by a large, healthy, reproducing population. That local population will be habitat-defined, as are all collections concentrated in favorable habitats. Nevertheless, the fact that this population is peripheral is not a matter of that population occupying the last piece of favorable habitat in that direction. For organisms as diverse as antelope or tree species, it is well known that, at the range, there is favorable habitat available close at hand but it happens not to be occupied. The range of a species is often a statistical matter. With no change in habitat at all, the range can move in at one point or out at another. The limit is set by the probability of local extinction in combination with the probability of accidental invasion of unoccupied, favorable habitat at the species periphery. The processes at this largest scale are different again from those that pertain to all the other more locally defined populations of antelope.

It is possible to erect climatic and topographic hierarchies that pass from local to global considerations. John Harper has analyzed climate and topography at scales so small that in his "microsites" seeds germinate or not depending on how they fit into local soil, water, and air relationships. At the other extreme, the climates of continents are caused by there being more land in the northern hemisphere. In between, climate and topography interact to produce environments of different scales, some of which relate to the scaled populations of the antelope.

Local mountain peaks make rain shadows that pertain to favorable habitat for discrete populations of antelope. At a larger scale, oceans, mountain ranges, and the jet stream outline the possible species range, although the actual range has the statistical property we mentioned above. At small scales, local texture of the terrain combined with microclimate determine the local resource base of the herd and guide the pattern of the conspecific cuing. Thus the antelope populations at their various scales can be related to their physical environment in several ways. The population biology of a single species has many scales and many distinctly scaled environments. The homogeneity within populations does not make them simple or untextured objects of study.

Equations for Population Dynamics

Much of the work on populations is concerned with dynamics of population numbers. The critical processes here are birth and death. Individual deaths and births are events and so, while fixed in time, they are not dynamic in themselves. However, collections of these events run together to give a pair of dynamics, the birthrate and the death rate. The balance between these processes gives the total population dynamics; equations with terms that are surrogates for birth and death rates appear to be useful descriptions.

A system is simple if it only requires one level of organization for its description; that is to say, it requires explanatory principles operating at commensurate reaction rates. The fact that members of a community are not all scaled equivalently means that, in a sense, members come from different levels of organization. Therefore, it is necessary to invoke a complex interference pattern for communities and ecosystems on the ground. By way of contrast, since all members of the population are scaled equivalently, they come from one level of organization, and this makes the population relatively simple. It can thus be mapped on the ground in a straightforward fashion.

Members of a population all belong to a single spatially and temporally defined level. This homogeneity allows populations to be described with something so simple as an explicit equation that plays birth against death. Note that nothing so simple can be used to describe the behavior of communities or ecosystems because the various parts of communities and ecosystems do not occupy commensurately scaled parcels of time and space. Single population-type equations for either ecosystems or communities would be unwieldy and certainly analytically intractable. Of course, there are models for ecosystems that consist of a large number of equations, but that is a different matter. Each equation there usually involves only one scale-defined relationship; that is what makes them manageable. In general we quantify and model communities not with analytical solutions to equations, but rather with the brute force of numerical computer-driven solutions.

Analytically tractable as many population equations may be, others yield only to numerical calculation on the computer. This is the clue to the fact that, despite the commensurate scale of all population members, there are still mismatches of scale in the functioning of populations as a whole. That is indeed what makes population dynamics such a challenge. If birth and death respond to external influences at different rates, they demonstrate different relaxation times. The interaction of disparate reaction rates generates complex behavior, and population equations are no exception. Even simple equations can yield very complex behavior if birth and death rates react over incompatible time frames. The differences in temporal responsiveness scale birth and death to different levels. Remember the general rule that complexity arises from the interaction of differently scaled processes, an interaction between different levels of organization.

Nothing in nature dictates that birth and death should be symmetric at all scales. Therefore, the different response times in birth as opposed to death rates in the equations can be perfectly realistic. True, birth cannot exceed death indefinitely. Even an animal as slow to reproduce as the elephant could not live for very long up to its full reproductive potential, if all individuals born live full life spans. Such a population very quickly would cover the entire surface of the globe with a skin of elephants standing shoulder to shoulder. If the birthrate continued unchecked a little longer, the earth would become a speck at the center of a sphere of elephants whose opposite surfaces would be moving apart at speeds that would require a correction for relativity. It is also true that deaths exceeding births leads quickly to extinction. A case in point is the now-extinct passenger pigeon whose flocks darkened the skies until as recently as the turn of the century. Nevertheless, birth or death can outstrip each other for short periods of time, and that can manifest itself as different degrees of responsiveness in birth and death rates.

There is no need to expect to find a balance between birth and death at any particular instant. This applies not only to waxing or waning populations, but also to stable populations going through seasons for birthing which are different from seasons in which death is common. Spring yields fawns while winter culls the herd. Death and birth being at different times of year, there is time for the carrying capacity to change between the seasons when birth and death prevail. An inclement spring could be followed by a mild winter. Furthermore, birthrates could be much more responsive to changes in carrying capacity than death rates, or vice versa. There is no reason to expect limited food to adjust reproductive effort on the same schedule as death occurs from starvation. Conversely, unusually abundant food is unlikely to affect litter or clutch size in temporal patterns which are mirrored in longevity. Birth and death are only symmetric as opposite ends of life; there need be no equivalence in their scaling. As a result, we can expect even simple cases to yield rich behavior, even in populations where birth and death relate simply to the carrying capacity, but with different degrees of responsiveness.

Despite the obvious differences between increasing birthrate and greater longevity, both increase the population. Conversely, decreased birthrate and increased death rate both decrease the population, although they are very different processes. Despite these differences between fecundity and mortality, population equations often are so simple as to combine death and birth into a synthetic growth rate which can be positive or negative. Unrealistic as it may be, this synthetic growth rate in population equations is often symmetric in its response to being over as opposed to under the carrying capacity. When death predominates during overpopulation, the population is modeled to fall at the same rate as it would rise were the population equivalently below carrying capacity and birth predominates.

Another wrinkle can be introduced in the application of two alternative strategies for modeling populations. Some equations acknowledge in their very structure the fact that changes in populations occur intermittently or seasonally rather than continuously. Instead of the instantaneous response of differential equations, difference equations are recalculated in discrete increments of time to give new numbers for the population intermittently. Differential equations give relatively smooth changes in numbers of individuals, while difference equations give a jagged course to numbers in the population. This discontinuous change in population numbers is often more realistic (figure 6.2a, b). The use of difference equations facilitates an integration of birth and death over some meaningful time period like a year. In this way not only is the explosive growth in the lambing season given credence, but the fate of the lambs is integrated into the change in numbers.

If difference equations are used to describe discontinuities in breeding or mortality, there is an implied lag in the system. The implied lag in the effect of resources on birth and death occurs between the times

A

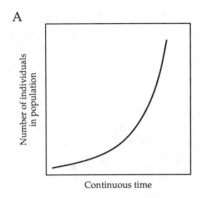

Continuous time

B

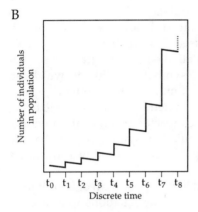

t_0 t_1 t_2 t_3 t_4 t_5 t_6 t_7 t_8
Discrete time

Figure 6.2. A. When a population is modeled with a differential equation, the change in population numbers is continuous and a graph of the population over time gives a smooth line in this case of exponential growth. B. When population is modeled with a difference equation the population numbers are updated intermittently and so a graph of numbers of individuals in a population over time follows a jagged line.

when the system is updated between time t_0 and time t_{+1}. It is possible to express the growth rate in terms of the time it would take the population to correct for values over or under the carrying capacity. The growth rate is conventionally given the term "r." The bigger the growth rate in terms of change in number of individuals, the shorter the time the population will take to reach numbers corresponding to the carrying capacity. Accordingly, the responsiveness of the equation can be well approximated by $1/r$, the relaxation time. This translates the capacity for change in numbers over time into a simple expression of time alone, namely the time to make the correction for a unit displacement of the population from the carrying capacity. Having translated population growth rate to time alone, it is easy to express that time in units which are the length of time between the update of the difference equations. We now have a dimensionless number (a relationship that involves only one kind of unit). That number is the time for a unit correction divided by the amount of time available for that correction before the equation is applied again. The power of dimensionless numbers comes from the way they reach unambiguously right to the heart of a relationship. In this case it is growth rate or population responsiveness against lag or time between updates. A high growth rate relative to the lag gives the population a long time to reach the carrying capacity.

We have in the relationship between lag and growth rate yet another example of the interaction of two different levels of organization-generating complexity. The lag fixes the scale at which the carrying capacity can exert its influence, while the growth rate fixes the scale of the population response. If the lag is short, because the intrinsic growth rate is relatively low, then the two factors affecting actual changes in numbers operate at similar rates. Therefore, responsiveness and lag would work on commensurate scales. In this case behavior is simple, for the population finds an equilibrium at the carrying capacity in an orderly fashion. If the population growth rate is somewhat but not excessively larger, exotic patterns of oscillation are found. Here the population regularly weaves above and below carrying capacity, never able to come to rest. It persistently over- or undershoots the equilibrium.

If the growth rate is very large, then the population has ample time to overshoot the carrying capacity even if the number in the population left by the previous application of the equation is very high or very low. Such model populations belong to a family called chaotic systems. In this case the effect of the carrying capacity on changes in population is the result of two very differently scaled forces, the growth rate and the lag. The result is very complex behavior which never repeats itself. Chaotic systems can be considered as having an equilibrium point in a fractal dimensional space. That is why such systems never repeat a state exactly. Theoretical populations of this sort are extremely responsive to initial conditions, such that the addition to a very large population of a single individual at the outset has a large final effect. The small change in initial numbers amplifies in a relatively short time to give population sizes very different than had the single individual not been added in the first place.

Chaotic behavior appears in many physical systems, although unequivocally chaotic population systems have not often been documented. Chaos in populations is quite hard to distinguish from random buffeting of systems from unseen outside influences. In terms of hierarchy theory and concepts like constraint, chaotic systems have some of the properties of middle number systems. The great sensitivity to initial conditions would indicate that upper-level constraints do not operate, and each individual has the capacity to alter the outcome of the entire population.

Despite the sensitivity to initial conditions, underlying chaos is veiled order. Initial conditions have large effects, but statistically defined order prevails in what are called strange attractors. The details of chaotic population states cannot be predicted from general principles like persistence at an equilibrium point or in a limit cycle. Nevertheless, there are regions to which the population is restricted. These regions make the strange attractor; the population visits only these regions, unless it is disturbed.

Population Interactions

The whole strategy of dealing with populations comes from their being composed of only one type of entity, usually individuals from the same species. The very homogeneity of populations allows us to describe them with such highly structured devices as analytical solutions to difference and differential equations. This homogeneity does not mean that populations are easy to study, for they can exhibit rich behavior very easily. Populations described with difference equations with even moderately high growth rates, as discussed above, are a case in point. Because members of a population are all the same for many practical purposes, they are reliable. If it were necessary to model in a single equation the complexity that might come from different response times of different individuals in the population, then all the complications of population behavior we have discussed so far would become second-order complexity, and the model systems become unworkable.

Nevertheless, some population studies do recognize a degree of heterogeneity in the population, but that requires a change in modeling strategy. The heterogeneity is usually introduced as a separation of different age or life stage classes. Instead of a single equation that models the entire population, each stage has two coefficients that determine its contribution to future population numbers. One is for survivorship which determines how many continue to the next stage. The other coefficient determines reproduction assigned to the life stage in question. By modeling death as connected to the next stage, and reproduction as connected to the youngest stage, the system of equations tends to decouple the parts of the system that manifest the different reaction rates responsible for the chaotic behavior discussed above. Much of the advantage of modeling with difference or differential equations is the insight gained by isolating various meaningful biological considerations as terms in the equations. Multiple-age or multiple-stage class models are often quite well behaved, but the biological significance of many of the terms and their effects are less than transparent. Multiple-age class models seem to be a good way of bookkeeping but leave something to be desired in generating biological insights and understanding.

Another type of inhomogeneity arises in population studies with interactions between species. As with multiple-age structured studies, there is a challenging amount of complexity in interactions between the different types of organism. A favored strategy in modeling species interactions is to simplify the equations for the species so that they are well behaved. There is ample complexity coming from the species interaction to keep the modeler occupied, without chaos or complicated oscillations coming from rich dynamics within species as separate entities. Accordingly, prey-predator interactions are often made tractable by using linear equations that are almost always well behaved. Linear equations behave as if the populations were always close to their respective equilibria. Interesting wrinkles of nonlinearity of the species existing alone are suppressed in favor of a tractable model of species interaction.

Prey-predator relationships are negative feedbacks; an increase in prey gives more food to the predator population which also increases in numbers, thus leading to increased predation and a correction in prey numbers back to the original level. Negative feedbacks are intrinsically stable. However, delays in the feedback produce oscillations. These delays allow the factors that move the population away from the equilibrium to continue to do their work. When the information that the population is too large or too small is eventually registered, the response of the correcting factor is likely to be strong enough to overshoot the equilibrium on the other side. If the delay is long enough, the correction comes at exactly the wrong time, and the overshoot beyond the equilibrium amplifies so as to make the system untenable.

The effect of release from predation causes an increase in prey numbers according to a schedule set by the intrinsic growth rate of the prey. There is no reason to suppose that this growth rate is on the same schedule as the positive response of the predator when it finds abundant food. Furthermore, the eating of prey instantly removes those individuals from the population, but the corresponding increase in predator numbers because of feeding takes time. Only eventually does good eating lead to greater fecundity. Greater longevity by definition cannot be instantaneous.

As lack of food begins to produce mortality, decline of the predator is on a different schedule than the decline of the prey from predation. Dying from starvation takes longer than dying from being eaten. The delay in the negative feedback of prey-predator systems comes from the parts of the system of equations (e.g., dying from predation vs. dying from starvation) responding at different speeds. Parts of the equations that behave slowly are coupled to the parts that give rapid changes in numbers of prey or predator. This is a prescription for rich, complex behavior.

Slowing down the faster part of the system is a way of reducing the lag, which in turn reduces the attendant system oscillations. In another biological system given to oscillation, the drunk driver cannot reduce the lag in his reaction time, but he can diminish the danger of fatal oscillation by slowing down the car to a speed commensurate with his impaired responsiveness. Lag is therefore less a matter of absolute time and more a matter of relative reaction rates. Different reaction rates belong to different levels. Therefore, lag which produces complex oscillation is a measure of the disparity of levels which are forced together in a description of a complex situation. Remember that complexity is not a matter of nature, but a matter of the description chosen for nature.

Prey-predator systems often oscillate for the reasons suggested above. Furthermore, the predator oscillation generally follows after the prey oscillation because: 1) it is the decline in prey that eventually starves predators, and 2) it is the reduction of the predator population below critical levels that immediately releases the prey. The demise of prey is the fastest coupling in the system. C. S. Holling and his colleagues have investigated the effects of manipulating system stability by changing the efficiency of the predation. They tinkered with the lag in the system by altering its fastest reaction rate. His particular findings are related to the form of the equation that he used to study prey-predator systems, but there appears also to be a significant generality to his results.

At the lowest level of abstraction, one graphs the numbers of the prey and predator over time. However, a more insightful image takes the same system states and graphs predator numbers against prey numbers at successive points in time: a trajectory over time in a two-dimensional prey-predator space. The trajectory of the system on the prey-predator plane is roughly circular. If the system starts with low numbers of prey and predator, increases down the prey axis eventually feed the predators and the trajectory turns to move up both axes following a diagonal track from five to four o'clock (figure 6.3). When predator numbers are high enough, prey are held in check. At that point prey numbers stop growing, but the predators have plenty to eat and so increase their numbers further. The system moves from four to two o'clock. Further increases in predators begin to overwhelm the prey and their numbers be-

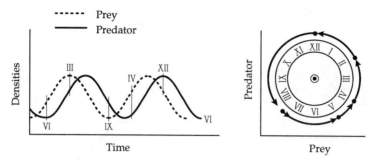

Figure 6.3. A graph of prey and predator over time oscillates, the line for the predator lagging behind that of the prey. Plotting prey numbers against predator numbers, the trajectory of prey/predator relationships follows a roughly circular pattern. Once around the circle brings the prey and predator densities back to their original values.

gin to fall. The system turns to twelve o'clock. As predators eat themselves out of house and home, both prey and predator numbers decline, taking the system back close to the origin at half past seven. The low predation pressure starts the cycle again.

In the most common linearizations of prey-predator equations, there is a neutral stability (figure 6.4). That is, however far from the from the equilibrium the system starts, it keeps oscillating to that degree indefinitely. In Holling's more sophisticated equations, there is the potential for damping of the oscillations until the system comes to rest at coincident equilibrium populations for both prey and predator (figure 6.5). However, in Holling's equations, if initial conditions involve very low or very high populations of either prey or predator, then the oscillations amplify until one or the other population, or both, is driven to extinction (figure 6.6). In terms of the prey-predator plane, there is a safe region in the middle of the plane, and any initial condition inside that region leads to damped oscillations and equilibrium. There is, however, a region where either or both populations are extremely large or small, and initial or perturbed conditions that take the system to these peripheral regions lead to amplifying oscillations and extinction (figure 6.7, 6.8).

As Holling altered the efficiency (k in figure 6.8) of his predator, that is, as he manipulated the implied lag in the system, the shape and size of the region that led to equilibrium changed in interesting ways. This can be best illustrated by introducing yet a higher level of abstraction in presenting Holling's results. Consider the prey-predator surface as an undulating plane, and the state of the population as a ball on that surface. Consider the safe region as a cup such that, if the ball is placed within the cup at the initial conditions, then the ball spins around and

NEUTRAL STABILITY

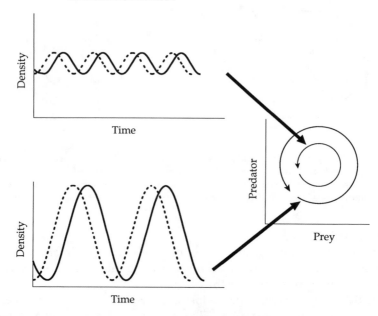

Figure 6.4. In the prey/predator equations, first applied 50 years ago, there was neutral stability. In this situation the initial deviation from equilibrium prey and predator numbers persists through time. The pair of prey/predator values never reach equilibrium together. They continuously oscillate the same amount above and below the unachievable equilibrium condition.

finally comes to rest at the equilibrium. Should the ball get outside the cup of the safe region, it spins away from the center until at least one population is extinct. Figure 6.9 is a section through the cup. The steep-

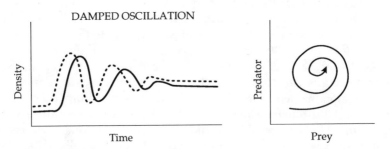

Figure 6.5. In Holling's more sophisticated equations, the prey and predator oscillations may diminish until the population values damp down to an equilibrium value.

AMPLIFYING OSCILLATION

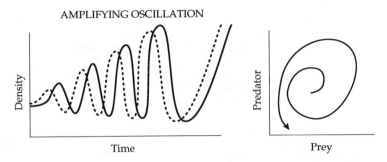

Figure 6.6. Again using Holling's equations, very large or very small prey or predator populations can produce amplifying oscillations until one or both of the prey and predator go extinct.

ness of the sides of the cup refer to the speed with which the ball moves to equilibrium. Once in a flat-bottomed cup the ball may roll around for a long time before reaching equilibrium. In a steep-sided cup, the ball plops into equilibrium in a single cycle of prey and predator.

A confused population literature uses words like "population resilience" (capacity to bounce back from a disturbance), "resistance" (the ability of the system to hold its state despite outside influence), and "stability" (the speed with which equilibrium is achieved). Inconsistent usage prevails; definitions we suggest above are switched between writers. Therefore, we will not press the cup analogy too far for fear of

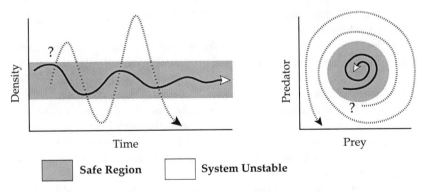

Safe Region   System Unstable

Figure 6.7. Inside a range of very high and very low values for prey and predator there is a stable region within which the population will spin down to an equilibrium condition. Only the prey is plotted on the two runs of the system. The "?" indicates the point on the trajectory where there has been a critical move outside the region of stability.

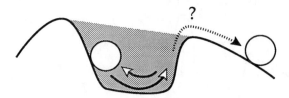

Figure 6.8. It is helpful to picture abstract the behavior of the pair of equations and their region of stability as a cup where the population is a ball that either spins around finally reaching the bottom of the cup as it achieves equilibrium or spins out of the top of the cup to some other condition. As in figure 6.7, the "?" indicates the point of no return.

adding to the confusion. Nevertheless, it does make an excellent vehicle for summarizing the sorts of things that Holling found as he made his theoretical predator more efficient.

If the predator was inefficient, then the cup was wide, flat-bottomed, and clearly bounded on all sides (figure 6.9). Holling suggests that natural ecological systems exhibit this form: a wide, well-defined range of states, across which it often passes, from which it is unlikely to be thrown by less than catastrophic disturbance. He notes that traditional tropical slash-and-burn agriculture of the Tsembaga in New Guinea follows this sort of track. They have a pig taboo which causes the pig population to rise unchecked, storing rich protein resources, until the tribe goes on a pig binge in times of strife. The advantage of inefficiency is that it leaves lots of slack in the system. The last thing the Tsembaga need is a well-meaning infusion of Western-style, efficient cropping, for that would reduce their slack.

If the equation is adjusted to make the predator more efficient, then the cup deepens and its sides become steep (figure 6.9). Equilibrium is achieved more quickly, but at the cost of the area of the safe region. So long as nothing is seriously wrong, such a system tracks the equilibrium closely, but at the cost of a certain fragility. This, Holling asserts, is like the Western industrial agribusiness, where efficiency leads from wood-lot to cornfield in one season, a system where government programs can reclaim or set aside vast tracts of marginal land with the stroke of a pen on an incentive program.

As Holling's predator becomes more efficient again, the sides of the cup become steeper, but one side begins to erode away. Eventually, further increases in efficiency remove one side of the cup altogether. At that stage, the system has no stable equilibrium, and extinction or un-

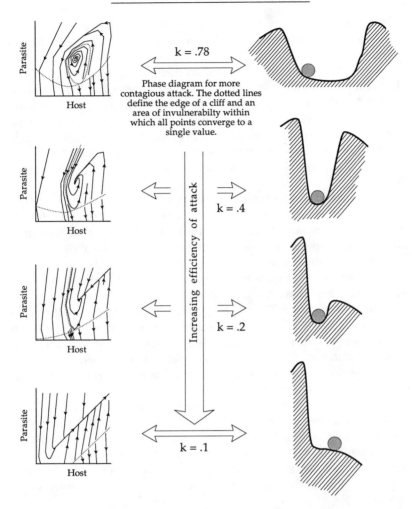

Figure 6.9. Holling and Ewing modeled host/parasite relationships changing the efficiency of assaults (k) upon the host. This produced a series of response surfaces ranging from a local stable attractor at the top to the loss of all stable equilibria at the bottom of the series. The response surfaces are paired with their respective analogous cups cut in section (after Holling and Ewing 1971).

controlled growth is the rule. Holling is anxious that the industrial global village is approaching that condition.

We would wish to modify Holling's assertion that natural systems have plenty of slack and suggest that natural systems can exhibit unstable architecture with no stable equilibrium. Indeed, we suggest that such is the source of much normal ecological change, so long as there is at least one scale at which the system exhibits slack and inefficiency. Consider Paul Colinvaux's observation that lions are inefficient predators which rarely strain the capacity of their prey to provide a food resource. On the other hand, Colinvaux notes that insect predators are deadly efficient. A single predator like a ladybird finds a prey population of aphids and proceeds to kill its prey as if they were sitting ducks. Furthermore, the predator reproduces quickly to generate a large, ravenous population of beetles that finally eat the aphid prey to local extinction.

Clearly the lion exists with lots of slack, even slack enough to ride out extended periods of drought on the Serengeti. They live in a relationship that has a broad flat cup, where the lion population is neither large nor quickly responsive to abundant prey as far as fecundity is concerned. The ladybird, on the other hand, is ruthlessly efficient and given to explosive population growth in the face of abundant food. With the arrival of one predator, the prey population is doomed to local extinction in a short time. There is no stable equilibrium because the predator is too efficient. If, however, we aggregate to a larger system, then the problem for the insect predator is in finding not items of prey, but prey populations. There we find the requisite inefficiency. When it comes to killing a single item of prey, the pride of lions is reliable and efficient. There is no means available to lions to be prudent and only half-kill a single prey item, for it is all or nothing and the big cats gorge themselves if they eat at all. Thus, the insect prey population corresponds to a single lion's kill, and the population of ungulates comprising the lion's food corresponds to a metapopulation of insect populations scattered across the landscape. Stability at all levels is not a requirement for persistence, for only one stable level of organization is necessary.

Catastrophic Reorganization of Population Systems

Too much detail in a model makes it a special case. In the extreme condition, a model containing all the details of what is modeled ceases to be a model and becomes the thing itself. What interests scientists is not the reporting of the ultimate details of nature. Rather we are interested in simplifications of the system or generalizations about it that are predictive, despite lack of detail. In fact, the less detailed is the predictive explanation, the better. A polynomial with enough terms can reproduce any sequence of states of nature, with no gain in insight at all.

A similar condition applies not only when we deal with nature directly but also when we work with models. An explanation of model behavior that comes only from tracking the detailed changes of state of all the variables is unhelpful, because such an explanation has no generality. It is not sufficient to assert that the modeler created the system and so can explain system behavior from investigating the details of the interactions of the parts of the equations. In population models, we do not wish to regress to a sort of natural historical description of the behavior of the equations in question. The insights we seek are predictions from simplifications, the minimal model.

Bearing this in mind, consider the unpublished work of Steve Bartell and his colleagues at Oak Ridge National Laboratory. They modeled trophic levels as a system of generic populations in an aquatic food chain. Aquatic ecologists have argued at length as to whether their systems are controlled from the resource base at the bottom as it stimulates algal growth, or whether it is controlled from above by big fish cropping the level below them, so releasing the level below that, and so on. Empirical evidence for each scenario has been accumulated and tossed into the argument. Theoreticians have managed to clear the air by showing that both processes are at work all the time, but either top-down or bottom-up control dominates at any given instant. Bartell's model showed the switch in control, even for a relatively simple set of equations. What is remarkable is that he could not develop a predictive strategy for when the control would switch (figure 6.10). More surprising, even though he was dealing with computer simulation instead of nature, it did him no good to make the ultimate sacrifice in generality by looking at all the numbers his model generated. He probed the finest details coming out of his system of equations, and still he could not identify the conditions that precipitated the change in the control hierarchy. Science advances fastest when it is met with complete failure, for then one finds the limits.

If it is not possible to predict from complete knowledge of system specification and detailed behavior, then what hope is there for understanding natural systems? The problem is that the answers to some questions we wish to ask are unknowable; the pathology is in the question not in nature. One of the boons of studying populations is that it is possible to identify unanswerable questions very quickly. When Bartell wished to know when the switch will occur, the pathology of the question is laid bare. One of the helpful reformulations of the question that can be answered is, "Will the system enter phases when its behavior is unpredictable?" This we can answer by looking for characteristics of chaos or catastrophe. Let us see what those characteristics are.

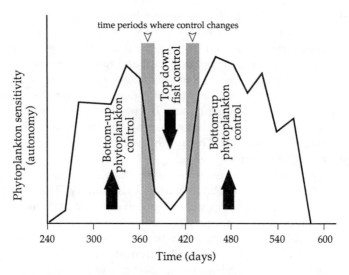

Figure 6.10. The output to Bartell's model for assessing control of the system by consumption as opposed to control by production. High sensitivity to changes in system structure indicates that the phytoplankton are their own master; low sensitivity indicates something else (fish predation) is controlling system dynamics. Bartell was unable to identify any general patterns that could predict the shift in primary control between fish and algae (courtesy S. M. Bartell).

We have established above that complex behavior will arise when very different rates are pressed together in the formulation of a population equation. At different times either fast or slow compartments in the model gain precedence. The complexity occurring in population models is akin to the unmanageable complex behavior that occurs in middle number systems. The difficulty in middle number systems comes from unpredictable changes in what constrains the system. An apparently minor component in the system gains control as the constraining entity. In population models the complexity can arise from the interaction of parts of the equation or system of equations that have different intrinsic reaction rates. Examples of these mismatches of scale could be instant death of a prey item coupled with a predator benefit that takes time to accrue, longevity. Complex unexpected behavior then emerges in difference equations with large growth rates because fast growth is coupled to long lag. As population models showing complex patterns of behavior move forward through time, different parts of the system of equations come to dominate. With a switch in dominance, the relaxation time of the output of the system of equations radically changes.

If one plots a fast variable against a slow variable, often the mismatch in relaxation times of the two variables causes the plot to curve so much that it actually folds. When the system moves past one of these folds, it goes through a phase when rapid change occurs. The details of the change are unpredictable from a knowledge of the relationship between the variables. After a short time the system finds the surface on the other side of the fold and a new relationship between the variables holds sway. Note that we can predict that such radical change will occur, but cannot predict its form. When the system is in transition between the folds of the surface, the inability to predict comes from the complete release of the system from the constraints of variables that make the folded surface. When the system passes over the fold, it is unstructured, and an unstructured system is not predictable (figure 6.11).

These folded surfaces have some general properties. If a change in the value of one of the variables takes the system over the cusp, then reversing the change does not move the system back down under the cusp to the original lower surface. Rather, it stays on the upper surface and exhibits a smooth continuous change that corresponds to the relationship between the variables on the upper, not the lower surface. Thus, in systems that show catastrophic changes of state, it is not possible to predict in the region of the fold, the state of one variable from the other, without knowing also which side of the fold is in use. This amounts to saying that to predict accurately, it is also necessary to know something of the history of the system. The state of the system, even at equilibrium, carries some information about the past in these systems. Usually the equilibrium in dynamic models destroys the influence of what happened in the past.

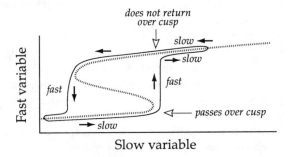

Figure 6.11. When the relationship between two variables is folded, the system moves rapidly and catastrophically to the other surface as it passes the edge of the fold. A reversal of the immediate change that took the system over the fold does not bring it back over the fold, but leaves it on the surface to which it has just moved.

In systems with folded relationships between variables, there is a degree of irreversibility in changes of system state. One is reminded of systems like supersaturated solutions or supercooled liquids where a single crystal precipitates a change of state. Withhold the crystal and the system remains liquid. From the relationship between the liquid and temperature alone one cannot predict system behavior. To predict freezing one also needs to know about the particle that acts as the nucleus of the freezing process. The local irreversibility of the change in state gives the local event lasting significance.

All too often the ecologist does not and cannot know about the small local event that precipitates the change. Knowledge of the general ecological process of change is as insufficient as is knowledge of the triangular packing of water molecules in predicting the form of snowflakes; it will predict that all snowflakes are hexagonal but not the detailed form of each or even any one snowflake. An equivalent situation applies to populations. There are all sorts of important determinants of population dynamics that cannot be known from even a detailed knowledge of what all individuals have in common. For example, sophisticated models in epidemiology cannot predict the start of the infection in the population, for that depends on some relatively unlikely event involving the arrival of the first infected individual. These awkward, irreversible aspects of systems which operate importantly far from equilibrium are relevant to population dynamics. In fact, the well-behavedness of populations in general, by virtue of their only consisting of one species, allows a clearer view of these difficulties than one might have in heterogeneous systems like communities.

Holling and his colleagues working on spruce budworm epidemics use their knowledge of the general characteristics of complex and catastrophic systems to work out what can and cannot be predicted. He has generalized his model to apply it not only to other insect pests but also to fire and grazing systems. We have already mentioned the budworm model in the introductory chapter, but it bears considering in the new context of catastrophe and entities wrestling for constraint.

The large-scale emergent property of the budworm system is the relatively long periods of low-grade infestation combined with occasional catastrophic outbreaks of budworm that kill almost all the fir trees in the forest. During the periods between outbreaks, the budworms are constrained by negative feedback involving predation by birds. Local increases in the budworm are met with increases in the bird population and increases in predation. The budworm depend on fir foliage as their resource base. As the trees grow taller, the negative feedback of birds and budworms oscillate around a gradually increasing equilibrium. The birds keep the insect in check, but the increasing foliage allows the insects to carry a larger predator load. The system is constrained by the slow growth of the trees.

There are, however, limits to how many birds can be packed into an area and how much food a bird can eat. These limits determine when the budworm population will escape the constraint of predation. The exact age of the trees when the packing of birds makes a difference depends on climatic factors that slow budworm growth potential. However, when the trees are between 30 and 50 years old, the birds are close to the limit of their capacity to control the insect (see figure 1.9). At this point the system is like a supersaturated solution or a supercooled liquid, for it is on the verge of a radical change of state. The exact time when the birds will saturate with budworm cannot be determined by the foliage or the bird population. The prevailing constraints are of no value in predicting when they will yield control to some other constraining factor.

At lower budworm populations, the influence of extrinsic factors, such as an influx of budworm from an infestation elsewhere, is attenuated by increases in predation that remove the excess in a timely fashion. However, when budworm densities are high and the system is close to the limit for bird density, a sudden influx of budworm is met with a sluggish response from the birds and constraint is lost. Once the population of budworms exceeds the birds' capacity for predation, the budworm escape the constraint of the negative feedback with the birds and a period of explosive increase in budworm begins. The exact time of the outbreak is unpredictable because it depends on accidents like wind direction coinciding with the right phase of the prey-predator oscillation. However, the fact that an outbreak will come within certain temporal limits is predictable, for that is based on knowledge that the system has a folded surface for the interaction between budworm and tree size.

Once the outbreak occurs, the constraining process in the system ceases to be the slow tree growth and becomes the rapid growth of the insect population. It is characteristic of switches in constraint that the timing is unpredictable. At the time of the switch, any of a large number of particular events can change the constraint. This makes the system a middle number system, with all the associated difficulties in prediction. With change in constraint, the dynamics of the forest change pace and character. The trees are not capable of a viable response to the new fast constraint, the whole strategy of "treeness" being slow behavior to keep hold of constraints. The trees are killed by the new regime, but this in turn wrests control from the budworm as they press against the new constraint of a diminished food supply. At this point, the epidemic declines and surviving or invading young trees establish the foliage and tree growth constraints once more.

The wrong question to ask is, "Exactly when will the budworm epidemic occur?" The answer to that question is unknowable. Better questions would ask about the texture of the epidemic cycle in a given region. Those more general predictions avoid middle number specifications of the budworm problem. The things most managers would really like to know generally imply the wrong sort of question, in that managers act to gain control only when the natural constraints on the budworm are losing control. Spraying insecticide is done when budworm are evident to human observers, and by that time the epidemic phase has begun. Spraying appears to hold the cycle in its least predictable and therefore least manageable phase. Essentially, contemporary management keeps the system in epidemic.

Holling recommends that management be focused at the larger scale of the whole cycle, the general form of which is predictable, and therefore controllable. He suggests that natural constraints on the budworm, like birds and disease, be bolstered. These constraints work during the phase of the cycle when managers are not aware of the presence of the budworm, although the birds can find budworm easily. By enhancing the phase where budworm is naturally constrained, the length of time to epidemic can be lengthened. In the knowledge that the forest is about to enter the phase where it is in danger of epidemic, the managers should impose a constraint like the budworm, before the budworm can reach its epidemic phase. Removal of the trees by cutting imitates the budworm and stops the epidemic from happening. Management needs to work at the level where the system is predictable, the level where constraints are reliable. It should work with the natural constraints.

Often management seems to act when the system is obviously undergoing change. It is much better to manipulate not the unstable phase of the cycle, but the phase when normal constraints can be employed as allies. Susan Riechert and her colleagues at the University of Tennessee have shown that generalist predators can keep a whole gamut of pests constrained. Until her work, it was generally held that spiders do not eat sufficiently voraciously to contain pestilence in horticultural settings. They eat with a moderation that is more like lion predation than the ruthless extermination meted out by insect specialists. The error in the conventional wisdom was to pay attention only to the outbreak phase of the nuisance. In fact, while spiders cannot deal with a pest outbreak, they can do a very good job of stopping the outbreak from occurring. Spiders gently but firmly press against small pest populations and represent a constraint that can be readily employed by merely offering them places to spin their webs in the garden.

Populations on Islands

We conclude this chapter with some sections that investigate the extent to which the functioning of communities can be explained by population phenomena. Conventional wisdom says that studying populations is a preferred path to an understanding of the mechanisms inside communities. Earlier in this book we noted that there are problems with that view which pertain to questions of spatial contiguity. Communities, as we conceive them, cannot be mapped onto a place, but populations can. There are two ways to tackle such incompatibility: deal with intangible populations; make communities tangible on the ground. It is their very tangibility that makes populations appealing, so the former seems unpromising. Tree populations are intangible because their individuality is masked by other forest species. In a forest, the tree population appears to be a conception that buys little and is usually not worth the effort. Either the population is unmanageable, as in a mixed-age, species-rich forest, or the community is trivial, as in a pure stand of a plantation.

The other solution, to force the community into tangible form, has its own problems. Trying to make the community tangible by putting it at places on a landscape does not help in the case of forests. The woodlot, as we noted, is a landscape entity where the spatially defined edge is the critical boundary. There is relatively little biology to be explained by the fence row at the margin, and so the woodlot becomes not just arbitrary, it is also biologically capricious. However, there are entities with many populations in them that are biologically meaningful and mappable on the ground. These are true islands, patches in an alien matrix where the boundary corresponds to a critical limit. Invasions are nontrivial and so extinction is meaningful.

Thus, a favorite type of community for population biologists who wish to widen their purview is the island. The study of islands has even been awarded its own named subdiscipline, island biogeography. The spatial boundary makes the island a peculiar type of community. Island communities resemble the tangible woodlot, but their boundaries are much more consequential. Very isolated woodlots could indeed be islands, but then their ecology is not that of the forest community that could cover the landscape if only humans would not cut it down. Isolation changes the scale of landscape occupancy and makes islands somewhat pathological, albeit interestingly tractable, communities.

If it is relatively impervious to invasion and emigration, then the boundary of the island puts upper limits on the scale of occupancy of the landscape. To the extent that the island forces occupancy of the landscape to just one scale, islands deny the interference pattern which has been the community of focus in this work heretofore. Since differences in scaling are commonly the explanation of coexistence in communities, curtailing scaling options by the bounds of an island is an invitation to competitive exclusion, which in turn denies the community its existence.

It is not surprising that population biologists should feel comfortable in an island setting because the island community allows the spatial mapping that is at least intuitively part of a population conception. It is also no coincidence that island biogeographic communities attract zoological ecologists over botanists. Remember that animal communities consist of populations, while plant communities have individual organisms as their attributes, at least according to the most prevalent community conceptions of botanists and zoologists. Apart from anything else, the patterns of plant versus animal data collection in the field lead us to this conclusion. Accordingly, it is no accident that island biogeographers use predominantly animal population invasion and extinction as their fundamental explanatory principles. Part of the agenda announced by many island biogeographers is the explanation of community structure and properties by building communities from their constituent populations. Often this is explicitly stated. Islands would seem to be the place where that could be achieved, if it can be done at all. The spatial tractability of islands simplifies the system by placing bounds on the upper-scale ranges of the community interference pattern. Animal studies do invoke the appropriate lower-level entities, namely populations.

Jonathan Roughgarden began his studies of the *Anolis* lizards on the Lesser Antilles with an explicit plan. His design was to show how complex communities work by studying simple island communities and extrapolating up to more normal communities with many species. His communities were so simple that they consisted of either one or two species. If the island had only one lizard it was always of medium size, 60 millimeters in length. In the relatively uncommon situation where there were two lizard species on the island, one was large, 100 millimeters or more, while the other was small, 45 to 50 millimeters. The conventional explanation is character displacement. Over time the lizards have apparently minimized competition by the selection against individuals of both species in the overlapping middle size range. The large individuals in the large species and the small individuals in the small species gain advantage by saving resources from avoiding competition. This displaces the character of lizard length away from the optimum manifested in the absence of competing species. If this conventional wisdom were correct, then indeed the workings of these simple communities would be examples of the beginning of the evolution of the complex interactions in species-rich communities.

Roughgarden has collected data from various sources that give the lie to competitive exclusion and the conventional view. His sources of evidence were as diverse as variations in tooth length over a short "fossil" record of a few hundred to a thousand years, and the unique patterns of local occupancy on a two-species island. Most islands are completely occupied by the lizard species present, but on one island he found the smaller lizard species excluded from the coastal area and inferred that it was fighting a losing battle in an upland enclave. The data suggest that there is little character displacement. Roughgarden's preferred scenario is that most islands only have one lizard because competitive exclusion has removed all rivals. In the case where there are two lizards, the larger lizard is the new arrival and is in process of exterminating the smaller species. The smaller species was medium size before the invasion of the large species, but it is selected to be smaller during the struggle. The smaller species always is driven to extinction in a relatively short time; hence the rarity of two-species islands. When the small lizard is driven to extinction, the larger lizard is selected so that it approaches the climatically defined, optimum, medium size, which is the size of all species living alone.

These results highlight the difficulty of using population studies to understand the important workings inside communities. Roughgarden came to realize that he was not studying simple communities beginning to form the complex interactions present in a mature community. The large invading lizards appear to come from large islands with many species in mature communities. What he was studying was not the emergence of community accommodations between species. Rather, he was studying parts of mature communities that had been excised from their context. What was missing from his small islands was exactly the constraints that his original agenda had stated to be the focus of the study. The problem with using multiple population studies as a means of understanding communities is the absence of the very constraints that constitute community structure. Population studies let ecologists study the raw material from which communities can be made, but they are inept at addressing the constraints that make the community more than a mere collection of species. Multiple-species autecology is not community ecology. Population studies can give some account of the naked, unconstrained parts of a potential community, but they cannot easily test community constraints.

Mutualism and Competition

Most of the population equations we have discussed so far have been prey-predator systems. These are negative feedbacks that involve constraint at the level of direct interactions between a small number of populations. Prey-predator equations are generally stable in themselves and need not involve invoking constraints from the community. It is possible to look at complex foodwebs as community structures but that is a different matter.

Mutualism, on the other hand, is a positive feedback relationship where different species benefit each other. There has been little work involving formal equations on mutualistic systems. One reason is that the mathematics of mutualism is not as transparent as the simple difference and differential equations of prey and predator. The other reason is that, being a positive feedback, mutualism is inherently unstable. Positive feedback systems are hard to study because the very relationship one seeks to study would be transient if it were left to its own devices.

Inside every negative feedback is a positive feedback trying to escape. Mutualism is that positive feedback. The negative feedback is the constraint that holds the mutualistic system in check, the context of the mutualists. By the time we find a mutualistic system, its dynamics are not principally determined by the mutualism. Rather, the positive feedback of mutual benefit is held in a context that limits the system state. If the constraint is unchanging, then so will be the state of the mutualistic association. This constancy would prevail even if the parties to the mutualism were changing their relationship by intensifying it or weakening it by flirting with an ambiguous parasitism or commensalism. Any change in state of the parties to the mutualism will be a reflection of a change in constraint that gives more resources to, or takes more resources from, the mutualistic system as a whole. The direct effect of the mutualism is not to produce change but to keep the system constantly pressing up against resource limitations. The same is not true for a prey-predator system because the prey-predator relationship itself provides its own constraint. Accordingly, changes in the state of the prey and predator can be independent of the context of the prey-predator system.

All this means that mutualisms, and not prey-predator systems, are the more suitable devices for relating communities to population considerations. The community is the accommodation between community members and it represents the set of constraints on the community members. Because they have their own built-in constraints, prey-predator systems are to an extent isolated from constraints at the community level. Certainly they represent a more complicated situation compared to mutualistic systems, whose dynamics are exposed naked up against the upper-level constraints of the community.

Mutualism is a positive feedback where all the interactions are positive. Competition is a relationship of mutual negative effects, where the success of any party has deleterious effects on the others. However, competition is still a positive feedback. Success in a competitive exchange delivers to the victor more resources with which to press the advantage. Therefore, for the same reasons that mutualism is a good tool for probing community-population relationships, competition is also a device of choice. The positive feedback of mutualism is constrained at the community level by resources. The constraint on competition is different; the community constraint on competition is the means whereby the loser of competition avoids extinction, while the competitive winner is denied a clean sweep of the spoils of victory. Simple resource limitation is not the key to understanding the role of competition in binding together community parts. Counterintuitively, the loser of the competitive exchange more than the winner gives the insights into community considerations, as we explain below.

This has implications for the protocol for investigating competition. It means that only some aspects of the competitive interaction relate to community constraints, while others are limited to only population considerations. This relevance of the loser to the community should influence our use of devices like the de Wit replacement series for investigating competition in our communities. In the de Wit experimental protocol there are extreme and intermediate conditions. At each extreme condition plants are grown in pure stands from a given density of propagules. The intermediate situations involve mixtures of propagules of the two species sown so that the propagules of both species combined have the same density as the pure single-species stands. The plants are then grown, and the effects of competition are assessed from the standing crops of the species in the mixed stands compared to the monocultures.

These de Wit replacement series offer a variety of insights into the competitive exchange, only some of which pertain to community processes. The method is capable of separating within- as opposed to between-species competition. In the intermediate case in the middle of the series, half the plants present in the pure stands are substituted with plants of the other species. This releases each species from half of its intraspecific competition and substitutes it with competition from the respective other species. If interspecific competition is indistinguishable from intraspecific competition, then the 50–50 percent mixture should have a total productivity halfway between the levels of primary production recorded for the species grown separately.

Arrange the replacement series as a sequence of bars on a bar graph, draw a straight line from the top of the bars for pure stand production, and we would expect the production bars of the mixtures in between to just touch that line, so long as competition is independent of species (figure 6.12). If the line drawn between the two pure stands along the tops of the intermediate bars bows upward, then production in mixed stands is greater than expected. In this case, intraspecific competition is more severe. One or both species can more than compensate for the lowering in productivity that should come from there being only half as many propagules in the 50–50 mixture. The lowering of the density of its own kind for each species in the mixture allows one or both species to grow more vigorously in mixed stands. Conversely, if the line for total productivity sags in the middle of the series, then total competition between the species is greater than that within species (figure 6.13). The introduction of individuals of another species does more harm to

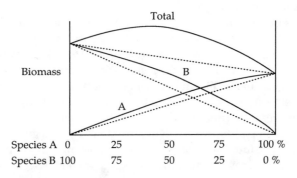

| Species A | 0 | 25 | 50 | 75 | 100 % |
| Species B | 100 | 75 | 50 | 25 | 0 % |

Figure 6.12. De Wit replacement series substitute competitors between species. At each end of the graph the organisms occur in pure culture. Halfway in between, each species is only at 50 percent density. Thus summed initial densities of both species are identical across the graph, but the proportions vary. If there is no difference between inter- and intraspecific competition, then a straight line connecting the final biomasses of the two pure cultures should be in line with the final biomasses of the various mixtures. The performance of the species separately can be judged relative to a straight line that goes from final biomass in pure culture of the species in question to the zero mark in the other pure culture. Values above expectation indicate that there is more interference within the species in question than between the two species.

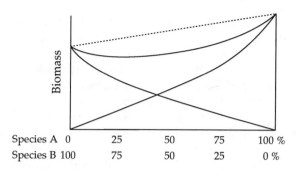

Species A	0	25	50	75	100 %
Species B	100	75	50	25	0 %

Figure 6.13. If in a De Wit series observed values fall below expected, then competition between species gives more interference than within species.

growth than keeping the density within a species as high as in the single-species stands.

As well as allowing insights into the balance between interspecific and intraspecific competition in the total system, the replacement series gives information on the species separately. In the same way that it is helpful to draw a line between the productivity levels of the pure stands, insights can also be gained from a straight line drawn from the productivity of the pure stand of one species to the zero mark of the pure stand bar for the other species. The line goes to zero because the species in question is by definition absent from the pure stand of the other species. If the productivity of the middle of the series for the species under consideration falls on that diagonal straight line, then the presence of the other species has no effect. The members of the other species are competitively indistinguishable from members of the species in question. If the productivity of the one species in mixed stands sags below the diagonal line, then that species is losing resources because of interspecific competition. Alternatively, if the middle of the replacement series for that same species bows up above the diagonal line, then that species suffers strong self-suppression when growing by itself. Strong self-suppression may occur in both species, in which case productivity of the middle of the series is higher than expectations (figure 6.12). Conversely, both species may suffer severe interference from the other species, in which case total productivity of the middle of the series is suppressed (figure 6.13). If one species has a diagonal productivity line that bows upward while the other species sags down, then the upward-bowing species is the clear winner of the competitive interaction.

The total productivity in the middle of the series is a reflection of the accommodation that the species have made toward each other. That is an aspect of species competition that might give insight into the way the species fit into the community. Identifying the winner of competition pertains to the aspect of competitive interaction that does not run its course in the community. It is a population and not a community matter. If the de Wit series told the whole story of the competitive feedback between species, then the loser should always be excluded from the community. However, the loser does persist by virtue of constraints on the feedback loop. Those constraints are aspects of the community above and beyond the sum of isolated population characteristics. If we wish to use competition between populations as a vehicle for understanding communities, we must focus on the way the loser survives despite competitive losses. The winner alone offers little to an understanding of the community.

Robert O'Neill thinks that the loser survives in the community by occupying the landscape, and competing on it, at a scale different from that of the common winning species. John Harper noted early in the modern work on plant demography that there were four species of poppy present in most weedy fields in England. What is remarkable is that the rarest of the four is present in the system with the same consistency as the ubiquitous common species. The inferior competitor is consistently rare but is just as consistently present. The question is, how does it lose competitively while avoiding exclusion?

The common species could be seen as competing for dominance of every square meter. It is successful in that regard and has high densities which translate to a small mean area per plant. By contrast, the rare species does not occupy the landscape at that scale. If it did work at the one-square-meter scale, it would be taking on the common species on its own terms and should be driven to extinction. The rare species has an ecology that works at a larger scale. It occurs at low density which translates to a large mean area per plant. That larger mean area is the scale of landscape occupancy that the rare species employs. The de Wit series as it is usually performed gives no insight into this community question. The alternative experiment to perform is one where a large number of replacement series are performed in a setting that includes the microsite variability present in the natural system. The common species will fairly consistently win in the various replacement series. However, if O'Neill is correct, the replacement series of interest will be the ones that give anomalous results, with the rare species winning the series in a few cases. The frequency of those unlikely events should occur in inverse proportion to its mean area per plant as compared to the mean area of the common species. Conversely, it would occur in direct proportion to the respective species densities. Counting the anomalies would be a community experiment and not a population experiment because it addresses a distinctly community characteristic. It is the different scales of occupancy of the landscape that generate the interference pattern which is the community intangible.

Conclusion

In this chapter we have shown that populations are not merely the attributes of communities. They are ecological entities in their own right that deserve their own approach. Only in some situations and some conceptions is the population the level below the community. Conversely, there are things about communities which cannot be readily laid bare by population approaches. Populations have a homogeneity of scale in their attributes, for all members are usually from the same species or are at least equivalently scaled for the ad hoc reason for erecting the population in question. This makes populations tangible and manageable in a way that communities are not. As a result, there is a depth of insight into the working of populations through formal representation with equations that is not available to the community ecologist. However, that very homogeneity of scale in populations is exactly why multispecies population work is not community ecology. Population ecologists are well advised to play to the strengths that their conceptualization allows. By all means, intellectual hybrids somewhere between community and population conceptions will be worthwhile. However, an insistence that the precision and tangibility of populations makes them a somehow intellectually superior conception is ill-advised. There is robustness in a diversity of approaches to complexity in ecology.

7. THE BIOME AND
BIOSPHERE CRITERIA

---◆-◆---

As we suggested in the introduction to our organizing scheme at the beginning of this book, biomes are conventionally considered to be spatially large, at a high level of organization, the major parts of which the biosphere is composed. Although we would prefer not to call the biome a level because we reserve the word "level" for scale-related definitions, in the case of the conventionally defined biome, there is indeed a scale-related component to it. However, advantage is to be gained from dissecting questions of scale away from other aspects of the biome, in the same manner as we did for other criteria. The biome criterion can be insightfully applied to systems that are in fact small in area and ephemeral.

Comparing Biomes to Other Criteria

Biomes belong in certain regions on the surface of the planet. Nevertheless, they are not simple landscape entities. Populations can also be mapped onto places on the ground, but they have their own distinctive characteristics that make their identity as a place only an incidental character. The same logic applies to biomes. There is a distinct causality for biomes which only secondarily circumscribes biomes in space. Physiographic features, a landscape consideration, are responsible for delimiting biomes in space, but they do it indirectly by defining the climate in regions of continents.

Biomes are distinctly climate-mediated. Despite the central involvement of the physical environment of the biota in defining a particular biome, the biome is not just a big ecosystem. This is because the physical environment is distinctly the context of the system rather than a part of it. An ecosystem, by our definition, includes the soil and the local atmosphere as being explicitly inside the system. The biome is defined primarily by its biota. If there is a general soil type that is associated with a particular biome, then that is seen, by our definition, as a consequence of the action of the contextual climate and the biota that identify the biome. It is not because the soil is part of the system. In our framework, large ecosystems would offer a better account of the soil and biota as an interaction over vast tracts of land. Let us emphasize again that we do not erect our framework as being singularly correct, but it does in this case allow certain distinctions that might otherwise be lost. Note how ecosystems by our definition are studied using mass balance and their components are pathways. Biomes do not generally yield a suitably rich picture under such a process-oriented approach.

The emphasis on a multispecies biota does not make biomes into large communities. While the biota of a biome define it, life forms and not species are the biological subunits employed. In communities, the emphasis is on an accommodation between different species. In biomes there is primarily an accommodation of the biota to the physical environment. Of course, a community must accommodate to its physical environment, but there is more slack in community responses to climate. The response of communities to microclimatic variation is manifested as interesting species differences, not as changes in life form. The environment of a community is the context in which one of several communities could emerge; the environment determining communities is now viewed by Roberts and other contemporary theorists as an error in the thinking of earlier community ecologists. By contrast, in biomes there does indeed seem to be an environmental determinism. According to David Roberts, the community is not a function of environment but is better described as a relation. Thus, environmental determinism is an inappropriate model for communities. In communities there is room for maneuver within the context of a physical environment. What makes a biotic collection a biome is the manner in which all members are pressed against certain constraints that dictate plant architecture of the dominant. The same vegetation can be seen as either an exemplar of a community or a biome. The difference is the type of environmental relationship that is considered and the causal chain which is given primacy.

A biome would be incomplete without its animals, whereas most community considerations are of either plants or animals, but not usually both at the same time. So different is the scaling of plants and animals that recognizing the multispecies interaction of plants and animals together makes for a very different entity than the plant or animal community considered separately. More than sheer size, the critical difference between the community and biome concept is the inclusion of animals wholesale in the life of plants. It is possible to consider particular plant-animal accommodations between species, but that is a separate issue discussed in a following section. For the moment, note that so differently scaled are plants and animals that the perception of each kingdom of the other is only in the most general terms, special plant-animal species accommodations notwithstanding. For most plants, almost all animals that matter are not so much distinguished by species as by some more general character like grazer or disturber. For most animals, plants that matter take on generic characteristics like food or perch rather than a species-specific identity. Thus biomes, with their required animal component, are defined by general animal activities like grazing or browsing. Thus the plants of a biome are not so much members of species as they might be in a community; rather they melt into some more general category defined by a generic service to animals. If we were looking for small biomes, local plant systems whose physiognomy was animal-mediated would be likely candidates. Climate, the other determinant of the physiognomy of conventional large biomes, cannot usually be so locally contained as to act as a unifying force for small "biomes."

Animals play a special role in holding biome components together and are important players in producing the remarkable homogeneity that we find across the vast tracts inside a single biome conventionally defined. C. S. Holling suggests that the movement of animals over large distances relates otherwise disjunct stands of vegetation. He notes that the spruce budworm moves several hundred miles, and in doing so produces the physiognomy of the boreal forest. Budworm move much more effectively than pollen to leave their mark on some distant part of the biome. In a population a balsam fir tree is a breeding member of the group; in a biome that same tree is more significantly a host providing food for an incipient epidemic. Certainly grazers play a large role in giving the grazed biomes their distinctive vistas. Thus individual grasses as constituents of biomes are less importantly embodiments of a genome and are more fuel for the grazers who beat back encroaching forest shade. Fire is the only other system attribute that can fashion vegetational physiognomy like animals. It is no accident that pestilence, grazers, and fire all move at about the same speed over the same sort of area. Only something with those general scaling characteristics can be the glue for a biome.

For a biome, the very essence of the situation is the manner in which the physical environment, mostly climate, determines what the biome shall be. The life form is all that the climate will allow. If there were more water there would be shrubs or trees. It is the quantity of water available or the critical limiting temperature in a given climatic regime that determines the biome which is found. There is a certain cost and benefit in being a tree. The costs are susceptibility to fire damage and grazing at the sapling stage, as well as respiratory load and critical water demand that must be met; the benefit is overtopping herbs and shrubs. For grasses, the costs of a tree are minimized and become the benefits of being able to live in a dry climate under ungulate grazing. The benefits of being a tree are exactly the cost of being a grass; they suffer shading if the climate permits trees. The physiognomies that characterize grasslands, shrublands, deserts, deciduous and evergreen forests are all direct reflections of what the climate will allow.

Consider now changes in climate and how they affect organisms. Individual organisms adapt to physical conditions with devices like closing stomates to conserve water in drought. However, there are limits to adaptive responses, particularly when the whole life cycle is considered. Trees may be able to withstand drought when seedlings cannot. It is the trees that give a forested biome its physiognomy, but without seedling trees that cannot sprout trees are not replaced. We raise again the issue of wave interference patterns. Think of the climatic regime not so much as an average condition but as a set of critical periodic events. For example, the longest period between showers in the wet season could define the site as one that can sustain reproduction of the dominant life form or not. Also the length of time that rain comes at a sufficiently high frequency to sustain seedlings determines if the wet season is long enough to allow establishment. Thus, in Ron Neilson's assessment, vegetation physiognomy is a stable wave interference pattern between climatic periodicity and tolerances of critical life stages of dominant life forms.

In wave interference patterns, very small shifts in the character of the component waves produce radically different patterns. For example, a continuous change in climate might gradually lengthen the longest period between rains; nothing happens to the vegetation until a critical threshold is suddenly crossed. Small differences in critical periods can set up the demise of a whole biome. More than that, the crippling blow is usually struck at the juvenile stages of the dominant vegetation type, so nothing appears to be wrong until some singular event, like fire, removes the adults that determine physiognomy. Changes in periodicity can have great effects quickly. Local newspapers have accounts of old-timers in southern Wisconsin puzzled as to why forests had grown up so fast since white settlement a few decades before. Change in fire frequency, releasing oak grubs, was the unrecognized cause of the shift in vegetation physiognomy.

Also related to sudden changes in vegetation is the relationship between land form and climate. Major topographic features which obviously relate to the placement of biomes remain constant over the time frame that it takes for major climatic shifts. Many aspects of climate change are continuous, like gradually increasing carbon dioxide. However, the major landscape features, like the Rockies, remain constant and represent a foil against which the jet stream plays to produce storm tracks or extended drought. Thus, regional climate is held constrained until it breaks through a set of physiographically set limits.

Even as the global climate changes, the major topographic structures on the earth remain in place casting a climatic shadow that is the context for the major biomes. Ron Neilson has data that suggest that the boundaries of entire biomes can be remarkably robust. He reports that the entire Great Basin can switch between shrubland and desert grassland with alternating extended wet and dry periods. The explanation is changes in wave interference, discussed above, set in an unchanging physiographic context.

When the climate shifts, the various woody plants can hang on so long as the drought is only for a decade or so. That is how woody biomes hold their territory. However, suppressed reproduction leaves the old biome vulnerable. Perhaps fire causes the catastrophic change, and grazers hold the ground so gained. While by no means do all climatically driven shifts in biomes occur at a persistent frontier, quite often there is a snap at which the boundaries between vegetation physiognomies move dramatically. In the drought of the 1930s the prairies started to break up wholesale. This surprise shook the faith of the orthodox Clementsians, who had always thought their climax vegetation was stable. In the case of the Great Basin, the snap can apparently be across the whole biome region at once. We have little reason to suppose that gradual encroachment of biomes into one another's territory is the rule.

Given the massive areal extent of biomes, they are remarkably homogeneous. The above discussion indicates why this is the case. A given climate sets the basic rules. Trees cannot persist in areas where the soil moisture is completely depleted for a month every year. In drought cycles trees will completely deplete the water supply and the weak will succumb. Without widespread catastrophe the trees could persist for their natural life span of several centuries. However, catastrophe through pestilence or fire causes wholesale change in physiognomy. At that point limits on establishment and the animals of the new biome maintain the new state of affairs. Physiography sets the critical limits that follow from the climate, and the forces for internal cohesion keep the response homogeneous across wide climatic regimes.

Plant-Animal Accommodations

A critical part of biomes is the relationship between plants and animals. We said that plants and animals do not form intimate relationships of community with each other because they are differently scaled. In biomes, plants and animals have generic models of each other. However, some readers at this point will be thinking of many exceptions to that general statement, and they are right. For completeness we turn briefly to some of those special relationships, noting that they are very different from the plant-animal relationships in biomes.

The exception to the rule of generic categorization across kingdoms would be the special relationships between plant and animal species one on one, where there is coevolution. It is no accident that plant-animal interactions as a subdiscipline attract population biologists with their proclivity for genetically and evolutionarily based definitions of groups of organisms. We confess that this type of plant-animal entity has received little attention in this book, partly because it lies unrecognized as a distinctive criterion in more conventional views. It has no name, although it probably deserves one. If we were to extend this book further, we could write a whole chapter on the "evolon," or some such term for entities with highly coevolved parts.

The evolon has its own distinctive relationship to time and space. Clearly a long time line is important here, because the tight accommodation between the species is the very essence of the structure. In communities there is an accommodation between entities that have done most of their evolution in the company of species other than those they encounter at the present time. The accommodation in communities is a fine adjustment to the species who happen to be there in this interglacial period. In the evolon, such as a highly specialized pollination system parasitism, or commensalism, the species involved owe their identity to the relationship: no interglacial happenstance collection here. The evolon must have a longer history than the community.

Another distinctive characteristic is that the evolon is usually held in a very narrow scale range. For example, in an evolon of pollinators of cacti, the animal ignores the plants as if they were rocks except when they have flowers. The critical relationship is held in the vise of a very particular time-and-space framework. This is probably required for the stability of an evolon. It could not persist long enough to form the intimate relationship between the parts if the scale of the interactions were not narrowly prescribed. For evolution to get a handle on something as canalized as a moth's tongue and a petal spur so that the two are the same length, the scaling of the parties to the interaction would have to be precise. If scales were different or various, signals would pass only with difficulty and with much noise. Slack in scale of the relationship of the parts would almost certainly break the chain of coevolution. Coevolution probably has to be a continuous process to produce the highly specialized structures that we commonly see in evolons.

Biomes in the Biosphere

In their book, *A Hierarchical Concept of Ecosystems*, Robert O'Neill and his colleagues suggested a pattern where physical constraints alternate with biological constraints in an ecological hierarchy. The argument is that when a biologically based advantage gains ascendancy, it is pressed by reproduction to a new limit that has a physical base; this is the same line of logic invoked by Darwin for the resource limitation underlying natural selection. The physical limit on some raw material is then broken by some economy of resource use or becomes irrelevant as some other biological advantage is evolved. The process is then repeated, as biological and physical constraints are interleaved while the systems evolve to higher levels that incorporate increasingly general limits. This alternation of biological and physical constraints is the organizing principle we will use to relate the biome criterion to the biosphere.

The four paradigms of contemporary biology are: organism, species, evolution, and mechanism. Together these paradigms fix the conventional view. The environment interacts with the genome to produce a phenotype. That phenotype performs variously well in the environment, and through an environmentally mediated reproductive success, the organism contributes to population evolution. In all this, the point to notice is that the physical environment is context that acts on the biology. This basic model has been the orthodox view of the functioning of the entire biosphere. However, in the early 1970s, Lovelock proposed the Gaia hypothesis. Enthusiastic followers have overextended Lovelock's hypothesis to levels of great mysticism which are entirely unnecessary. Simply, the Gaia hypothesis invokes a different causality from the conventional wisdom. Instead of life responding to and being held in the context of the physical environment, life is seen as the context of the physical environment in the vicinity of the earth's surface.

The need to invoke a switch in the patterns of control comes from the very unlikely scenario of the earth's atmosphere. Despite the fact that the sun has increased its energy output by 30 percent since the beginning of life, the temperature of the atmosphere at the surface has remained within a narrow range. For the last three and a half billion years, somewhere on the earth's surface has remained warmer than freezing and well below the boiling point of water. This is far too unlikely were physical forces driving the system. It would seem that life has been controlling the atmosphere at a biospheric scale, and not vice versa. Life is more than complicated physics and chemistry, it can only be understood as a goal-directed system. For example, pathology can only be understood in the context of proper functioning in the pursuit of a purposeful healthy state; camouflage and mimicry patently serve a purpose and cannot be well understood in other terms. If life is the governor of the atmosphere, then life's purposiveness pervades any adequate account of long-term atmospheric behavior. This argument does not propose a consciousness of the atmosphere. Such a suggestion fails to understand that purposiveness invoked in the understanding of anything comes from the observer.

Given the model of O'Neill et al., if the entire biosphere is under biological control, we might expect large ecological subsystems immediately below the biosphere to be under physical constraints: remember that biological and physical constraints appear to alternate. The biological control of the entire biosphere does indeed appear to be in stark contrast to the control of the biomes. Biomes are held in the vise of physical constraints. When there is not enough water to support the demands of woody vegetation, then the biome changes to a shrub or grassland type. If O'Neill and his colleagues' assertion is general, then all very large but sub-biospheric entities should be best seen as held in the context of a physical environment.

Biospheric Scale Applied Ecology

Very large-scale ecology has only just become an object of study and has been left for meteorologists and paleobiologists to impose their own particular point of view. However, industrial might is now able to exert such force on the biosphere that ecologists are being forced into taking a large-scale point of view. With these new endeavors we discover some remarkably general principles. This section lays out one such set of principles.

The destruction of the ozone layer has serious consequences. It does appear to be something new, for we have measurements of ozone in the stratosphere that go back far enough to indicate that there was no hole twenty years ago. The most important destroyer of the ozone layer is freon. It is a harmless chemical at our layer of the atmosphere, the troposphere, such that it is used as the propellant in spray cans. It is also put right next to our food, for it is used in refrigerators. Freon is a class of compounds that resemble the smaller hydrocarbons, like methane, but they have fluorine and chlorine in place of some of the hydrogens. The class of compounds is functionally inert in most circumstances, although it is hazardous to smoke cigarettes around freon. We used to use ammonia as the principal refrigerant, but that is highly reactive and not the sort of material one wants around the house in concentrated form.

Ozone is the unstable molecule consisting of three oxygen atoms. Even isolated from other reactive molecules up in the stratosphere, ozone is so highly reactive that it has always degenerated to ordinary oxygen molecules with only two atoms. Ozone is replenished by the sun's ionizing radiation which creates ozone from oxygen, reversing the natural degenerative process. In the stratosphere, freon is broken down by the sun's rays to release its chlorine. Chlorine from freon is a highly reactive arrival to the stratosphere, and it reacts with the ozone to make oxygen, reverting back to chlorine gas again after it has done its destruction. Chlorine therefore acts as a catalyst in the destruction of ozone and is not itself consumed in the reaction. It would appear that the damage will continue for at least a century or so, until the chlorine is lost into space.

If ozone is so reactive, one might ask why other chlorine-containing gaseous material does not wreak the same havoc in the ozone layer. After all, there have always been chlorine-containing gases in the atmosphere. The answer is that chlorine itself is so highly reactive that it has reacted with something else and is immobilized in some large molecule before it can reach the stratosphere. The estimates are under some revision, but it takes between two and ten years for tropospheric freon to reach the stratosphere. Almost any gas that exists in the atmosphere for that long is going to react with something so as to immobilize the chlorine it contains. And there's the rub! For a molecule of its size, freon is remarkably inert, so it does not allow the chlorine to be trapped in some large molecule, one too large to reach the stratosphere. The inertness of freon makes it a vehicle like no other for transporting chlorine up into the stratosphere.

Underlying the above chronicle of events leading to ozone destruction are some general principles about relative scaling. Together they form a predictive framework that indicates far-reaching strategies for human design in an ecological setting:

1) Predictions are not that something will change, but are that nothing will happen—we can only predict that whatever orders the system will remain in place.
2) The effect of nothing happening is likely to be the same across a wide range of situations, e.g., if something is inert, its behavior can be predicted to be nothing for a large number of conditions, whereas if something is reactive, one reaction is a poor indicator of the effect of other reactions.
3) Inert material accumulates. This explains a lot about the way we find the world; the composition of air, for example. Run the global system for a few billion years and something as inert as nitrogen gas must accumulate, for everything else has reacted and been removed in the process.
4) "Inert" is a scale-relative term. Being inert now means that only the passage of time, albeit a long time, is needed for there to be a reaction.

Let us expand upon and weave together these statements. The outcome is counterintuitive, for in the final analysis it appears that biological material can radically alter its global environment precisely by stabilizing its local environment.

One very effective way for the human social system to choose elements that will not harm us is to select inert materials. If a substance is inert, it is scaled so as not to be capable of involvement in our own chemistry or the chemistry of anything else that is dear and, literally, near to us: for example, it does not interact capriciously with our house plants or house paint. That is what makes freon so appealing as a household material. At an immeasurable rate, it probably does become involved in chemical reactions in our bodies, but one in a quadrillion or so is also undetectable by our biochemistry at any level. So freon appears completely safe.

For a student of relative scales, "inert" is a relative term. The human industrial universe in which freon is created sees it as inert, but over a long time (several years) and in a larger universe (the total atmospheric system) freon is far from inert. By temporally scaling freon away from the scale of increments in seconds for our biochemistry, we have scaled it up to the decades and centuries upon which total biospheric time proceeds.

The general principle for human design in ecological systems is as follows: things which are inert at a small scale can often be expected to be critically reactive at a large scale. By virtue of their being inert, stable

Figure 7.1. The production of stable material over time necessarily leads to an accumulation. Eventually this will lead to a change in an undefined upper level producing a surprise.

materials can accumulate such that the quantities created over decades, centuries, millennia, or increments of geological time are large enough to have large-scale effects (figure 7.1). Being inert has a generic effect that leaves unspecified the universe to be perturbed. With such a large target, the principle is predictive. As predictions go, saying that something that has absolutely no effect today will have an effect some time later is a robust forecast. At first it appears an uninteresting prediction, but with examples like freon destroying the ozone layer, it appears to pertain to unexpected and important situations. Generally inert material behaves, or rather does not behave, the same way with respect to a large number of situations and agents. Inert material is predictable in a general way. We use freon in our homes because we can predict its safety with respect to many aspects of home life. Therefore, we do not need to consider these aspects of human existence one at a time.

Scaling Strategy in Natural Systems

In human industrial systems, there is an active decision-making process. Premeditated action is a standard operating principle in industry. In natural systems that is no so, but evolution does produce systems that can be described as goal-directed, even though evolution itself is not a system with its own cognition and innate purposefulness. We might expect living systems to find inert materials helpful for the same reasons that industrial systems actively choose them. For organisms, using inert materials so as to scale their immediate context away from their own level of behavior has very general advantages. First, the material will be decoupled from the general level of functioning of the organism. Second, if it is inert for one critical process, it will probably be inert with respect to all critical biochemical processes. Accordingly, many organisms use inert materials in their homes. That is the reason diatoms make the frustules in which they live out of silicon; sand is inert stuff.

Another inert material used by microscopic organisms is calcium carbonate. Calcium bicarbonate is soluble and reactive, and it is the vehicle that delivers calcium to microscopic homemakers. Respiration and photosynthesis involve carbon dioxide, which changes the local pH up and down depending on the light conditions. This causes the precipitation of calcium carbonate, a relatively insoluble form of calcium. It has been a favorite building material for microbes ever since the blue-green bacteria started forming calcium deposits about three billion years ago. Many aquatic organisms now use calcium carbonate as a protective covering: for example, molluscan seashells and microscopic rotifers. It is inert with respect to their physiology and so does not interfere with bodily processes, while it plays a crucial structural role in bodily protection. It is predictably safe with respect to biochemistry, and keeps doing the same good job throughout the organism's entire life.

What is particularly interesting in the present discussion is the way that fossil accumulation of calcium and silicon does indeed have an effect on the functioning of the present biosphere. This would be a prediction that we might make from the way freon has biospherical-level properties. Diatomaceous earths may appear inert, but they may well be players in vulcanism. Diatom production is large in shallow seas, so there is an accumulation of this material on coastal shelves. The sheer mass of the sediment presses the edge of continental plates down, encouraging plate subduction that leads to extensive volcanic activity. Volcanoes are a significant source of greenhouse gases. Thus the inert shells of diatoms can play a significant role.

Fossil calcium deposits are even larger players in the modern biosphere. As with diatomaceous deposits, it takes millions of years for calcium to become an active player. In the end everything inert becomes active, and in the case of calcium it takes changes in sea level and a rising of the sea bed to become land. Calcium gives lime-rich soils that determine the flora of entire regions. Calcium influences the solubility of most other plant nutrients, and so the inert homes of ancient microbes set the ground rules for all contemporary terrestrial plant communities by being abundant or otherwise in a given locale. Given its effect on other nutrients, calcium is directly and indirectly responsible for the concentrations of productivity in modern oceans close at hand. At the level of nutrient cycling over millions of years, and the recycling of continents in contrasting periods of flooding and upthrusting, the inert homes of sea animals play a crucial role. The sedimentary rocks react to water erosion differently from igneous rocks, and these differences create physiographic structures of ecological importance, for example Niagara Falls. Landform in general can have the placement of calcium sediments as one of its causal agents. Formerly inert calcium is a big player in the modern biosphere.

Consider in the same light the contemporary crisis in rising levels of carbon dioxide and the projected global warming. One of the main drivers of change in this scenario is deforestation. It fits our model perfectly: trees use cellulose, a relatively inert material for building; carbon trapped in forests represents an important part of the global carbon budget; the present levels of carbon dioxide are homeostatic with immobilized gigatons of wood carbon as part of the equation; humans harvest wood because it is relatively inert and so is good for building; if we use wood for building only, then all will be well; however, synchronous destruction of large numbers of trees and the burning or rotting of their wood products would seem likely to precipitate global warming, a major biospheric effect.

Scaling Principles in Application and Design

We might have predicted the general reactivity of ancient inert material in the modern biosphere, in the lesson learned from the role of freon in the stratosphere. Now we can generalize to other human influences. The accumulation of plastic is a serious civil engineering problem. The very reason we use plastics is the reason why they are such a nuisance. They are inert, that is what makes them useful, and that is what makes them a waste disposal problem. This we could have predicted.

Perhaps the solution is to use reactive materials instead, but that gives rise to unpredictable situations. Inactivity is predictive at the scale of real-time biology, but activity is not. Unlike inactivity which is generalizable, reactivity can take many forms, some of which will give unpleasant surprises. A beneficial reaction only portends other reactions in the humanmade material; it does not indicate the favorability of the unknown reactions.

Consider what happens when one tries to make plastics exhibit rapid behavior after they have been used and are ready to be discarded. Note that under combustion, plastics immediately lose their desirable properties of existing close to human beings without doing damage. We can drink coffee from polystyrene on a camping trip, but try to burn the cup in the campfire afterward and the local environs become a very unhealthy place. The inert stain-resistant materials used for airplane interiors become sources of deadly gases in the fire following a crash.

Biodegradable plastics amount to a rescaling so that the materials still have the low level of reactivity to make them useful, while making them sufficiently reactive to avoid the present waste disposal problems. However, that low level of activity makes such plastics less generally useful for they can influence the flavor of foods stored in them. We cannot expect biodegradable plastics to be a panacea. As they degrade in landfills, we can only hope that the end products that percolate to the water table are harmless. It is not that we think that biodegradation of plastics is pointless. Rather, we raise the issue that we have been surprised before the long-term effects of apparently nontoxic or action-specific chemicals like DDT. Therefore, caution should replace optimism.

Plastics and freon raise a class of problems that emerges with a vengeance in pesticides. In terms of the ecological design principles that have emerged in the foregoing discussion, pesticides represent the worst of all worlds. From the point of view of the pesticide user, he wants it stable enough to complete its work without the expense of reapplication. It is almost impossible to engineer an exact and moderate period of persistence. The price paid for the pesticide being even moderately stable is that accumulation is bound to happen. Note that being a pesticide involves some biological activity in the real time of the life of plants and insects, so they are persistent but not exactly inert. Accordingly, the deleterious upper-level effects come more quickly than the effects of plastic or freon because there is a biologically significant channel of activity through which the accumulated pesticide can be expected to wreak havoc. We have now come to expect these mid-term effects of pesticides and are apt to forget that we were very surprised by bioaccumulation when it first manifested its effects high in the food chain. Of one thing we can be certain, new classes of compounds are going to surprise us again. Lest we become lulled into a sense of security by our newfound understanding of bioaccumulation, the general lesson is: moderately persistent materials that appear benign in their focused activity can be expected to cause ecosystemic problems almost immediately (figure 7.2).

The ecological problems of nuclear power appear as related models. In conventional nuclear power in the United States, the material at the end of the process is very active in obviously deleterious ways. Reprocessing those materials can alleviate the problem, but there are implications for nuclear weapon proliferation if we go that route. The es-

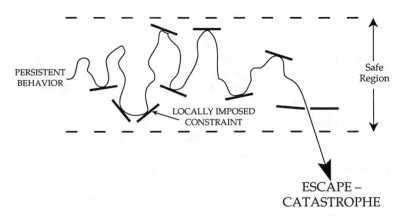

PERSISTENT BEHAVIOR

LOCALLY IMPOSED CONSTRAINT

Safe Region

ESCAPE – CATASTROPHE

Figure 7.2. Production of material which is persistently reactive requires constant monitoring and preemptive action to avoid undesirable effects. It is to be expected, however, that eventually some preemptive action fails and an undesirable consequence follows.

sential problem is that nuclear waste is persistent in its potential to do harm. The permanent storage of nuclear waste is probably tractable. The strategy is to immobilize the waste, thereby making it predictable for the hundreds rather than thousands of years it takes to decay to safe levels: worrying but probably manageable so long as we are cognizant of the scales that matter.

Consider the large number of materials made by the chemical industry that are persistently dangerous. The general principles raised here place a heavy burden of proof on industrial engineers. The only thing needed for disaster is time; and that is not a speculation, for we can predict it from the basic principles laid out above. The planners have to answer all possible scenarios, while the problem is free to take any unlikely course. We have enough faith in engineers to say that the disaster, when it happens, will take a very unlikely course; but it will happen. We cannot escape the general condition with an economically expedient list of special controls; implementing the list will bankrupt us.

All this highlights the genius of recycling. It is no accident that the metabolism of healthy organisms and the functioning of natural ecosystems both use recycling. It is a strategy both for escaping the constraints of scarcity, and for regulation. Organisms control the whole oxidative process by limiting the amount of ADP, the low-energy form of fuel in the ADP/ATP cycle. ADP accepts energy changing to ATP. ATP then acts as an energy supplier and regenerates ADP. Biological systems use recycling to avoid both the problems of supply and of waste disposal.

In a cycle, the cycle itself gives the properties of predictability through persistence that are required for a humanmade material to do its job reliably and safely. On the other hand, the cycle is a persistent pattern of reactivity, so there is no accumulation of end products; there are no truly *end* products in a cycle. Since there is reactivity in a cycle, we might be concerned about the generally unpredictable nature of reactivity. However, the positive feedback that would generate the unpredictable behavior is held constrained by the upper-level negative feedbacks in the cycle. The cycle is always in place to out-compete any undesirable, unpredictable reactions. Thus the reactivity is predictably favorable.

Only when the cyclical activity is suspended can other reactions take over, but then the supply of the reactive starting point is also suspended. Furthermore, because of the cycle there has been no accumulation, so the unpredictable side effects are limited. In terms of the problems of accurate prediction raised here, the advantage of recycling is that it contains most of the effects at one level of organization. This

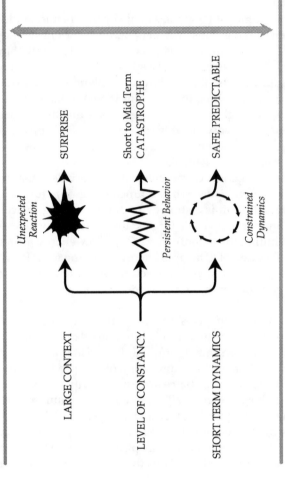

Figure 7.3. If the middle level is defined by some constancy or structure, then the general characteristics of system predictability depend upon the level at which the important dynamics occur. Production of stable material leads to accumulation and the expectation of upper-level, surprising dynamics. Constant reactivity at the middle level can be expected to produce undesirable effects in the mid-term. The particulars of these effects will be difficult to predict. If the constancy is a cycle, then the reactivity occurs at a lower level and the system is predictable in its behavior.

254

should limit the number of surprises coming from effects at other levels. The cost is that recycling is a delicate matter to the engineer, which we know from the difficulties in enacting effective legislation on the matter. Even so, a recycling option that might consume as fuel the waste of conventional nuclear plants in the U.S. appears tractable within a few decades.

With new humanmade materials, we can expect the old dilemma between reactivity and inactivity, unless we plan ahead for it. Even then, there is no guaranteed solution. On the positive side, we are now at least appraised of this class of problem. The advantage of expressing the problem in terms of only relative turnover times is that we are not locked into the details of particular scenarios (figure 7.3). Therefore, experience with apparently unrelated materials can be used to steepen the learning curve in an arena where all seems different, but where we cannot continue to make mistakes at the rates we have in the past. The Babylonians knew about particular right triangles for laying out fields with square corners. The contribution of Pythagoras was to generalize the problem so that it applied to all right triangles in principle. All that we have said above about freon, plastics, and pesticides was already known, even if it was not common knowledge. What is new here is the general statement of principles with broad application in unfamiliar circumstances.

Conclusion

In this chapter it has been profitable to see the biome independent of the large size which is usually ascribed to it. From this treatment it has emerged that the biome is an autonomous intellectual device that is more than a large-scale loose compromise between other criteria. A deal of this chapter has been concerned with basic science issues.

Despite the advantages of considering the biome independent of the large scale usually associated with it, we have spent effort in discussing large-scale questions. Some of the most pressing issues for ecology are large-scale. We have therefore used the biome concept to address large-scale questions of an applied sort. In addressing global problems, this chapter has made a suitable prelude to the next, which is centrally concerned with applied ecology at all scales.

8. MANAGEMENT OF
ECOLOGICAL SYSTEMS

———◆◆◆———

The intention in this chapter is to make what has gone before useful for the manager. By the end of this chapter we will be in a position to suggest a strategy for ecological management in both management planning and application. To get to the point where such recommendations make sense, we will first develop a clear view of ecological management and its relationship to both pure and applied ecology.

We will start by contrasting pure and applied ecology. They have drifted too far apart and we will make some preliminary suggestions as to the advantages of reuniting them. Then we will move on to look at both applied and basic ecology in terms of both hard systems of the type used by engineers and soft systems as employed by social scientists. Ecology emerges as a firm science at best—with a soft center. Next we attempt to unify the two halves of ecology. At this stage, restoration ecology becomes a conceptual hinge between pure and applied ecology. Before a unified discipline can be turned on contemporary ecological problems, we proceed to the last step, which offers a fresh conception of the contemporary biosphere. We need a clear view of the system that the manager must address. Humans are a disturbance that has been incorporated within a new structure, the anthropogenic biosphere. This notion has been explored by Tier de Chardin under the name "noosphere." It is a world where the major subsystems are forcibly integrated, an earth that will remain best described as out of equilibrium, but not necessarily unhealthily so.

With a unified ecology at hand, and a realistically defined world as its object of attention, we can make our final recommendations. We propose a family of protocols that can use basic science ecology and applied ecological insights to address ecological questions scaled from local sites to the level of national policy and beyond. In the final analysis we suggest modeling as if we were the natural context of the system. Instead of attempting to wrestle down all the details inside the managed system, we advise working with the system constraints from above. However, we have to develop fully a line of argument in this chapter before those final recommendations become meaningful.

The Two Subcultures in Ecology

Ecologists usually identify with one or several of the major subdisciplines, but not with all parts of ecology at large. At professional meetings, groups of ecologists concentrate their attendance on particular sessions. The same faces are seen at the landscape meetings or the population sessions, but not usually both, although attendees of either group may go to sessions in a third area, such as community ecology. This political structure within the discipline seems innocuous enough. However, a price is paid when one group ignores the work of another as irrelevant, when a cross-fertilization of ideas would, in fact, be helpful. A particularly wasteful schism is the one between pure and applied ecology. As we explain below, there is some contact across the divide, but there is also enough disjunction to waste opportunities for cooperation.

The Ecological Society of America has recently started a new journal for practical application of ecology. The British Ecological Society has published the *Journal of Applied Ecology* separately from the pure science counterpart, the *Journal of Ecology*, for a quarter of a century. Thus, the division receives official sanction, albeit inherited from times when pure science was more confident and autonomous. As the economic and political climate changes and funding is more for solving ecological management problems, there is less self-conscious identification of those who work on questions of pure or basic, as opposed to applied, ecology. Even so, basic scientists working on applied questions have a different worldview from those employed by government agencies with a mission in ecological management, like the USDA Forest Service.

Wildlife and fisheries resource managers are distinctly applied in their work but still have ties to population theory generated by basic science ecologists. Furthermore, the basic theory itself has come, as often as not, from observations of prey-predator relationships or single-species dynamics of commercially significant populations. Even so, academics in zoology departments view themselves as very distinct from wildlife and fisheries colleagues in the local college of applied life sciences. The demarcation lines between pure and applied ecology are just as clearly drawn in plant sciences; this is reflected in academic degrees and departments of forestry, horticulture, plant pathology, and agronomy, as opposed to botany or biology. Sometimes managed forests are, indeed, the focus of ecosystem studies, conducted from forestry departments. Nevertheless, explicit management recommendations about growing trees come mostly from silviculturalists who often do not think of themselves as ecosystem or community ecologists per se, but rather as applied forest ecologists. Thus the divisions are based less on antipathy than upon mutual lack of interest. There is some overlap, but also enough separation for the two academic subcultures of pure and applied ecology to have evolved each in their own way. The applied scientists and managers do pay attention to the basic science literature, but pure science ecologists do not read reciprocally in applied ecological journals.

The independent development of scientific subcultures gives each its own special insights, as well as each its own blind spots. Thus strengths and weaknesses are not matched, leaving the possibility of encompassing the best of both subdisciplines, while avoiding the respective limitations of each. It is time to hybridize applied and basic science ecology and select the strong traits in both. Our attempt at this hybridization is by no means the first, for in European landscape ecology there is a strong interaction. Our effort is more broadly based, for we wish to effect the cross-fertilization across all aspects of ecology.

Schools of ecological thought have speciated before, only to interbreed later with intellectual hybrid vigor. Certainly the British intellectual descendants of Darwin made relatively little progress in the latter half of the last century, for they became preoccupied with lines of descent and filling in the missing links. By contrast, vigorous schools of plant science in Germany with an adaptationist bent took Darwinian evolution into the field and grounded it in observation of living material. Eugene Cittadino, in *Nature as the Laboratory,* has just shown how much modern English-speaking ecologists owe those central European plant biologists.

The intellectual development of ecology follows the analogy of a braided stream, the image used by Cooper to capture his conception of the evolution of plant communities. The German schools of natural history mentioned above became the fountainhead of American ecology through their influence on Clements and others. Channeled by the shared English language, American ideas trickled across to Britain.

Eventually this produced a flood of natural historical studies of a derivative type, but with a distinctive Anglo-Saxon empiricism. Meanwhile, the rationalist intellectual tradition on the Continent stagnated in a set of stereotypical, formal classifications of vegetation. However, successes in quantitative vegetation analysis in America in the middle of this century have, in the last twenty years, cut a channel back to Europe. From Scandinavia to Spain, vegetation science is now turning to the use of computers and multivariate analysis of plant communities. Europe is presently providing a refuge for American vegetation scientists overtaken by a spate of population studies in the American journals. We contend that the time is right for a confluence of pure and applied ecology. That conviction is in fact the raison d'être of this book.

Hard and Soft Systems

An ungenerous view would say that anything which puts "science" in its title is not science at all: social science, political science, library science, domestic science, and so on. All these "sciences" deal with messy systems where controlled experiment is often impossible. In the soft sciences the critical problem is to find a powerful way of looking at the system. The subject matter of the above soft sciences, and soft science in general, does not offer self-evident entities; finding workable structures and explanatory principles is the main task. Accordingly, measurement has to wait for a definition of what is to be quantified.

In hard science the form of the problem is often clear, and the job of the hard scientist is to measure with precision, quantify appropriately, and build predictive models. It may require hard work, ingenuity, and facility, but results are very satisfying. The behavior of some physical systems in hard science is so predictable that engineers can build complicated structures from elaborate mathematical models with remarkable reliability. True, engineers do build in a large safety margin, but even so, one is impressed with what can be calculated and then built with confidence. Would we could build plant communities with the reliability of bridges.

Scientists are defined as soft or hard by the nature of the material they choose to study. The entities invoked by the important questions determine the appropriate mode of operation: hard or soft. The concrete achievements of the hard sciences and their applied partners are so great that our judgment has become distorted. Like it or not, the social standing of scientists roughly corresponds to a ranking from hard to soft. Accordingly, the temptation is to move to the hard-science end of any given discipline; sometimes productive, this can degenerate to mere calibration. Worse, a hard-science approach forced on soft subject matter quantifies the details before the system has been defined in a meaningful way. In ecology it is manifested as measurement for its own sake, an old problem across ecology, but one exacerbated by modern gadgetry.

Hard-system scientists work with invariants like tensile strength; that is what characterizes the hardness of the science. Despite the hardness of the material they study, physicists and engineers still have some choices as to style of model building. Thus, hard scientists do make some arbitrary decisions, although soft scientists must devote much more of their effort to making judgments. Accordingly, the difference between hard and soft science is a matter of degree. Ironically, the subjective and creative part of science emphasized in soft-system approaches is highly prized in the hard sciences. The Einsteins of this world make their contributions exactly by choosing another point of view.

In physics and chemistry, the observations are hard won, but once the complicated detection devices are working, they appear to reflect very general situations. One does not make allowances for the differences between electrons, for most purposes. At the soft end of the axis, human individuality dogs the investigator every step of the way. In more positive terms, it is exactly those human quirks that make the soft sciences fascinating.

Soft science is not hard science with more slack, rather soft science has its own distinctive subject matter. Hard science allows humans to act predictably on the material world. Cutting and removing a forest changes the material world in obvious ways. More interesting for the soft scientist, humans acting on or in a material world also change attitudes and values with respect to the material system. In economics those changes may be in price; a resource continuously consumed might first be hard to process, then be readily exploited and finally scarce. Thus, a monotonic change in the material world leads to alternating economic value: expensive, cheap, expensive. Soft science addresses changes in values, a subject matter for which the hard scientist typically abdicates professional responsibility. Hard scientists do soft science as amateurs, if at all. In soft systems, the entities are generally not fixed, and if they are, there are many alternative ways to address them. The social implications of the bridge, not its loading factors, embody a soft-system problem.

The methods of the soft science disciplines above may hold the key to dealing with the more awkward aspects of ecological systems, like developing methods for their management. As we shall explain, this liaison with the soft disciplines does not mean that ecology will be driven to be less scientific. There is a way to be rigorous and address larger issues, but it will be with a different sort of rigor. It will emerge that ecology already has a lot in common with the soft disciplines, so we need to put an end to ecologists' studious neglect of soft-science tools.

Clearly, ecological management is of soft systems, particularly if multiple use is envisaged. However, even academic, pure science ecology is by no means a hard-systems endeavor. A tree almost imposes itself on our senses, for it takes little effort to see it standing out from its background. Easily won data come directly through unmodified human senses, and each one of us has a style of personal observation full of values which cannot be suppressed completely. The sheer number of reasonable approaches and questions that ecological material can generate means that investigator preference is a major contributor to the results. For example, is it better to consider this or that organism member or a population member in this particular situation? All but the most incremental ecologists work in a relatively soft-systems mode. The most contributive spend a deal of effort choosing which of the myriad possible paths they actually take in a given study. Most of the work goes into finding a good way to look at things and then making measurements in that context.

The other side of the coin in a contrast between hard and soft sciences involves the limited dependence of soft sciences on the invariants that are the stock in trade of the hard scientist. Even in the softest system, what can happen depends on physical possibilities. The politics of a public building project does not have to entertain every and any lunatic scheme, like houses in the fifth dimension to save space. Soft systems are not completely a matter of human creation, independent of the realms of possibility.

In ecological systems, physical limits do apply, of course, as they do to all life sciences. Ecological systems may afford a large number of alternative views, but this does not mean that ecology can violate physical principles. The physical underpinning of ecological soft systems comes from the scale dependence of the processes that work inside communities, population, organisms, or the other types of organizer discussed here.

We may choose to call the endemic collections of mites that clean the pores on our skin a community in balance. We also choose to emphasize that our bodies, the community substrates, are organisms in their own right, rather than some undefined resource base. If the system of interest was the epidemiology of a scabies mite outbreak we might choose to name the system parts an exploding population. All this is the observer's decision. However, what we cannot change is the physical size of the mites, and the scale-dependent particulars that come from being that small. Diffusion, for example, has very different consequences for mites than it does for us; surface tension of water is a great force for them, whereas for us it is not. The important processes whereby material ecological systems work are scale-dependent in a way that relies on physical laws, not ecological concepts. Concepts are crucial, but are still predicated on material, scale-dependent possibilities.

Management Units as Devices for Conceptual Unity

A significant part of the field manipulation conducted the U.S. Forest Service is imposed not on areas defined by community type or ecosystem function, like a watershed, but is applied to management units (figure 8.1). There is sometimes some compromise between management units and terrain or a general community type, but management units are formally delimited to work as homogeneous production systems. They may be defined on purely geographic convenience. Some cultural demarcation, such as a road, might cut across the middle of a homogeneous example of community or ecosystem, and yet for the management unit, the road might provide a very workable boundary.

When it is to be seen through the eyes of the manager, a landscape falls into pieces whose identity turns on production of resources in a homogeneous fashion; heterogeneous production calls for subdivision into smaller units more homogeneous in their potential output. Only incidentally might this production map onto a homogeneous community, or an area of homogeneous ecosystem function. The homogeneity of the production is based on several criteria: geology, soils, topography, or vegetation type. By contrast, tree age is usually only a secondary consideration, even though it could be exactly the organizer for a homogeneous community at a particular successional stage. Management units are a mixture of community, ecosystem, and landscape entities, ranging in scale from ten acres to several hundred acres, depending on the heterogeneity in the major factors listed above. Legislation requiring consideration of joint production of public lands has increased the degree to which modern management units are multifaceted entities.

Once the units are established, management actions, such as clearcut or selective cutting, are usually applied to the unit in a relatively homogeneous way. Details of past management actions do introduce some heterogeneity on the ground; this can lead to some variation in the application of the management action across the area at a given time. The actual management actions on the ground come from an annual work

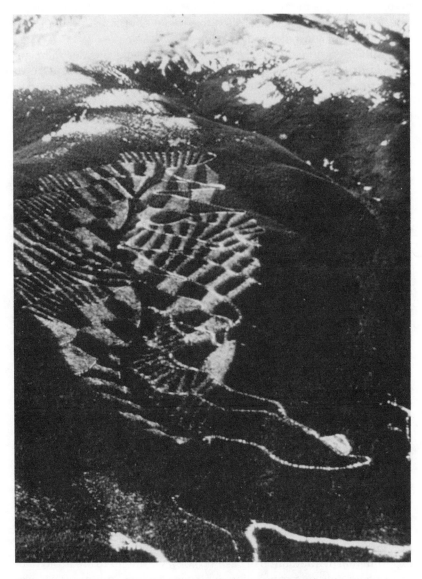

Figure 8.1. The Fraser experimental forest has been subjected to various experimental patterns of management units, seen here in an aerial photograph (photo U.S. Forest Service).

plan, which is distinctly tactical. It is created in the context of a strategic forest plan, which is itself selected from a set of options, perhaps created by a linear computer program like FORPLAN.

On the occasions when basic-science ecologists consider management units, they emphasize that the boundaries of landscape units, ecosystem watersheds, and community types are unlikely to correspond to tracts circumscribed by management action; fragments of community types will be lumped, and watersheds will be incompletely included. Even so, it is mistaken to dismiss studies arising from management goals as unnatural and so less likely than communities, ecosystems, or landscapes to yield ecological understanding with general application. Resource use is no more arbitrary and anthropocentric than are the academic ecological entities; a nitrogen atom cares not if it is in the leaf of a community dominant, part of an ecosystemic nitrogen retention pathway, or is located in a landscape entity like a hedgerow. Like management units, communities, ecosystems, landscapes, and biomes are reflections of human ways of thinking, not reflections of ecological reality beyond perception and conception. As a way to deal with the undefined fluxes of matter, energy, and information in ecology, management units can be seen as just another conceptual tool to be used alongside academic ecological criteria.

There is a tradition of pragmatism in applied ecology. Academic ecological criteria are often applied in a manner that is just as ad hoc as management criteria. Therefore, it seems a pity that, since academic ecologists also pay the price of pragmatism, they do not often fully avail themselves of its utility. The power of using management questions like "How do we get an additional 10,000 board feet out of here without decreasing water quality and elk habitat?" is that it forces a simultaneous application of the conventional academic criteria.

Although nature functions simultaneously as communities, ecosystems, and landscapes, even dual, let alone tripartite, descriptions are rare. Since they introduce an ecological entity new to many students of natural systems, management units may be used as a helpful device to pry open new intellectual possibilities. By considering management not only as a tool to achieve effective resource use, but also as a tool for experimental manipulation in basic ecology, we intend to break old habits of using only one conception at a time. Neither communities, ecosystems, nor landscapes deserve primacy; any one of the three may sometimes be allowed a somewhat larger role than the others, but all three can simultaneously make their different contributions to the composite understanding.

Using management units applies a perturbation across the major academic criteria. In management we have a wealth of experience with many "experiments" already reported in the literature. A set of given management practices (experimental signals) impact upon a part of nature differently depending on whether it is conceived as community, ecosystem, or landscape. Employing management action as an experimental manipulation, we can build a composite description using the power of the three conventional principles without being limited to any one of them. In fact, environmental impact statements legally require this rich conception of the manager. Management activities typically do not conform to academically defined entities, for a composite is usually involved. Basic-science ecologists have not previously made much effort to work such composites, although managers nearly always do.

If we are able to identify community criteria as well as ecosystem criteria that both map onto a given management practice, we may well be able to identify circumstances where a community entity is also a functional ecosystem flux. In tropical agriculture, the landscape mosaic of slash and burn allows community recovery. However, agricultural return time which is too short, or fields that are too extensive, both interfere with mycorrhizae and nutrient cycling. The degradation of communities can be, therefore, an ecosystem-related problem. We hope that our approach will uncover similar cross-links that were unsuspected until recently; it is designed to do so.

Management applies across a wide range of scales. Management practices come in all sizes, from the harvesting of a solitary tree to the recreational use of a county. Resource management is an ideal tool for investigating the effects of scaling and hierarchical arrangement. Using basic research concepts of community, ecosystem, and landscape to address resource management questions will allow management practice to take full and explicit advantage of ecological insight coming from outside the applied research literature.

Restoration: Simplified Management and a Test of Theory

Restoration ecology has, until recently, been seen as a sort of gardening with wild species in natural mosaics. The emerging academic specialty of restoration ecology hopes to use restoration as a device to test ecological theory. A new society of restoration ecologists has emerged, based at the arboretum of the University of Wisconsin. The agenda of its founders appears to include bottom-line testing of ecological theory. If we do indeed understand prairies and forest communities, then we should be able to translate that into management action which produces communities that appear natural, as far as the scale of the project will allow.

Restoration can be the acid test of our understanding. It would appear that, after all, it is possible to perform community and ecosystem studies through direct manipulation.

Like a restoration ecologist, the silviculturalist works with the express aim of generating a forest stand with a desired mix of species and physical characteristics in management units. The object is not to sell the trees, unless that would help achieve the silvicultural ends. The principal concern is with the supply side of wood production; the silviculturalist emphasizes joint production. By contrast, a forest management decision has the silviculturalist's objective, but now the decision is based on the costs and benefits. The forest manager and the forest economist introduce the consumption side of wood production and work to balance it with the silviculturalist's supply; the manager emphasizes multiple use, not joint production. The management decision asks how much of the silvicultural agenda should be carried through to completion. Outside the realm of resource production ecology, the restoration ecologist only goes as far as the silviculturalist, and does not finally implement the economic plan of the forest manager.

Restoration has to bear in mind not just the scale of the restoration but also the type of ecological system to be restored. Restoring a community may well not restore the associated ecosystemic functions. Sometimes there is a head-on conflict between restoring one type of system as opposed to another. For example, restoration of a habitat of desired rare species may well demand halting natural processes of the community. If a species is by nature a refugee, moving from one habitat to another as succession closes its opportunities, then its ecology demands a very large mosaic wherein at least some areas, at any time, are prime habitat. Often a species is rare exactly because large areas are removed from the staggered cycling process. The effect of this is to divert the development of next-phase prime sites, while the few places left in their prime phase are moving naturally on to the next stage in the cycle with nothing to replace them. Without the option of restoring the large mosaic, the alternative stratagem is to take a prime site and decouple the cyclical processes. Unnatural as it may be, the safe site is held in prime condition, because it is the only site, and others are not going to move into prime condition to replace it should the local natural cycle be allowed to proceed. The decoupling of the cyclical processes of succession causes havoc in the community. Often a population restoration is at odds with maintenance of a healthy community.

In a sense, the restorationist is working on a simpler problem than the manager. The restorationist has the luxury of taking management action with a highly focused, unambiguous goal. By restoring for one explicit purpose, the restorationist avoids conflicting aims. The manager must also execute a plan, but with the doctrine of multiple use, he has to find the best compromise between different desired conditions which may have mutually exclusive elements.

Ironically, the manager works closer to nature than the restorationist. The very focus of the restoration is likely to be completely artificial, by virtue of a one-scale/one-criterion agenda. The manager, on the other hand, deals with the richness of a world working according to population, community, ecosystem, landscape, and organism criteria all at the same time. A holistic solution is required of the manager. It is only through understanding the relationship between all the criteria that the manager can find a satisfactory answer, if it can be found within the set of actions available inside the management unit. The multiple criteria required of the manager invoke systems that may not fit neatly inside the management unit. Thus, although the scale of the management unit is fixed, multiple scales are implied. The restorationist can ignore all that occurs outside the unit to be restored. The manager can only involve all the required multiple criteria by at least acknowledging the context with its larger scale as seen through the criteria that invoke that context. Even so, the manager cannot act directly on the context while remaining within the bounds of the management unit at hand. We offer a solution to the manager's dilemma at the end of this chapter but raise the problem now as a point of tension.

The restoration ecologist, like the ecological manager, usually has a prescribed universe to restore. This dictate presses the restoration ecologist to work at scales he may not otherwise have chosen, and obviates the usual strategy of most field ecologists who define the scale of their system to their convenience. The convenient scale may, for example, lead to a study circumscribed by the edge of a watershed because nutrient budgets are easy to assess inside those bounds. However, with the scale of their system taken out of their hands, restoration ecologists are likely to work at scales where inconvenience has excluded other researchers.

The explicit scale of a restoration means that only some phenomena can be part of the restored system. Consider prairie restoration. In the Middle West of the United States there are many examples of restored prairie situated on suburban lots, as substitutes for conventional front lawns. Fire is not usually a management option there. In the Curtis Prairie of the University of Wisconsin Arboretum, for example, there is not enough room to restore the buffalo to the system without destroying other parts of the restoration. In the virgin prairies before white settlement, there was a natural pulse of heavy grazing as the herd passed through, with times in between for vegetation recovery. Barring an un-likely traveling circus of bison run by philanthropic conservationists, this cannot be accomplished in most restorations. However, the buffalo are not missing from small restorations. Because they are too large to be contained by small restorations, large entities like buffalo do not belong there. This is characteristic of animal restoration problems, because movement of large animals sets the scale too large for most restorations of moderate size.

A perfect restoration is not a small version of a pristine biosphere, for even the primeval world was not made of parts that were microcosms of the whole. It is a mistake to imagine that a restoration is somehow in-complete because something in the global pristine system is missing. Even if we could restore the entire biosphere, we would probably not want to do so. The essential scale dependence of all particular ecological systems is brought home by the explicit scaling of restorations.

The Distinctive Character of the Managed World

At a workshop held in Santa Fe, New Mexico in the autumn of 1988 on the topic of ecology for a changing earth, one working group focused on the human component. The group made three critical observations which can be woven together to give a broad picture of the world under human influence. This is the world upon the parts of which managers focus. The first point was that food webs containing humans have very indistinct boundaries. The second point was that the larger the human presence, the more leaky the ecological system. The third point was that, relative to the historical and prehistorical past, even the major eco-logical subsystems in the biosphere are now out of equilibrium.

FOOD WEBS

When the working group started to consider what was missing in the data base to address their charge, Joel Cohen pointed out that very few published food webs have humans as one of the nodes. Human food webs are distinctive because of certain qualities pertaining to the system boundary. All food webs have an arbitrary boundary which reflects the interests and time investment of the investigator. Bigger food webs take longer to create. Some published food webs have no primary producers at all, although there has to be a basic reduced carbon source some-where to support the consumers. Others may variously exclude whole classes of organisms, because they are not responsible for the move-ment of much carbon, or because they do not pertain to the goals of the study at hand. All food webs have an arbitrary veil line that excludes species too rare to be encountered in the sampling of the system, even though they are regular contributors to carbon flow. Nevertheless, the inclusions in most webs are both purposeful and reasonable, and the exclusions are defensible. This is particularly so for a favorite system, namely aquatic food webs. One of the margins of that habitat is usually taken to be the water's edge and surface, a robust boundary by almost any standard.

Not so the human systems; the critical feature of humans and their food consumption is that there is almost no place to draw the food web boundary that is arguably better than any of a very large number of alternatives. Perhaps an Andean village would present fewer problems than many human systems of consumption. However, even a village of equivalent size, relatively isolated on the western plains of the United States, would have a very leaky boundary, no matter when one drew it. For example, although the village might harvest wheat, that eaten by the farming family would, for the most part, be bread made from flour processed somewhere else. The mind boggles at the thought of a food web that might describe the patterns of consumption of New York City. Clearly the irrigated central valley of California is very much part of the New York food web. Seville oranges from Spain are processed into marmalade in Britain before they grace cosmopolitan New York breakfast tables. Probably little enough alligator meat from Florida is consumed in the metropolis for us to exclude it from the web as rounding error, but even that is a judgment call. We offer no solution to this problem, but we do note that human systems of food consumption have very indistinct boundaries and significant amounts of material come from geographically far-flung places.

The Leakiness of Human Systems

The second critical observation of the working group was that pollution problems appear to be more deeply rooted than we first thought. In particular, managers have identified most of the point sources of nutrients on the major rivers of the United States. Even so, the rivers remain heavily impacted by nutrient load. Part of the problem is agricultural land use, where fertilizer gets into the aquatic system. Less expected, but very important, is the nutrient input that sheets off suburban areas. This suburban runoff, let us emphasize, is not collected sewage or point discharge; rather, it is a reflection of the nutrient leakiness of suburbia as a whole.

The points of nutrient enrichment on the Hudson River have been identified and mostly shut down; now unmasked, the full impact of suburban runoff is apparent. In the Lake Wingra project of the International Biological Program, the scientist found, to their surprise, that the kick start of the eutrophic cycle in the spring came from the simultaneous melting of accumulated dog feces on suburban lawns, frozen through the winter. By contrast, the forested lands on the opposite side of the lake contributed almost no nutrient load, and certainly not the pulse that was felt from the spring thaw of front yards (figure 8.2). Predominantly natural ecological systems retain their nutrient material and forests are masters at the game. It is by holding nutrients inside the forest through cycling that woodlands escape the pressing constraint of low nutrient input from the air. A nitrogen atom entering a forest system can expect to be held for well over a thousand years, even though it is likely to be mobilized for new growth at the beginning of most growing seasons. Lacking a carefully orchestrated capture-and-recycle program, the suburbs leak material profusely. This characteristic of the suburban landscape applies to all other intensive human uses of the landscape. Human-dominated systems leak material.

Figure 8.2. An aerial photograph of Lake Wingra showing the suburban areas in contrast to the forested vegetation of the University of Wisconsin Arboretum (photo Wisconsin Department of Transportation).

BIOSPHERE OUT OF EQUILIBRIUM

The third observation of the working group was that human-dominated systems, even at the scale of the whole biosphere, are undergoing radical changes of state, such that the old equilibria are dysfunctional. We should hasten to add that equilibrium and nonequilibrium are not properties of nature so much as they are modes of description of nature; that is a very different matter. The nonequilibrial nature of the modern world is something that emerges when we try to describe ecological systems in terms of the old parts of the biosphere that existed before the human population explosion.

The changes of which we speak are so profound that pristine systems and heavily managed systems alike are all casualties. Even situations that have already been fully impacted by humans, like the tall-grass prairie region become the American corn belt, appear not to be stabilizing in a new configuration. The continuous production of corn masks radical changes which are still happening on the farm. Marginal lands come in and out of production depending on federal government incentive programs. Furthermore, the prairie soil is still deep, but is eroding at a rapid rate by geological standards. Should the predicted, anthropogenic, global climate warming and attendant climatic shifts occur, then the whole corn belt will move north to Canada.

In forested areas the changes are even more apparent. Vast areas of forest are being converted to grasslands across the tropics. In managed forests, the removal of minerals in logs, and the erosion that follows even careful logging operations, both remove soil elements faster than processes of soil regeneration can replace the loss. Even areas that felt the heavy hand of humans long ago show this global pattern of change. The long-deforested lands of Atlantic Europe are accumulating organic material in bogs in a way that did not happen in the ancient forests. This is evidenced by the monoliths of the first agriculturalists found resting on soil not greatly different from that in the primeval forest, but whose stones are now buried under tons of peat. Almost ironically, there are protests that the increasing mining of this peat for fuel is destroying the Irish landscape. Whether the changes are good or bad, the modern biosphere appears to be in a state of flux as never before.

Small, fast changes might have been within the capacity of the old system under the rules of assembly that prescribed old relationships between parts. Certainly, very large changes have occurred in the biosphere before, but they have generally been so slow in coming that the major subsystems of the biosphere could accommodate them only by moving to a new latitude, not by undergoing structural change. The important thing about the present changes around the world wrought by human beings is that they are both large-scale in area and magnitude of change, while also happening very fast. The speed of the change is so great that there is not enough time for natural systems to move with them without going out of equilibrium and losing integrity.

At first the working group couched this observation in terms of an uncertain future. They were not prepared to predict the future of large-scale ecological systems because humans are changing the world so fast and so extensively. A view was expressed that we cannot predict because the system will be out of equilibrium for some time. However, it appears to the present writers that the future envisaged by the working group is already here. There is no point in waiting for equilibrium because it will never come. At least it will be postponed indefinitely, until the human race is broken on the wheel of its own deeds. That turn of fortune may not come in less than geological time. It is the way of the new age to be in disequilibrium, by the standards of past biosphere. Let us hasten to add that this is not an expression that all is well with the world, or that present styles of ecological management are adequate.

As we considered the development of complex systems, we recognized the phenomenon of incorporation of disturbance as one of the ideas that hierarchy theory uses to address the development of complex systems. If the modern world is running true to the form of a complex system, it should have incorporated this human disturbance. In doing so, it will have collapsed to a higher level of organization. Before the human disturbance, the major ecological systems were relatively loosely constrained in a manner that allowed a high degree of autonomy for the biomes, major communities, principal ecosystems, floristic regions, and faunal realms. However, with the expansion of human influence, these natural systems suffered disturbance. After repeated disturbance, it is characteristic of robust complex systems to accommodate to disturbance by forming a new entity that includes the ameliorated disturbance as a working part. This is what has happened to humans; they have become incorporated inside a new entity that takes large-scale, cohesive human activity as a given (figure 8.3a, b).

This new entity, which we call the anthropogenic biosphere, is much smaller-scale than the entire global system, and is constrained by it. Nevertheless, the anthropogenic biosphere is larger-scale than the major natural systems that it contains. The emergence of this new entity explains: 1) human food webs do not have discernible boundaries; 2) human-dominated systems are leaky, and 3) few, if any, major ecological systems can be adequately described with equilibrium models that would have applied before human intrusion.

A. THE PRISTINE BIOSPHERE

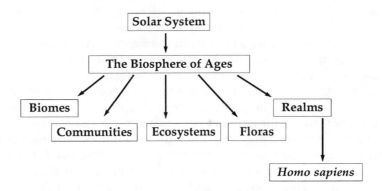

B. THE HUMAN IMPACTED WORLD

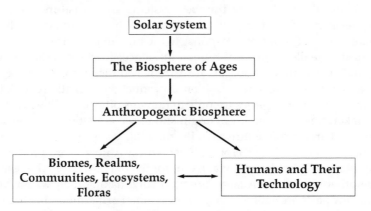

Figure 8.3. Before the imposition of modern humans upon the biosphere the major ecological subsystems were held under a relatively loose set of constraints. However, with the arrival of the anthropogenic biosphere the major ecological subsystems are held under tight constraints, well away from their old equilibria.

The biosphere is large enough to leave plenty of slack for the old natural systems to relax close to some sort of equilibrium, but the anthropogenic biosphere works faster, imposing tighter constraints. That explains the apparent lack of equilibrium in the quasi-natural ecological systems in a human-impacted world. Hierarchical organization involves sets of upper-level constraints that hold system parts out of equilibrium. It is in the nature of the new anthropogenic upper-level structure that the parts will remain held out of equilibrium indefinitely. The anthropogenic biosphere is a local phenomenon relative to the long-term biosphere driven by vulcanism, catastrophic meteors, and ice ages. Ice ages will come again, no matter what humankind wants or tries to control. There have been much warmer epochs than this. Therefore, despite anthropogenic global warming, we should not overestimate the significance of human activity at the scale of eons.

The major subsystems of the globe have not changed scale since the coming of the industrial age. Being a smaller context, the anthropogenic biosphere is scaled much closer to the major ecological subsystems than was the primeval biosphere. Accordingly, the anthropogenic biosphere forces a much greater degree of integrity on the globe's biota. In the new biosphere, matter, energy, and biota are moving around the globe as never before. These fluxes are the connections that reflect the greater integrity. One of the manifestations of this forced global integrity is the essential leakiness of the habitats in which humans concentrate. The nutrient-leaky suburbs and agroecosystems are a case in point. As smaller parts of the biosphere variously feel human impact, each becomes variously open as a response to the human presence.

In the smaller context of human influence, food webs that significantly involve our own species are extremely difficult to delimit, except on a whim, in an arbitrary fashion. Introduction of exotic species is a normal condition. Species friendly to humans often do well when they are taken from their points of origin and raised elsewhere. It is no accident that arabic coffee thrives in Brazil, while cocoa of the New World Indians is a critical cash crop in West Africa. A new, more productive equilibrium is, however, not at hand, for the old pests from the respective other side of the Atlantic are gradually making their way across, and new pests are evolving in the new realm. Furthermore, the hand of man will probably introduce the respective predators and diseases shortly after the pests either arrive or arise. With transported cultivars, chased by old pests, who in turn suffer their parasites sent on behind by humans, an ever more convoluted chain of consumption is created by human actions. The anthropogenic biosphere seems no more ready to settle down to a new steady state than are the ways of humanity in a strictly cultural sense. Integrity through flux is the order not just of the day, but of the foreseeable age.

Managing Systems Away from Equilibrium

To recap the main points made in this chapter thus far: 1) basic and applied ecology need to cross-fertilize; 2) restoration, having a single-minded agenda, can force the system to work at the scale prescribed by the area to be restored, but since management involves many criteria, a more holistic approach is required; 3) management must take account of scales including those of the context of the system to be managed; and 4) in the human-centered world of management, the context of the system to be managed has been importantly altered. Let us now weave these ideas together to give ecological management a new style. The bottom line for this new style, as we shall explain, is for the manager to maximize the natural contributions of energy to the functioning of the managed system, while minimizing artificial energy subsidies.

Having recognized the systems we are to manage will function away from equilibrium, we are now in position to use an explicitly hierarchical approach to management. It is in the nature of hierarchical control that lower-level entities are held away from equilibrium by the constraints imposed by upper-level contexts. Effective management needs to be particularly mindful of two points that follow from an explicitly hierarchical approach: 1) what it manages will be out of equilibrium, all the more so because it is being managed, and 2) as a higher-level context, the management practice must offer a viable context for the system under its charge.

In the anthropogenic biosphere, great tracts of land have been altered: forests cleared; grasslands fenced, grazed, and seeded; wetlands drained; riverbanks urbanized, and so on. This means that large and small ecological systems, including those at the scale of management units, have lost their context. Surviving prairies exist as islands, and woodlots occur as fragments torn from the whole forest. For animals which migrate or occupy large home ranges, the absence of the larger context is felt acutely.

Thus out of context, the fragments degrade because the support and protection afforded by the vanished context goes missing too. Over time, species diversity of woodlots must decline because the forest is not there to keep up the pressure of fresh propagules. Although we couch the disappearance of large animals in terms of whole states or regions, in fact the mechanics of the retreat eliminates one pocket of survival at a time. Sometimes firearms play an active role, but for the most part individual local extinctions are not exceptional or unnatural in any way. Local extinction is normal for many species; it is the absence of surrounding populations for reestablishment that drives endangered animals out of regions. More than man the hunter, it is humans as changers of habitats—the woodsman, the agriculturalist, and the industrialist—that forces the retreat of wildlife.

The central management principle we wish to erect is: *the most effec-*

tive management will recognize the manner in which the context is missing, it will identify the services that the context would have offered to the managed unit, and it will subsidize the managed unit to as close to that extent as possible. Effective management pays for other human activities having removed the context.

Before management, the unit to be managed lies orphaned from its context. Management fosters healthy development in the absence of a natural context. If management can achieve what we prescribe, then the managed unit should be serviced as if it were in context. Accordingly, well-managed units should behave as if the context was indeed there, although it is not. The managed unit should be free to function without deprivations; what the extant context cannot offer, the human management system provides instead.

This leads to a second principle of management which at present is only a hypothesis that may in time be verified: *if the management regime is effective, the managed unit will offer a maximum subsidy to the management effort.* If the managed unit is being provided with all it might expect from a natural context, then it can function to full effect. For example, the animals in the unit have suitable habitat within the unit, and anything they would have gone outside the unit to obtain is provided by the manager. Effective management would provide infusions of genetic diversity from a zoo or a distant wild population. Normally the context would do that in the form of outcasts or strays from neighboring populations, but in their absence, human contrivance plays that role instead. The effect of these infusions should be to maintain the managed population as a vigorous unit, making it appear as close to self-sustaining as possible. Breeding in the managed population subsidizes the management effort.

If a system is poorly managed such that it wants for things that would have been provided by the natural context, then the internal working of the managed unit becomes dysfunctional. At that point, quite the opposite of providing a subsidy to the human effort, the managed system will demand all the more, this time for assistance in its internal working. Neglecting to provide small infusions of genetic diversity could cause breeding to become ineffective. Instead of the occasional cost of a solitary individual from outside, the only course is the wholesale importation of many individuals to buoy population numbers.

The difference in system demands comes from the poorly managed system failing to subsidize the human effort. Game reserves that are too small to sustain elephants become degraded; the internal processes of vegetation regeneration collapse. If increasing the size of the reserve is not an option, then vegetation has to be subsidized by trucking resources onto the reserve. It is important to do whatever can be done to avoid stress of the natural vegetation in a reserve because native plants can provide a diversity of plant resource species and a staggering of timing which might be exceedingly difficult to replicate. Providing adequate resources allows the managed system to perform for itself some of the more difficult tasks of resource supply. Failure on all fronts receives judgment from the elephants themselves whose numbers decline.

Management of ecological crisis through preservation in zoos and seeds in gene banks is a poor substitute for habitat regeneration or game reserves. The inadequacy of such last-ditch stands can be counted in terms of the high cost of maintenance of zoos and germplasm centers. The high cost comes from the system to be preserved offering little subsidy to the enterprise.

When the context of an ecological system is in place, there exists a homeostasis. The ecological constancy of this homeostasis can take many forms, from actual stasis, a constant cycling of successional states, or constant recovery from repeated disturbance. Across all these possibilities there is a constant input of resource material from the system's context, as well as a continuous leakage or expulsion of material to the context. When the context is removed, that exchange comes to a halt, and there follows a period of degradation as the system adjusts to the loss of support. This is recognized in island biogeography as the period when a newly formed island loses species. It has been suggested that there are still adjustments in progress that are responses to the last major change in sea level.

If management steps into the breach, then the process of degradation is arrested. By behaving like the missing context, the manager holds the isolated system away from the depauperate, equilibrial state that would otherwise be its fate. In this way the manager works as a classical upper level. Higher levels in a hierarchy both hold lower levels away from equilibrium as well as offering lower levels a context that protects the lower-level entity from outside interference. Management both guides system function and guards the system from outside intrusion.

Ecological management need not be an attempt to return the system to some pristine condition. Nevertheless, good management will create situations that are sustainable. Sustainable solutions can only be achieved if the manager works with the underlying processes in the system to be managed, not against them. There is a difference between managing the system from outside, which is what we recommend, and forcing the system to perform in some focused, prescribed way. Insisting on a given outcome before one has monitored the relationship between internal functioning and external context amounts to a shot in the dark. The more the management regime can align with processes that are already in place, the happier will be the outcome. Fighting the processes that are in place instead of using them to carry activity forward will consume all the resources that the manager has available. The project will be ineffective at best, and actively disastrous at worst.

Our gentler style of ecological management is the one prescribed by Holling in his recommendations for control of budworm. He suggests that fighting the epidemic is the worst thing to do, for it holds the cycle in its most destructive phase. Furthermore, as our principles would predict, fighting the natural process of the outbreak is both very expensive and environmentally destructive. Since budworm infestation is a fact of life, the best strategy is to mimic budworm behavior; arrive with chainsaws ahead of the epidemic. The gentler farming style of tropical gardeners similarly aligns human activity with processes already in place. The garden is about the size of a blowdown and the natural processes of soil regeneration are employed by the gardener. Furthermore, woody plants that will offer resources to humans while simultaneously playing a role in soil recovery are either left alone in the clearing phase or are encouraged to invade while the garden is in place. The phase after abandonment of a slash-and-burn patch can offer abundant resources for gathering.

The above examples of working with natural processes are distinctive so as to make our main point forcefully. However, we do not see the style of management we recommend as being a great departure from many ordinary and successful management efforts that are in use at present. Our task is not to introduce such exotic things as slash-and-burn techniques into northern climes; our agenda is the elimination of poor management, not the introduction of radical or alien methods. We recommend a strategy for introducing techniques and practices that might appear as nothing out of the ordinary, but which come from a rational grand plan. Exactly in line with this recommendation is the strategy in place for the reintroduction of wolves to the western lake states. Note that there is nothing striking about it, except that it seems to be working. Rather than dramatic reintroductions far from extant populations, the preferred method of the state and federal agencies is to allow and encourage healthy populations in the boundary waters of Ontario and Minnesota to expand into the forests south of Lake Superior. As a method it is both cheap and effective because it only slightly redirects natural processes.

One of the difficulties in ecological accounting is the lack of a suitable unit. Our scheme alleviates part of that problem, for it allows an accounting scheme to measure management success. If management is inept, the system offers little subsidy to the human effort. A change in the management scheme for the better should elicit some collaboration from the managed unit. The subsidy that the managed system offers is often in the same units as the cost of management. Well-managed wolf populations generate a dividend in wolves. Wolves are what the natural context would offer, if it were still intact; wolves to supplement the population are what the manager must provide; and wolves are what the well-managed system spontaneously generates. If a large area for forage of elk is the missing factor in the distressed context, then forage transported from farmers' fields is the cost of good management; the subsidy from the manager is in the same units as what is missing in the context. The benefits of effective management are in unequivocal elk. True, different management problems will be given account in different terms, for there is no universal translation between, for example, tons of corn and increases in genetic diversity. However, a need to translate costs between disjunct managed situations is not often what is needed. More important is an accounting system across different management actions inside a single management task at a specified site. Efforts to find some generalized unit of ecological accounting, like energy, are probably misguided. It is better to do accounting in terms that pertain to a particular question and explicit goals couched in tangible ecological terms. That is the sort of accounting problem that our scheme facilitates.

The U.S. Forest Service uses linear programming as a tool to optimize management activities; a main model is FORPLAN. Given a set of explicit requirements and a set of explicit a priori limitations on what can be done to achieve the desired output, the model can deliver explicit management prescriptions. Nevertheless, limits to computation are very real, since each run is time-consuming, and therefore expensive, and requires much human effort dealing with input and output of the model. The linear program needs to be run many times so that the consequences of changing preset limits become clear. The weak link in the whole process is identifying what should be the limits to possible activity prior to the run. Thus, quantification is not really the critical constraint, for the computer can deliver on that front. Final success or failure turns on qualitative decisions as to acceptable limits on the run. By approaching the problem in terms of subsidies, the manager now has a tool that can assist in making those qualitative judgments. Subsidy offers an intellectual framework within which rational decisions can be made; then the computerized linear program can do its job effectively. With unreasonable, ineptly specified, or contradictory prohibitions, the linear program works as all computer analyses on the principle: nonsense in, worthless results out.

By recognizing that the critical factor is the changed context of the system to be managed, the human creature is brought into the center of ecological management. It is human activity which has changed the situation before the management even begins. In the management of the Great Lakes Ecosystem Basin, the International Joint Commission has adopted an approach that, like ours, explicitly includes humans and human activities. The "Ecosystem Approach," as it is called in the international agreements, argues that ecological, sociopolitical, and economic systems all coexist in the functioning of the basin. The "Ecosystem Approach" requires an integration of all three sectors in the search for management solutions. The emphasis is only slightly different from ours, for the absence of the natural context is just another way of pointing to the presence of the human sociopolitical and economic sectors. Ours is not a management scheme for a pristine world, but one for a world full of human activity, one where even the fragments that resemble the primeval condition are artificial islands. If we manage them as pristine wilderness, all will be lost.

One of the main statements of this book has been that one cannot know the upper-level constraints by only studying the smaller-scale lower level. The upper level only emerges as what it does not permit the lower level to do. However, by taking a contextual approach to management we have a device for addressing upper levels from the lower level. Consider once more the two principles: the role of management is to stand as a substitute for the missing context; the managed system offers a larger subsidy as the role of the missing context is more effectively substituted. If these principles hold, then we have a device for knowing the missing context by the management actions that are most heavily subsidized by the managed system. The greater the subsidy of the management by processes inside the managed unit, the more the management mirrors the missing context. To know the missing context, all we have to do is understand what has been done in the successful management action.

Conclusion

In conclusion, let us summarize the last sections of the chapter first and work back to the opening premises. In the discussion of the nature of the world that includes modern humanity, the critical feature to emerge is displacement of the major ecological subsystems away from primeval equilibria. From this it follows that management is not only of systems that are out of balance, but management itself explicitly holds the system away from equilibrium. Out of this emerges a management strategy that is explicitly hierarchical, where management is a substitute for defunct natural constraints. The central concept is subsidy: subsidy of the managed system in recompense for the destroyed context (absent forest, grasslands, and wetlands); subsidy of the human management activity by the managed unit, so long as the management is appropriate.

The central theme of this whole book has been a contrast between different ways of looking at ecological systems. In this chapter we have emphasized that the manager is forced to look at nature using several criteria simultaneously. The contrast of the manager with the restorationist is helpful because the latter generally deals with one ecological category at a time. The multiple criteria demanded of the manager at once presents difficulty and exposes the richness of ecological material. We have analyzed what this means for management in contrast to basic research. Further implications for basic research of a multifaceted view of ecology will be explored fully in the next chapter.

The different facets of ecology are not so much a matter of nature as a matter of choices in human perception; the material system does not function discretely as a community or ecosystem, or any other conceptual categorized entity. By being forced to deal with several ecological categories at once, the manager comes face to face with the human subjectivity that makes ecology more a soft than a hard science. Lynton Caldwell once noted that when professionals manage an ecological system, they do not manage the ecological system itself; rather they manage the people who act on the system.

Ecology is a fairly soft science. We have tried to show that the softness of a scientific endeavor is related to the changes in human value systems that occur when the object of study is raised. A hard-science ecology would not only be impotent when it comes to management, it would also be intellectually sterile. The essential beauty of ecological material can be seen with remarkable clarity through the eyes of the manager. The naturalist and the preservationist do not have a monopoly on the joy that is to be had from being an ecologist in the woods. Management is a very esthetic matter. In the modern biosphere, human activity is part of the system in a new dynamic interplay. Our species has the next dance with nature, and it is the ecological managers who should be the dancing masters and the orchestra leaders.

9. A UNIFIED APPROACH TO BASIC RESEARCH

————◆◆————

The Distinctive Scaling Styles of Basic and Applied Ecologists

The research that underlies management issues can be conducted using applied or basic questions. However, basic research in ecology involves a style of scaling the question very different from that in strictly applied research. The original questions in basic-science ecology do not usually involve a consciously held scale of operation. The scale only becomes fixed as the questioning becomes operationalized. By the end of the project, the scale is fixed, but it may not be stated explicitly. The original research question is not conceived as being scaled, but the answer to the question often has a convenient scale at which it can be resolved with reasonable effort. The researcher seeks that scale as part of the practical matter of conducting the investigation. The testing and empirical part of an ecological investigation must be scaled compatibly with the theory or central concepts involved in a research project.

The above would seem obvious, but what is less obvious is that there is no simple way to achieve that commensurate scaling a priori. There are so many subtle facets to the scale of an ecological question that trial and error is the usual method whereby the match is achieved. If the original research design begins to manifest inconvenient scaling properties, then the sensible researcher rescales the design until it possesses the desired qualities (figure 9.1). For example, ecosystems are often studied using entire watersheds because the nutrient and water bud-

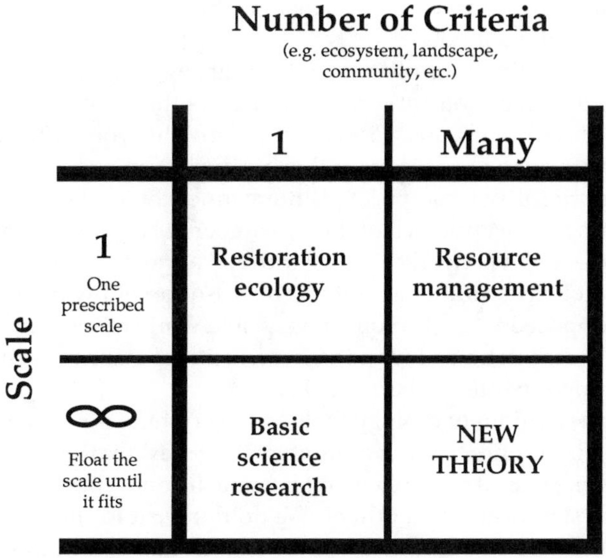

Number of Criteria

(e.g. ecosystem, landscape,
community, etc.)

	1	Many
1 One prescribed scale	Restoration ecology	Resource management
∞ Float the scale until it fits	Basic science research	NEW THEORY

Scale

Figure 9.1. Managers must use multiple criteria and are forced to use the scale of the management unit. Restoration ecologists have one fixed scale at which they perform a particular restoration on one criterion. Basic ecologists use one criterion but float the scale until it is convenient. New theory is needed to link criteria unfettered by particular scales.

gets are easily calculated at that spatial scale. Community studies will often include only certain classes of organism, so that there is a certain homogeneity in the working of the system; the decisions as to which taxa to include and which to overlook are in fact scaling operations. This rescaling of the research protocol is not dishonest or bad in any way, for the original conception of the problem is only implicitly scaled, and the readjustment of protocol just brings the mechanics of the field or laboratory work in line with the original ideas. Basic-science ecologists rescale as is convenient.

Another characteristic of the basic scientist in ecology is the number of approaches and conceptions that he uses. Usually there is only one. The basic scientist works like the restoration ecologist, who restores either a community or a rare population, but not usually both. This is indicated by the way that most academics in basic research would have little difficulty in classifying themselves as having one overriding interest: plant community, or ecosystem, or animal behavioral ecology, but usually not more than one. So not only is the scale of the protocol free to float to a convenient level of inclusion, there is usually a very small number of criteria involved.

A system defined simply to avoid conflicting criteria combined with rescaling to make protocol convenient allows the basic ecologist to ask highly focused questions. However, one might worry that the basic ecologist's system goes too far in the direction of engineered simplicity, ignoring crucial but complicated interconnections. The one-criterion and floating-scale approach of the researcher of basic questions may be productive, but it is clearly a very contrived approach. This applies not only to ecologists in the laboratory, but also to field workers probing basic as opposed to applied questions. Field workers sometimes accuse laboratory researchers of working in an artificial manner, so it is ironic that the same charge can be leveled at them.

The folk wisdom of ecology that says everything is connected to everything else is only true in an uninteresting way, for the whole reason for doing ecological research is to find which connections are stronger and more significant than others. We do not wish to show that everything is connected, but rather to show which minimal number of connections that we can measure may be used as a surrogate for the whole system in a predictive model. That is the strategy of the basic scientist in ecology. Nevertheless, more things are connected to more things in significant ways, and at more scales, than the protocol of the basic researcher would indicate. Therefore, there is something to be said for combining power of floating the scale, as does the basic researcher, while considering how the multiple criteria of the manager might impinge on a general research problem. The applied ecologist has it pressed upon him that communities do not work independently of ecosystems, even though it is conceptually tidy for the basic ecologist to work as if they did. In this chapter, we examine a possible basic/applied hybrid, with an eye on how the basic researcher might conduct richer investigations.

The Cost of One-Scale/One-Criterion Research Protocols

In floating the scale to fit the empirical protocol, and choosing one ecological criterion, the benefits are clear. The sharp focus of the question that comes with such a research strategy is important. Less obvious are the costs of this conventional approach to basic research. The very pointedness of the question can hide the fact that the scaling of the final model might be an awkward hybrid. Some examples might illuminate this point. In all these examples there is a rescaling such that the measurement system loses power. Of course one can always rescale by taking a different point of view. The point here is that in each case the ecologist probably intends to measure with regard to the new scale we raise not the constraints that scale the original measurement protocol.

1) All the animals in a study might visit a convenient study site of defined area, in an orderly sequence. However, there is no guarantee that they move across a larger area in an equally orderly fashion. Some visitors may be local, while others may be wide-ranging. Thus the complex entity studied in this scale-focused fashion does not pertain to any definable area, except the trivial special case of the particular study site.

2) By normalizing data collection to damage done by herbivores over a given area, over a defined period of time, one produces a neatly scale-defined study. However, differences in leaf longevity and replacement rates between species means that the measured damage is in differently scaled units for neighbors in the apparently scale-controlled data base.

3) A community mosaic may be studied in a well-defined area, giving an apparently homogeneously scaled study. However, the two phases of the mosaic might have completely different turnover times. Also, the scale at which each mosaic phase exhibits equilibrium patch dynamics may well be very different. Thus the system could be importantly confounded between scales of functioning, even though the study site is explicitly scaled.

4) Professor Grime of Sheffield puts competition, disturbance, and stress on one triangular graph to order plant habitats (figure 9.2). Without entering the fracas that surrounds the "proper" definitions of stress and competition, let us give Professor Grime his definitions. His graphs are insightful. Even so, it would be most surprising if competitive mechanisms operated at the same turnover times as the mechanisms for stress. Thus there are mixed scaling complications in what appear to be comparable studies.

5) The cornerstone of the conventional view of Darwinian evolution is competition applied to variations in form and function. One of the problems with Darwinian evolution is that it is capable of explaining almost anything that we observe by invoking some mutation or contrived competitive contest. Anything that explains everything explains nothing. There has emerged a fixation amongst adaptationist ecologists to put everything in terms of competition. Sometimes other processes come to the fore, but the situation is rescaled until what is happening can be expressed in competitive terms. The simplest case might be mutualism; it is possible to re-

High Disturbance

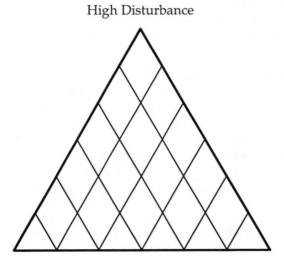

High Stress   High Competition

Figure 9.2. The triangular graph form used by Grime to contrast stress, disturbance, and competition strategies in plants.

scale the relationship such that it is expressed as giving a competitive advantage to all parties to the mutualism over anything that is not in the association. In this case the rescaling seems innocent enough, and it makes intuitive sense. However, an insistence on always rescaling so that the context is always a competitor is in danger of organizing the world experience around the concept, instead of the concept around the experience. The fact that one can make it fit the data is beside the point, for there is a hidden cost. The neat fit is deceptive, for one thinks one is finding out something about nature, when in fact all that is happening is the study of an arbitrary human-scaling decision. What if one knew the answer to the question "Who was the first human?" Of one thing we can be certain; he was like his uncle. In other words, knowing the first human only reveals the details of one's criteria in one's definition for being human. It tells little of human nature. Similarly, fixation upon competition as an organizing criterion tells little about nature and everything about the arbitrary details of what we choose to have "competition" mean.

The predictive structures posited by the one-criterion/one-scale approach can work spectacularly well at that scale and using that criterion. However, the tidiness of the answers they give can be deceptive, suggesting more generality than is warranted. Apply the findings of basic ecology at a slightly different scale, or try to use them under some new criterion, and what was powerful becomes impotent. From being a helpful framework against which the researcher can gain purchase, the one scale and one criterion suddenly changes into a limiting context that boxes the results into a very special case. One might expect tight, consistent results to represent something fundamental, and therefore generalizable. However, quite to the contrary, the more focused a finding, the more specialized is its utility. If the investigation that gave the community insights was couched in terms compatible with landscape or ecosystem considerations, then the translation might be possible, but predictions from pure community ecology can be expected to be inept when they are applied out of context, under the other criteria.

One of the generalizations that has been touted as a great success has been the application of island-biogeographical principles to the creation of nature preserves. Certainly the theory is highly focused and might be an archetype of the one-criterion approach. The generalization of island biogeography to the landscape patterning of nature preserves is a fairly natural extension, in that preserves do have many of the properties of islands, and preservation is the obverse of extinction, one of the building bricks of island-biogeographic predictions. Therefore, the extension into conservation is a modest one, and so is more likely to be successful than some more grandiose application. Certainly there is some promise, but the jury is still out as to how successful the more general applied predictions have been.

That island biogeography is an equilibrium-centered theory need not be a problem. However, many aspects of preserves will clearly require nonequilibrium description. For example, consider preserves that are large with respect to the biology of the organisms of interest. For many aspects of an insect's biology, an acre can be large, and a square mile might as well be a continent. Continental biotic fluxes cannot be addressed with a scheme that uses discrete islands to calibrate the system. On the other hand, more mobile insects or birds could visit between preserves, but have their migration rates be buffeted by external conditions. Neither of these cases can be adequately described by islands at, or even heading for, equilibria between invasion and local extinction. Furthermore, island-biogeographic predictions are taxon-specific, but preserves often have to serve diverse types of organisms. One might expect thresholds to play an important role in nature preserves, and island biogeography does not address such questions directly.

The efforts of island biogeographers here are laudable, and we do not wish to criticize or discourage them. Rather, our purpose is to identify that: 1) island biogeography is one of the few areas of academic ecology that can boast hard unequivocal predictions; 2) the extension of this pure science theory is modest; but 3) the generalizations in even this case are tenuous.

Scale Changes and the Coincidence of Different Criteria

Whatever we understand will only scratch the surface of what would be tractable even if we knew how to deploy research resources to best effect. Then there is the much larger set of things we very much want to know, but are intractable no matter how we deploy resources: despair! Therefore, we do not complain that the narrow focus of basic science in ecology has achieved much less than what is knowable in principle. However, the narrow focus of academic ecologists has meant that we have not started to deal with any sort of broad view of how the various parts of ecology might fit together.

Even taking a provincially human perspective, our environment influences us in important and obvious ways at many scales and according to many criteria. A tree plays many roles all at the same time, always within the bounds of physical laws. Ecology has only just begun to make modest proposals as to how we might try to put disparate perspectives of just the commonplace together, let alone achieve a multifaceted view of what is less than obvious. It is not that ecologists have been idle or unintelligent, for meshing our complex of concepts with the undefined fluxes of material in nature is a formidable task. A simple example of difficulties of mapping between ecological criteria will show exactly the convolutions we envisage. Let us watch the concepts of ecosystem and landscape fall in and out of step in the physical movement of mineral nutrients across an increasing extent of time and space.

1) Landscape and ecosystem at a small scale: plants placed on a landscape, say a hedgerow, lose nutrients dynamically in the pattern of the spatial arrangement of the plants (figure 9.3a, b).

2) Ecosystem fails to map onto landscape: the related process of recycling, an ecosystem property, does not map simply onto the landscape. Absorption of those same nutrients after their loss by the plant in leaf fall is the simple next step for the ecosystem, although it may be diffuse and follow a convoluted trajectory across the

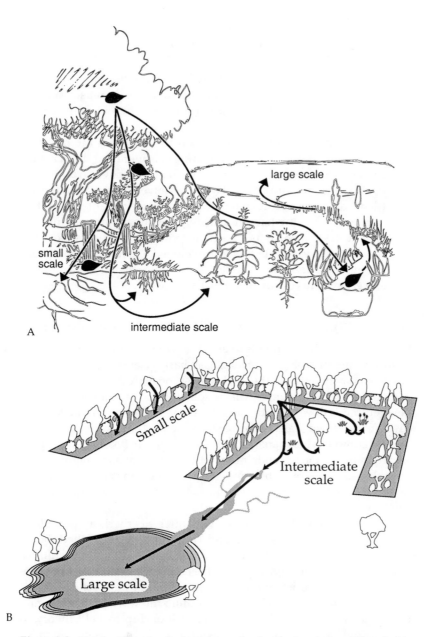

Figure 9.3. Nutrients lost to a hedgerow may be absorbed again in a diffuse fashion that does not easily correspond to a place on the map. However, water leaving the terrestrial ecosystem in streams will pass into a lacustrine ecosystem thus restoring correspondence between a landscape feature and an ecosystem process. A. Pictured with small scale in foreground, B. diagrammed from outside the largest scale to emphasize relative scaling of the three levels using a single scale of projection.

landscape. In a particular case, surface runoff may move nutrients a short distance from the hedgerow so that they are absorbed back into the ecosystem biota at an arbitrarily different point on the landscape.

3) Ecosystem and landscape again concur upscale: if the nutrients enter a channel of water, a landscape consideration, then a scale change is likely to occur that takes us to a level above on both criteria. Having entered a channel while dissolved in the water, the nutrients may be reabsorbed by the biota at some distant point. They may well leave the terrestrial ecosystem altogether and enter an aquatic system, such as a lake. Accordingly, they have not only changed ecosystem, but have also moved to a fundamentally different type of landscape. At this larger scale, changes in landscape and ecosystem concur, albeit by happenstance.

Awakening Interest in Multiple Criteria for Basic Ecology

Our intention in this section is to encourage the use of some aspects of the approaches of ecological managers in basic-science, academic ecology. The manager works in the final analysis on a prescribed site with its scale on the ground defined. This is something of a straitjacket for the manager, but it does have the advantage of anchoring prediction and activity to a defined scale. Now we suggest something more daring: the multiple criteria of the manager combined with the floating scale of the basic researcher. Our agenda may not be as radical as it might at first appear because others are beginning to make the same move toward the melding of multiple criteria. The present section draws attention to some of that work.

There have been attempts to link two different criteria at the scale of the second step in the ecosystem-landscape three-step described above. Remember how local ecosystemic recycling was confounded by diffuse reshuffling across the landscape. At this relatively small scale, the mapping of ecosystem function onto the spatial matrix was first developed by Peter Sharpe and his colleagues continue the work in an emerging body of ideas called field theory. The relationship to field theory in physics is only metaphorical. The theory suggests that the performance of plants at a given site can be predicted by knowing the size and position of neighboring plants, and calculating the field of influence of those neighbors (figure 9.4). A "field" of suppressing influence could, for example, be the effect that one plant has on another, at a distance, by casting a shadow in that direction. Fields of facilitation could be action at a distance as a windbreak. To this point, field theory has been used by Joe

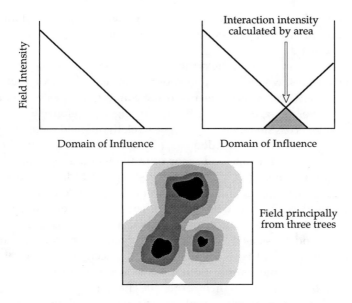

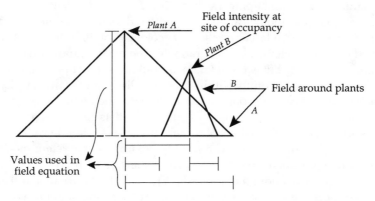

Figure 9.4. In field theory each place on the map is identified as falling under the influence of a field of a certain force. This can be related to patterns of growth on the ground. The height of the triangles is the influence of the plant in question on the spot it occupies. The base of the triangles defines the width of the plants' domains of influence, while the sides indicate decreasing influence toward the domain margin. The form of the triangles change with changing environmental resources leading to different maps of field intensity between say wet and dry seasons. Maps of fields are hypothetical here. Walker et al. (1989) have calculated various scenarios of domain interaction.

Walker as a vehicle to integrate the complex patterns of measured interference and facilitation between trees, shrubs, and grasses in an Australian woodland.

The fields of influence in field theory are explicitly generated by organismal entities, which act as points of reference. Field theory before now has dealt with the mechanisms inside communities, as we define them. However, Sharpe was recently attempting to generalize the theory, to consider ecosystem fields. The closest relative of field theory is the point source influence in plant demography that uses nearest neighbors as the predictor of the size and vigor of individual plants. Note that nearest neighbor theories are tied to the distances to individual points and are not generalized to integrate the influences of several neighbors through a model using a field of influence. Thus, field theory takes population insights and generalizes them so that they are at last of use to community and ecosystem ecologists. Both Sharpe's community and ecosystem fields are an explicit statement of the mechanisms that relate organisms to the interference patterns that we used to define communities and ecosystems in their respective chapters.

The problems of fixation on single criteria in basic ecology have been recognized by those shaping the discipline. In the early 1970s the American International Biological Program (IBP) did focus on a limited number of sites at which large numbers of ecologists of different types collected data in tandem. The integration of the effort at individual sites was on the agenda but was not taken to fruition. At that time, American ecologists were so bound within their subdisciplines that the physiological ecologists, the population biologists, the community ecologists, and the nutrient-cycling-ecosystem scientists all worked on their own problems largely as if the other researchers were not there. The integration was done in the creation of large, untidy ecosystem models that were often created with no particular question in mind.

We were naive enough in those days to think that if we took full advantage of the burgeoning computer technology the solution was at hand. The intention was to put everything we knew about the site and its functioning into the computer and so capture a piece of nature and ask any question we wanted, later. There was a style of modeling that was prevalent in those days that we now disparagingly call "garbage pail models." Underlying such models is a naive realist philosophy that says if we know it happens in nature, we must have it in the model, otherwise the model will be a less complete approximation of nature than we could achieve by putting everything into the code. The notion that models were descriptions of a purposeful conception of nature, not approximations, was not generally understood. We behaved as if the model mysteriously became nature itself, to a degree. The departure from nature was seen as a matter of degree that could be minimized by taking nothing out of the model once it had been coded in. The conception was distinctly large-scale reductionist, not the holistic modeling exercise that we thought we were performing. Some sites were more sophisticated than others, but the above description certainly fits the Lake Wingra site, in the eastern deciduous biome, at which Allen was stationed at the time. We learned a lot about rounding error and the importance of having a question. Some of the aquatic modeling did give helpful insights to field biologists, but the integration of approaches that was envisioned never happened. Those fighting for the resource for the program should certainly be credited with vision. Without having gone through that exercise, American ecology would be the poorer now.

The latest effort to integrate the activities of different types of ecologists is showing more promise. It is no accident that it comes as a remedy to a scaling problem. Presumably an integration of effort from different ecological subdisciplines will take longer than the usual three-year granting period, or the time it takes to earn a Ph.D. Recognizing this, enlightened forces at the National Science Foundation have created a network of Long-Term Ecological Research centers (LTER sites). As well as providing longer-term funding, the granting agency encourages intersite comparative work. The jury is still out as to whether the approach will move the discipline as a whole into a more integrative mode. Certainly, at the Northern Lakes LTER site, Tim Kratz is boldly putting together a synthesis of landscape ecology and biophysical characteristics of his lakes.

Kratz is attempting a meta-analysis that takes variables of many sorts and classifies them into biological, physico-chemical, and a special class he calls oxygen-related. He uses a ranking of his seven lakes as a means of calibrating the variability of a large set of disparate variables. The pH bears no simple relationship to species diversity, but the two can be compared by ranking the lakes according to their pH variation over time and the variation of species diversity over time. Thus the rank of lakes by pH variation is in the same units as the rank of lakes by species diversity variation.

At the outset one might not expect landscape position to be related to broad classes of variables. Nevertheless, they do show an almost perfect ranking in variability with a difference in altitude for the whole suite of lakes of only fifty feet. Better than that, preliminary results make intuitive sense, but only once the results are in hand: the parts of the total landscape system that are closer to the external forces driving variation are more variable. In the case of Kratz's lakes, it is those higher on the landscape, the up-landscape lakes closer to the naked influence of rain, that are more variable. A similar pattern with interesting differences emerges for another LTER site in a coastal zone of South Carolina. In the case of the brackish North Inlet LTER site, it is the ocean that drives system variability. The sites closer to the open sea are more variable. Interestingly, only some of Kratz's classes of variables show pattern. Furthermore, the driest Jornada LTER site in the New Mexico desert shows no pattern with position on the landscape. The full meaning of these findings is not yet clear, but it is a new approach that is trying to make connections across different types of ecological criteria. It points the way to a unified ecology.

Fuzzy Criteria: A Problem and a Solution

Particularly with requirements for multiple-use planning, management problems involve multiple criteria. Different management actions are often differently scaled, and so despite the fixed area of the management unit, the manager invokes multiple scales when it comes to implementation. The different criteria recognize their own types of system parts. Often a given entity may be recognized by several different criteria.

For example, a tree can be a population member, a community member, or an ecosystem storage unit. Each conception emphasizes its own aspects of form and function. As a population member the tree relates to other population members through reproduction and genetics. With the tree as an ecosystem part, reproduction and genetics are usually taken appropriately for granted. Ignoring reproduction and genetics might offend the sensitivities of population biologists, but population considerations have little relevance for most questions an ecosystem ecologist might ask. The ecosystem ecologist would see the tree as possessing biomass, or rhizospheres for nutrient cycling. Clearly both the roots and seeds are crucial components without which everything else about the tree could not exist. However, we are concerned here not with the tree itself, but with it as an object of study by an ecologist, and that is a very different matter.

The aim of this chapter is to engender a more catholic worldview in ecological specialists, such that the tree might be given attributes in more combinations, so that it is studied through alternative conceptions that may not fit neatly inside the present academic specialties. Unlike the pure scientist, the applied ecologist considers as many aspects of the tree as pertain to his multiple-use criteria. This breadth of vision, with all its problems, is what we recommend for extending basic ecology to make it a unified science. Accordingly, the basic scientist will come to share the manager's classification dilemmas in ordering criteria for observation and analysis. The fundamental problem is that no single classification of the tree on a single criterion will serve all pertinent aspects of the system, so somehow an explicit link must be made between the criteria.

There is a formal algebra for dealing with situations that only belong in part to a defined criterion. It is fuzzy set theory. The notion of fuzzy sets is not in any way vague, despite the name. Sets have been conventionally considered as crisp and quite concrete; that is to say, an entity is either a member of a set or not. In fuzzy sets, membership is not all or nothing, but is a matter of degree. The notion of fuzziness is helpful for sets such as the set of "tall people" to which one belongs usually only to a degree. The archetype of the set might be a professional basketball center or a Watusi tribesman, while the obverse would be a dwarf or pygmy. Most of us are somewhere in between. Wilt Chamberlain would score a value of 1, in that he completely in the set, while Tom Thumb would have scored 0, because he was definitely not a tall person. Allen at 5'9" might score about .5, whereas Hoekstra, at 6'1" would score higher, say .8.

If we were assigning elements formally to a particular fuzzy set, then there would be rules that would place the two authors unambiguously and quantitatively in the set of tall people to their various but explicit degrees; Hoekstra and everyone else of his height would have a particular value associated with their degree of membership. Given the definition of the criterion for the set and the degree of assignment to it for all members, all else is firm and concrete. Fuzzy does not mean ambiguous. Once there is a clear criterion for membership, there is a well-defined algebra of fuzzy sets which is as particular as that which applies to discrete or crisp sets. Polar ordination of vegetation is a rigorously defined technique, and is performed by the not-only-but-also operation of fuzzy sets.

A given action or experimental stimulus may relate to several different criteria. The assignment of an ecological action to a given ecological criterion is a fuzzy problem. The fuzziness comes about by virtue of a specified action having impacts that are variously split between ecological systems defined on alternative criteria. For example, removal of a tree or two, namely the practice of selection cutting, is a community consideration, for it shifts the balance of dominance. On the other hand, a clearcut impacts the landscape. In practice there is a continuum between the two extremes, such that forest management removes trees in various sized mosaics, so impacting both community and landscape to various degrees.

It is clear that effects on landscapes would increase with size of the cut. The effects of increasing size of the units removed has a less straightforward effect on communities. Increasing the size of the units that are cut might increase the homogeneity of tree removal, in that larger cuts will take everything independent of size and species. While more community members would be cut in a larger logging operation, the impact would be move evenhanded between species. The old dominant trees taken in a selection cut may be ones that are big because they moved into the site quickly after it was last open. A clearcut would open the site, and so leave dominance relationships alone, even though biomass is reset close to zero. Thus a linear monotonic increase in landscape effects, namely an increase in size and completeness of the cut, might bear a complicated relationship to community effects.

We have more experience in using multiple-criterion approaches in managed systems so, although the ultimate interest here is basic-science ecology, management systems make a helpful vehicle to advance the argument. This management-oriented definition of the problem transfers easily to basic-science questions. Instead of a management action, the event to be evaluated could be a hurricane, a fire, or an insect outbreak in a species of no commercial value. An experiment need not just focus on the impact of the test signal on reproductive effort that would interest a plant demographer, but could include nutrient status, with an eye toward coupling population and ecosystem effects. A clearcut also has ecosystem consequences. This was the focus of the Hubbard Brook study, a basic-science endeavor that looked at community composition and ecosystem nutrient cycling. It emerged that the seeds of the pin cherry, a pioneer species, lay dormant for almost a century. When the stand was catastrophically opened, the cherry as a population moved into the site immediately, saving much of the ecosystem nutrients. A selection cut would pass unnoticed on the ecosystem criterion, so the ecosystem would represent yet another way to contrast the various effects of a given action.

Multiple-resource management practices are the easiest example to explain the fuzzy description of ecological complexity. Multiple-resource management directed at recreation use is principally a landscape consideration (people mostly go to places and look across vistas, i.e., landscapes). Let us say that it is .8 a landscape question. However, it is also a community consideration, in that vegetation physiognomy is of recreation management importance; say it is .3 a community consideration. Degradation of land through recreational abuse, or degradation making it less useful for recreation both make recreation an ecosystem question in some small way; say it is .1 an ecosystem consideration.

The objective of a multiple-use plan is to develop a systematic suite of management actions in a management area which involves a spectrum of community, ecosystem, and landscape aspects. The above recreational demand would position at .3 on the community axis, .1 on the ecosystem axis, and .8 on the landscape axis. It would be possible to define a particular management action in a three-dimensional community/ecosystem/landscape space. This procedure of fuzzy set assignment of different ecological criteria to a given action will permit a systematic series of fuzzy classifications for a management area, ordered on an increasing scale of the management action.

We have already suggested in another case that selection cutting of an individual tree is principally a community consideration, so the management action would be positioned down in the zero-zero corner of the landscape/ecosystem plane, but high on the community axis (figure 9.5). We have also noted that there is a continuum of cutting activities all the way up to clearcut of an entire watershed. As we increase the size of the cut for small patches, there is rapid increase in the landscape impact. The management action influence point would move up the landscape axis as patch size increases. Only as we approach large enough clearcuts such that the patches abut streams, would the point turn in the three-dimensional space to move down the ecosystem axis, as the management practice begins to alter system hydrology. Thus there would be a scaling trajectory through the fuzzy criterion space that would allow the ecologist to identify the scales where there is a crossover of influence between criteria.

This scale-of-action trajectory in a criterion-dimensional space is at least an instructive exercise for gaining insights into the relationships between the alternative criteria and ecological scale. Tools for only thinking about multiple scales and multiple criteria are uncommon. We are already aware of shared community and ecosystem effects, as in the case of the salmon populations acting as both fish community members and as a reverse nutrient pump to counter the movement of nutrients down to the ocean. However, even with concrete examples of that sort, there exists no systematic scheme for netting relationships from a knowledge of general principles of the scaled relationships. The fuzzy criterion space with a scale trajectory will generate a map of the likely spots for criterion correspondence, like the tree as a structure relevant to several criteria. We hope the fuzzy criterion space identifies not only the tree as a point of articulation between criteria, but also other less obvi-

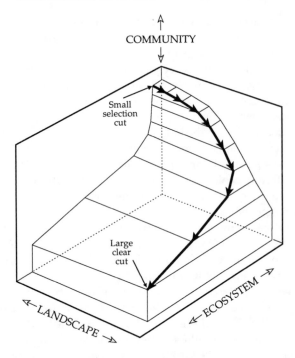

Figure 9.5. A hypothetical trajectory of the fuzzy influence of tree removal at various scales. Small-scale action influences communities but has only a little influence on the ecosystem and landscape criteria. Large-scale clearcuts exert influence on the system according to both those criteria, but perhaps little effect on the community.

ous or tangible cross-criterion structures. Whether or not it does that in practice, we have found it a worthwhile mind experiment to open up insights into the general problem of the relationship across both criteria and scale.

Ecologists make many types of measurements, but expend little effort relating across types. In communities we collect species lists, diversity estimates, relative size classes for major species, recruitment and regeneration estimates, seed rain, mortality, and wildlife foraging data. Detailed measures of ecosystems include fluxes in energy or matter: runoff, snow accumulation, evapotranspiration savings, nutrient and sediment flushing in streams, local temperature and humidity, and productivity. Finally, detailed measures of landscapes might be proportional area of land use types, composition of land use types, cover types, corridors, strips, intersections, edges and fragmentation effects, interiors, texture (evergreen-broadleaf mosaics), vistas, waterways, etc. All of these categories of data were in fact collected in the Fraser Experimental Forest in Colorado, and we found papers and reports published on them all. Such a wealth of data from a particular forest should be inter-collated. The scheme for the fuzzy assignment of major ecological criteria to management actions offers one of the few intellectual frames for the unified treatment of that wealth of information. Huchins, one of the most distinguished presidents of the University of Chicago, suggested that in contrast to the Middle Ages, which were rich in concepts and poor in data, we are now awash with data but have little unifying conceptual structure.

Cycles and Their Points of Contact

If ecologists are to pull the loose ends of their discipline together, then they must seek natural structure, structure which is predictive and reliable. This is another way of saying that science should seek that which is robust to transformation, that which persists when viewed on several criteria. The several criteria each reveal a new aspect of the structure, an emergent property. Each new emergent property or phenomenon requires new explanatory principles. In terms of surfaces, predictive power is more general for entities defined inside surfaces that have high dimensionality, surfaces where there is a coincident attenuation of many processes. Inside these surfaces there is a closure of processes, a containment of cycles of behavior. Sometimes the cycle might be a life cycle, in which various stages take their turn. Note that the tadpole in the life cycle is itself composed of a cycle of anabolism and catabolism, which itself can be broken down into smaller cyclical biochemical pathways. The cycle of interest could be a successional cycle or a nutrient cycle. Structure can be equated with the closure of cycles.

In the immediately preceding section we discussed the importance of ecological entities, like trees, that are players in populations, communities, ecosystems, and landscapes. Remember that populations, communities, ecosystems, and landscapes can be considered as the embodiment of cycles. Players in more than one ecological structure occur in more than ecological cycle. It is those players in more than one game that embody the places where the cyclical processes that pertain to several criteria come together. Those structures are the places where the various cycles of nature kiss. Those will be the instruments for generality that are central to the unifying scheme that might pull ecology into a cohesive whole.

Much is already known about the behavior and evolution of cycles. Manfred Eigen has built a body of theory about cycles that identifies how upper-level structure emerges. The new level results from a cycle of behavior that locks together a set of lower-level cycles. His ideas are related to Prigogine's notions of the emergence of upper-level structures through the amplification of fluctuations. In Prigogine's dissipative structures, when a system is away from equilibrium, a random fluctuation starts a set of coordinated fluctuations. This coordination sets up a new structure that persists away from equilibrium.

The details of the new nonequilibrium system are very sensitive to initial conditions. For example, the form of a mass of crystals depends on the exact form of the seed crystal, and on when and where it was put into the supersaturated solution. The importance of the unique initial conditions of Prigogine corresponds to the critical phase in the formation of Eigen's cycles. An accident may set up the first upper-level cycle, but once it has emerged, the new cycle outcompetes all other potential cycles. The first cycle to emerge at a given level gives the form of the large-scale, long-term dynamics thereafter, exactly as does the seed crystal in a Prigoginian system reorganization. The seed crystal imprints itself on subsequent patterns of crystal formation. The seed crystal grows through cycles of patterned growth leading to patterned growth, which preempts the resources of other patterns of crystal growth. By the time the seed crystal has finished growing, there is not enough solute left to form any other crystalline pattern.

Another example of the first cycle outcompeting all later comers is the way that microbial life preempts the resources of any other system that might be the origins of a different form of life. As a result, all life that we know uses not only a limited set of fundamental compounds, but it uses predominantly only one isomer of several major classes of molecule, namely amino acids and sugars. There is an echo of early crystalline structure at the base of life. One cycle of organic reaction in the primeval soup outcompeted another, so fixing that cycle over the other. For example, the cycle of ATP to ADP and back again is what fuels all living systems. However, GTP could do the same job, and indeed is part of the fueling system for protein synthesis. The "A" and the "G" above refer to adenosine and guanine. The cycle is from a phosphate chain of length two to a chain of length three, and back again: hence diphosphate and triphosphate. At some stage adenosine gained the ascendancy and that fixed the rules for the future.

In Eigen's scheme, hierarchies of cycles build ever higher, each stage depending on an accident outcompeting all later accidents, so fixing a stable hierarchical structure. The mechanism of the competition is simple. Various separate structures become involved each as one component in a cycle of behavior. Presumably these structures that have become components of the first cycle could become involved in other cycles. They could, but they do not, because the first cycle is already extant, and so uses its components with an efficiency that cannot be matched by an incipient cycle. Subsequent cycles never get started. Thus, in his scheme a lower-level structure is held in just one upper-level cycle, for it relates only to the next member in the chain. The track of the dominant cycle does not allow any other relationships to become as important as those between its own parts.

At first sight, our assignment of individual structures to several ecological criteria that imply several cycles would appear to be at odds with Eigen's evolution of cycles. In our scheme, the tree is part of a cycle of succession in the community. It is also a part of the carbon and nitrogen cycles under the ecosystem criterion. The cycles of the landscape are less obviously cyclical, but there are cycles there all the same. In Georgia the landscape is now in its fourth cycle from a forest to field and back again. The coincidence of these cycles in involving a given tree is entirely intuitively reasonable, so we must be talking about something different from Eigen's primary cycles, and indeed we are. The competition between Eigen's cycles whereby one reaction outcompetes another only involves one criterion for both competitors, by our standards. The reason why it is one base or the other in the fuel cycles of cells is that the ATP/ADP and the GTP/GDP cycles are fundamentally the same. They work under the same set of biochemical constraints, really under the same criterion. By contrast, the cycle of succession compared to the nitrogen cycle operates in such different terms that there is no basis upon which they can compete.

Thus we envisage a set of cycles that are free to share given components, because each cycle uses the entity as a part in such different ways that there is no competition. True, one of our coexisting cycles might have arisen before the others. That first cycle will have come into existence by outcompeting some potential set of cycles for the resource base, and so it was the one that persisted. Once that first cycle was extant, then other cycles could use structural parts from the first cycle, so long as they did not compete for the resource captured in the structures of the first cycle.

Once evolution has created a tree with its role in succession, then the same death event that moves succession of the community incrementally forward also releases the nitrogen in that dead tree for an ecosystemic cycle. It is the same material, doing the same thing, whether we view it as an ecosystem event or a community event. The different cycles are only conceptually different, not physically different. Two biochemical cycles might indeed compete for a given molecule, for if it is converted into the next stage of one cycle, it is necessarily not converted to the next stage in the competing cycle. Community and ecosystem do not compete with each other for resources because the tree dies in the same way for an increment in both cycles. Quite to the contrary of competing, they coexist one with the other, and so form an accommodation to each other over time. At first a commensurate ecosystem and community cycle might be independent of each other, but later they will form a mutualistic structure. It is these mutualistic structures that we seek in the meshing of cycles based on different ecological criteria.

In biochemistry, various cycles have become very stably fixed. Presumably, each of these cycles, like the Krebs cycle, will have come into existence by setting themselves up as a structural solution to a biochemical circumstance far from equilibrium. What is remarkable is that these cycles do share chemical resources. On the face of it, this would seem surprising because the various cycles in the cell do use the shared molecules in exactly the same terms, such that they must be competing. When glucose phosphate is used in one cycle, it is necessarily not used in another. The solution to this apparent dilemma is, as Eigen points out, that there is an upper-level cycle which is using those quasi-autonomous lower-level cycles as its parts. The whole forces an integration on the parts. The whole is an upper-level cycle, which could be either metabolism, the cell, or the organism, depending on your point of view. The cycles are forced to use the same materials in a forced integration. Presumably, the different low-level cycles in cell metabolism arose somehow in isolation, and were only later pressed together as members of the higher-level cycles of whole cell metabolism.

This highlights what is missing in a conventional ecology that is divided along subdisciplinary lines. At any given scale of time and space there is a physical world that is locked into ecological configurations. At a given scale, there are populations, communities, ecosystems, landscapes, and organisms, the structural devices that we use to do ecology. The conventional way to study ecology is to focus on one, or maybe two, and hardly ever three of those types of structures. It is reasonable to expect that there are critical structures which, like the shared molecules that hold metabolism together, hold ecological structures together. These structures, mostly unidentified as yet in ecology, are the points of articulation inside a unified ecology. Ecologists might grudgingly give biochemists credit for their undeniable achievements, but might denigrate them for not having put the cell together. The biochemist is not holist enough to put his functioning whole together, a defensive ecologist might say. The irony is that biochemists have done a much better job of treating their system as a working whole than have ecologists. Ecologists, for the most part, study their material using only one of the types of cycles. Unity is sought by studying differently scaled populations, or ecosystems, or communities, but never the entity that presses those different devices together at a given scale.

In our unified conception, lower-level ecological structures are shared by cycles of very different nature, but all of them ecological cycles. Imagine a given ecological structure, like a tree, as a bar magnet (figure 9.6). There is a field of cycles around it, each cycle being a line of magnetic force cycling from the north to the south pole. An organism can be a part of many ecological cycles, each based on a different criterion. Furthermore, each of these cycles passes through many other structures, each one of them with its own field of cycles. The tree in the nitrogen cycle releases minerals in leaf drop. Before the local cycle returns the nitrogen to the roots, it has passed through many other structures. Some of them might be organismal, like earthworms, and others might involve whole guilds of species, like the rhizosphere. At any given stage in this thought excursion, we could change criteria in one of the intermediate structures. At that point we could head off into an evolutionary or genetic cycle, perhaps via one of the fungi in the rhizosphere. The physical material of which ecological systems are made does exactly this. It is not committed to being in a successional entity, or any other type of ecological structure. A carbon atom in a successional entity like a tree trunk could either pass next into the atmospheric part of an ecosystem or become incorporated in the DNA of a millipede.

It is important to recognize that it is not structures in the ecological

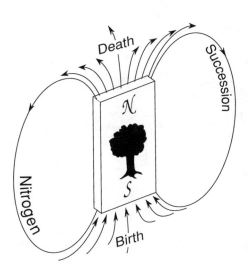

Figure 9.6. Consider an ecological tangible (e.g., a tree) as a bar magnet with lines of magnetic connection between the poles. The magnetic field represents a set of cycles of which the present tangible is only one stage (e.g., succession, birth/death, nitrogen cycling).

cycles that are the focus of attention. They are only the devices that we need to follow the lines of connection in the ecological cycling of material. Perhaps the analogy of a magnet is inappropriate because it puts emphasis on the solid magnets instead of the lines of connection. More appropriate would be a torus, for there the hole in the middle is the ecological structure of reference (figure 9.7). Touching tori would be even

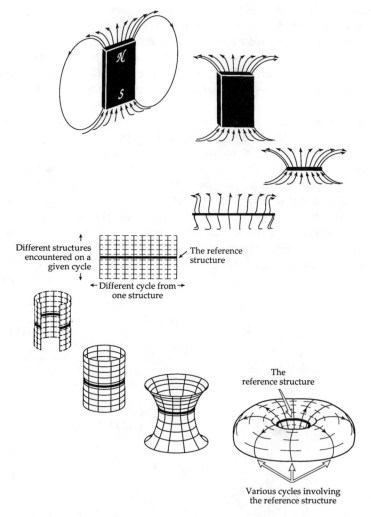

Different structures encountered on a given cycle

The reference structure

← Different cycle from → one structure

The reference structure

Various cycles involving the reference structure

Figure 9.7. Simple topological manipulation of the cycles and the tangible allow a change in emphasis where the tangible becomes the hole in a torus and the intangible cycles come to the fore as the torus itself.

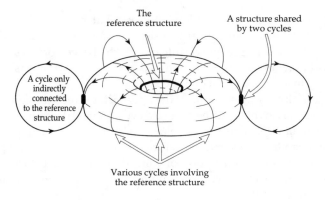

Figure 9.8. Ecology seen as a set of cycles with tangible structures relegated to the status of holes and points of contract.

better, so that the thought experiment can circle around from one torus to the next. In this sense, ecology can be imagined as a box of bagels or doughnuts (figure 9.8).

The value of this approach is that suddenly our ecological concepts will be working for us, instead of us working for them. We are free to use an ecosystemic nutrient cycle as long as it serves our purpose. When a community conception becomes appropriate, then there is nothing to stop the change in criterion as desired. By emphasizing the arbitrary nature of the criteria that ecologists erect to further their ends, we can use the criteria and structures without being fettered by them. Only when ecologists have a clear distinction between what is physically necessary and what follows from conceptual devices can they avoid confusion. Then we can build large, powerful, and predictive ecological structures at a level that subsumes ecosystems, populations, organisms, or communities, in any mixture of our choosing.

There is already an example of the approach we recommend, a study of prairies that involves three separate cycles focused on one ecological entity. The tangible entity is big bluestem, the grass which dominated the American tall grass prairies before the coming of the white people. Independent of our conception of coincident ecological cycles, David Wedin recognizes three cycles pertaining to three different system criteria, all focused on the tall grass prairie-dominant species.

In the tall grass prairies of Minnesota, Wedin and colleagues from the long-term ecological research site at Cedar Creek have been performing a set of highly focused studies on nutrients and competition of the dominant grasses. Wedin is now in process of standing back from the detailed results of that work to include several different criteria with respect to the processes maintaining prairies. He researches with David Tilman, who has established a paradigm of nutrient status and competition using the population criterion. In that paradigm, the species that can draw nutrient levels lower than any other and still hold the site are the superior competitors. This population-oriented approach is reminiscent of the positive feedback hollows in Roberts' conception of community. The dominant grasses of the tall grass prairies have an optimum that is close to the environmental condition that they create.

According to the results of garden experiments, the Tilman paradigm appears to characterize the dominant grasses of the tall grass prairies. Wedin uses the observed capacity for drawing down nutrients as the centerpiece of a model that links grazing disturbance, fire, and community structure. With it he explains how the tall grass prairies can occur in such widely different climates, as well as why they appear to be so fragile in the face of disturbance.

If the dominant grasses can establish a population, they draw down the nutrient level, particularly nitrogen, so that no other species can perform well. Their dead, above-ground parts, which might decay and contribute to soil nutrients, in fact have so little minerals in them that they do not effectively recycle minerals. As a result, the community dominant becomes established in its position. The dominant grasses manipulate the nutrient cycle in a positive feedback where success leads to lower nutrient levels which lead to further dominance. By breaking the nutrient cycle, the dominants set up a population competition cycle that continues tall grass dominance. This positive feedback of community dominants creating a favorable competitive regime is the first of the critical cycles.

As is common with positive feedbacks, this system can feedback in the other direction, allowing those weedy species which can improve the soil nutrients to gain control by winning competition for light; high nitrogen leads to dense vegetation and more intense competition for light in which the prairie dominants lose. Thus the danger for the dominants is the conversion of the soil into a high nutrient status. Should that happen, say by manuring or agriculture, the positive feedback of dominants and low nutrient status is broken.

Wild grazing animals, by urination and defecation, have the potential to reverse the direction of the feedback. Grazing and nutrient cycling are related to distinctly biome and ecosystem criteria. It is surprising to find population and ecosystem criteria applying together to the workings of a single system so transparently. The dominant grasses do offer a large green biomass for the bison early in the season, but the nutritive value of the plant to the animals drops precipitously after about six weeks into the growing season. Accordingly, the animals might be expected to be forced to move on before they can change the critical low nitrogen situation. This grazing cycle of returning bison being driven onward is the second critical cycle.

Wedin notes that the bison were exterminated from the tall grass prairie by 1840 or so. The serious, academic students of prairies like Clements did not work until some forty years after. There are reports in the formal literature that prairies were enormously tall. It is conventional wisdom that grazing by huge bison herds had occurred, although they were long gone before the formal reports were made. Wedin's work might call that wisdom into question. Either the prairie was a lot shorter than the ungrazed examples Clements reported and we see now, or the bison were occasional and not persistent grazers.

The last cycle is the fire cycle. The prairie is a fire-adapted community, so this last cycle belongs to the community criterion. Although fire is used in tropical gardens to release nutrients from the biomass, the benefits are to potassium and phosphorus, not nitrogen. Fire can burn off more nitrogen than it releases to the soil. Although the conventional wisdom is that prairie fires benefit soil nutrient status, a closer examination suggests that the dominant grasses burn readily to reduce nitrogen status as much as they pass on minerals to the surviving root culm. The fire cycle is self-perpetuating by virtue of fire giving the competitive advantage to the tall, readily burning dominant grasses. Thus there are three cycle: population dominance through reduction of soil nitrogen; an ecosystem grazing cycle; and a community fire cycle. The cycles appear at first to be unconnected, but they are united as they pass through the tangible dominant species.

Soft Systems Methodology for Managing Natural Systems

The underlying physical dependence of ecological systems appears to have misled ecologists into thinking that, if they only work hard enough, they will have hard systems and become hard scientists; not so! Invariant physical laws behind the scale-dependent parts of the discipline do make ecology a firmer science than some, but it does not make it a hard science. This is so, no matter how precise and sophisticated are the data collectors at the physiological end of ecology.

Because of its rich conceptual base, ecology is a relatively soft-system science. There is an explicit methodology for dealing with soft systems. By accepting point of view as the very substance of the discourse, soft-system methodology has given structure to what would otherwise be idle, capricious opinion. By now it should be clear that there are enough decision points in an ecological investigation to require some formalization of the decision-making protocol. An explicit science of soft systems has proved very valuable in applying what social psychologists and sociologists know about human behavior. It finds effective uses for resources in social contexts, exactly what the applied ecologist needs to know. It should also prove helpful in channeling basic research resources.

There are a number of protocols for dealing with soft systems. The scheme laid out here has been developed by Peter Checkland, an industrial chemist become business and private sector administrator, who is now professor of systems science at Lancaster University. It is particularly helpful for our purposes, because it amounts to a translation of the central theme of this book into a series of steps for action. Checkland's protocol is called "soft-systems methodology." It is a scheme for problem solving in messy situations where there are too many competing points of view for simple trial and error to prevail. This is exactly the situation that pertains in writing environmental impact statements or in developing a plan for multiple use in an ecological management unit. It will also be useful in identifying questions of importance in the conduct of basic ecological research. The translation to basic-science questions will require some modification of the scheme, for it is distinctly a resource-use approach coming from business applications.

There are seven steps to the process of problem solving in the scheme. *First is feeling the disequilibrium, recognizing that there is a problem, even if it is not yet expressed.* In ecology, this could be seeing that there is a problem in the sense that an environmental activist might raise an issue. Something is changing, but the source of the problem is unknown. In ecological management it might take the form of a major landowner who remembers when he was a boy that the streams were clean and the lakes were clear. Apparently there have been changes in land use without anyone noticing the fact.

At this early stage, the point of tension might only amount to the troubled feeling that an ecosystem, community, or population ecologist might have that the problem he has chosen could perhaps be solved by a different sort of ecological specialist. The new problem then becomes how to entertain those alternative system specifications. Another problem only felt intuitively could be one where joint production of water, vegetation, and wildlife is not adequately given account at a single temporal and spatial scale. Somehow, but in an as yet unspecified way, water, vegetation, and wildlife all occupy the same general area, but in ways that do not mesh. How can the basic scientist study these resources in a unified way? A last example might be in working an interface between basic research into biological diversity and the maintenance of biological diversity by management. Diversity takes various forms depending on the objectives; different objectives may lead to mutually exclusive definitions of diversity, and so indicate mutually exclusive courses of action. All of these applied problems will require the execution of basic-science ecology before the management issues can be directly addressed (figure 9.9).

"Mess" is a technical word here that couches the situation in terms that recognize conflicting interests. Multiple use in a management unit is a mess, as would be a basic research question that mixed ecological subdisciplines. After intuiting as above that there is a mess, *the second step is to generate actively as many points of view for the system as possible.* Checkland calls this stage "painting the rich picture," or the "problem situation expressed." The distinctive feature here is not the building of a model that has a particular point of view, but rather a taking into account of as many explicitly conflicting perspectives as possible. This second stage generates the system as described after deliberation, rather than the system as given at the outset. It is the richness of the picture which is important at this stage, not the restricted mental categories one might create to deal with it. We are still considering the system as it is directly observed, not the system as it is intellectualized later. At this stage it is important for the physiological ecologist to consider the view of other sorts of ecologists, or for other specialists, say community ecologists, to

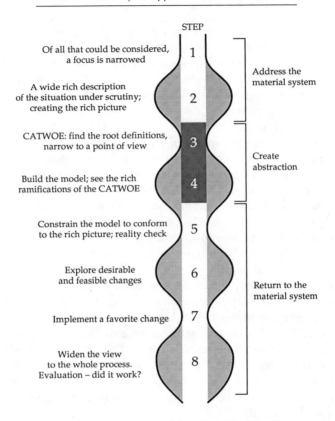

Figure 9.9. A schematic representation of the whole Checklandian process. Note a successive narrowing and widening of focus. Note also the distinction between the modeled, abstract realm (step 3, 4) as opposed to that which deals with the material structure.

consider the physiological aspects of the problem. The danger at this stage is finding a particularly appealing angle and allowing that to curtail the generation of ideas.

The third stage is the most critical, and involves the explicit development of abstractions. It puts restrictions on the rich picture in the hope of finding a workable solution. Checkland calls this stage finding the root definitions. There are certain aspects of the system that need to be identified at this point. They are crucial to making the system manageable, for it is talking at cross-purposes with respect to these system attributes that produces the conflict and confusion that was the problem as felt at the outset. The particular physical parts of the system corresponding to the various system attributes will change as the scale of the system is changed, or as the point of view is altered. The point of view relates fairly directly to the sort of ecologist that the researcher or manager admit to being. This, Checkland's third step, is an explicit locking together of a particular scale to a particular type of study. It amounts to actually using the abstract scheme we have erected, the scheme where we separate conceptual levels from scale-dependent levels. When the implicit scale is fixed, and the important things have been identified, then all the critical system attributes fall into place. Checkland's genius has been to identify the system attributes that link scale and structure with phenomenon, not just in a tidy intellectual scheme like ours, but as a working, problem-solving engine. With these critical system attributes identified and linked, ambiguity disappears. The researcher or manager still has to find a workable solution, but at least by getting the root definitions, he can avoid confusion.

Explicitly, the root definitions can be remembered by the acronym CATWOE. "C" is the client of the system and analysis; for whom does the system work? Sometimes the "client" is the person for whom the system does not work, namely the victim. "A" refers to the actors in the system. These could be the client or victim as well, but often the actors are separate entities. In the scheme that we have used to this point, these are the critical structures. In human social problems these are likely to be actual people, whose scale depends on their scope of influence. However, the actor could be a forest in an ecological system. Implicitly, the actors set the scale (figure 9.10a, b).

Even in an apparently well-defined problem there can be several candidates for the actor. Consider the quandary of DNA that repeats a meaningless signal many times over. The problem can be considered at several scales. If the actor is the whole genome, one will focus upon selection for adaptive advantage at the level of conventional evolution, the organism. It happens that the redundant mass of DNA has no meaning at that level and does not relate to organismal fitness. However, a selfish gene model would have to posit the DNA sequences as the actors, so scaling down the level of the explanation. That appears to be the appropriate level, for DNA that can selfishly replicate occurs in greater quantity. All this is clear enough with hindsight, but in less obvious situations it is important to be open to new levels of analysis and explicit as to the level in use. Choosing an actor achieves that end.

"T" are the transformations or underlying processes. What does the system do? What are the critical changes? These critical transformations are generally performed by the actors. "W" identifies the implicit worldview invoked when the system is viewed in this particular manner. In

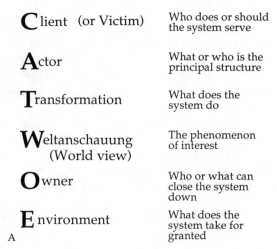

C lient (or Victim)   Who does or should
                            the system serve

A ctor   What or who is the
                            principal structure

T ransformation   What does the
                            system do

W eltanschauung   The phenomenon
(World view)   of interest

O wner   Who or what can
                            close the system
                            down

E nvironment   What does the
                            system take for
A   granted

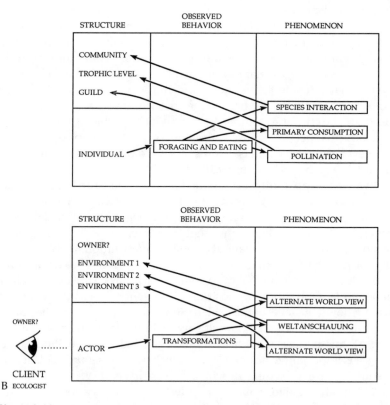

Figure 9.10. The CATWOE acronym translated into the criteria of Allen, O'Neill, and Hoekstra's scheme for moving between levels of organization. A. The acronym, B. the translation.

the scheme that we raised in the opening chapter, "T" identifies at the least our "differentials," the naked, measured changes of state. In a larger view, "T" pertains to the subjectively identified significant differences, our "phenomena." "W" isolates the subjectivity embedded in the choice of the phenomenon from the transformation that is implied in the critical change that embodies the phenomenon. Remember how in our earlier scheme for moving between levels, the phenomenon fixes the type of entity that is either found as a context, or as a mechanism. These would be the upper-level actors, whose identity defines either a reason or role when one moves upscale, or an explanation and mechanism as one moves downscale. Thus "A," "T," and "W" together identify whether it is a community study, an ecosystem study, or whatever else, and the scale of the investigation.

"O" refers to the owners of the system, who can pull the plug on the whole thing. Like the actors, the owners could be the client or victim of the system, but usually the owner is someone else. The scaling issues of grain and extent emerge here. With power to terminate the system, the owner defines the extent aspects of the scaling of the study. By contrast, the actors will usually define the coarsest grain that can be involved in scaling the system because they have to be discernible at the level of resolution associated with the specification of the system.

While owners fall out readily in an analysis of business or social management, it appears less important to identify them in strictly ecological settings. However, it is helpful to know the ultimate limits to the functioning of an ecological system, and the concept of owner might be of service there. For example, the fact that ice ages have pulled the plug on plant community associations in the past indicates the extent to which communities start as ad hoc entities. As an owner, global climatic shift puts limits on the evolved accommodation that is embodied in community structure. In managed or restored ecological systems, the owner can apply in very literal terms.

Last, "E" identifies the environment, that is, what the system takes as given. Anything longer term and slower moving than the whole system is a context in which the system has to live. By default, the environment defines the scale of the system extent by being everything that matters which is too large to be differentiated.

It is important to realize that the several different sets of root definitions are not only possible, but desirable. The actors in one set of definitions will be different from those in another. That presents no problem, but it is mandatory that the actors in question only act in the model for which they have been identified, and are not mistaken for actors performing at some other scale on a different set of assumptions. In fact, that error is exactly the sort of confusion which arises if the formal scheme recommended here is not followed. Mistakes are easy to make if there is not a formal framework to keep track of all the relationships. That error of sliding the scale or change of worldview is a favorite device for vested interests to confuse the issue when they know that their own position is inconsistent. Lawyers representing either the company in an environmental litigation, or an environmental action group bringing suit, can confuse an issue in this way, if they are in danger of losing.

Having defined the CATWOE, one is ready to *build the model, which is the fourth of the seven stages.* There will need to be a different model for each set of alternative root definitions. *The fifth stage returns to observation of the world, and the model is checked against what happens.* If the actors are people, then one can ask them their opinion of the model and modify it to be consistent with their special knowledge. *At the sixth stage one identifies desirable and feasible changes for the system.* In an applied eco- logical setting the plan for management is generated here. In basic research, the hypotheses and the protocol are married at this point. Note that desirable may not be feasible or vice versa. If desirable and feasible changes cannot be matched, then it is possible to return to stage three and look for new root definitions. Sometimes there is no feasible and desirable course of action, in which case there is nothing to be done, but at least one knows that more clearly than before.

The seventh stage is implementation of changes. In the case of an ecologi- cally managed system, that would be the implementation of the forest plan. In basic research, the tests are made, the data are collected, and the findings are studied. We should emphasize that this is not a fixed sequence, for one can jump from stage to stage, modifying as appropri- ate. Anyone who has done basic research knows that it is never done the way one writes it up for the journal article. By allowing steps to be repeated or taken out of order, the scheme acknowledges how re- searchers really operate. It does not try to force the process into the myth of the scientific method which emerges in the sanitized account that is finally published.

Consider the difficulty that the USDA Forest Service has in managing to "maintain biological diversity," as indeed they are mandated to do by an act of Congress. Managing for diversity is clearly a mess, in a techni- cal Checklandian sense. The problem is that "maintaining biological di- versity" is an undefined action leaving room for many points of view, each with its own agenda. The original Congressional action might be step one in Checkland's scheme, a general recognition that something is wrong. Step two, the problem situation expressed, in this case shows how complicated it is: 1) hunters desiring diversity of animals for their sport and unhappy about closed growth that offers few resources for animals of choice; 2) lumber men may not care about diversity, but they change diversity through logging old growth, so making a young growth more common than in the presettlement forest; 3) preservationists concerned with maintaining old growth in large blocks so as to provide for animals that require habitat that would otherwise become dissected; 4) motorists who desire a diverse vista; and so on. Different CATWOEs can tease apart those conflicting views to allow development of pertinent models so that a plan can be made without ambiguity. With the scale, structure, and function of the forest defined, compromise can be sought in a meaningful way, and management action will be enlightened by an understanding of what is intended, and how it is to be achieved. Rhetoric and hidden agendas can be exposed and those that are disingenuous can be rejected.

In basic research, complex problems can be made explicit. They may not become simple, but at least the researcher is cognizant that he has a point of view and knows what it is. Explicit entertaining of other scales and points of view, whether or not they are finally used in the investigation, highlights what has been done and sets it apart from what would have been interesting, but has not been researched.

The literature of herbivory is a case in point that would be greatly improved by the application of the explicit scaling that Checkland recommends. We referred earlier to the work of Becky Brown in this regard. The compensation that some plants make for being grazed can be seen in very different lights, depending on the researcher's point of view. At one level the plants suffer capital depletion and damage to the means of production. At a higher level they may replace losses in foliage from another plant part like the roots. In doing this they may replace moribund productive tissue and so, in the mid-term, increase biomass of the entire plant above what would have occurred without grazing. Alternatively, a community point of view might find that increases in production come from a change in the competitive regime making more resources available. In all these ways and many more, plants respond to grazing so as to make it appear an advantage to be in a grazing regime. A CATWOE would tease these points of view apart.

Note that even when the measurements are made of material flow and hard percentages of plant material consumed, all is meaningless until that material system is seen in the light of a point of view that recognizes biological purpose. Checkland's system acknowledges the subjectivity of the scientist, and so frees him of delusions of objectivity. It will force the contrast between critical differences in measured material flow, as opposed to differences in point of view applied to the same material findings. At this point the literature does not emphasize that distinction and the groomed accounts in the journals do not help the reader to recognize the tussle between points of view that were, in fact, part of the research process. Research is messy, but that may not present a problem unless we pretend that it is neat, objective, and based on clean measurements of hard systems.

The Difficulty of Seeking Ecological Scales of Nature

When we make an observation at a given grain in a universe of a certain extent, we experience something that corresponds to an aspect of the material system. We can reasonably presume that what we see does have some correspondence to something in the world beyond the observer. Change the scale, by changing the grain and extent, and appearances are different; we can presume that those changes are also some reflection of nature beyond the observer. However, those cautious statements are very different from saying that nature itself has a certain scale and that simple measurement will reveal that scale. A scale of nature is meaningless unless there are fundamental units to which the scale applies.

Photons may indeed be such fundamental units; perhaps we have to acknowledge the speed of light as a scale of nature. However, in ecology we do not have such unequivocally fundamental units, not even in organisms, units of selection, ramets, or any other ecological entity. The scale of a given ecological entity, or any process associated with that entity, cannot be readily untangled from the arbitrariness that is part of all ecological discourses. Therefore, ecological scales of nature are not going to be those nice clean scales that occur in physics. As an example of this ecological messiness, the "scale of allelopathy" has no meaning in general. The scale depends upon the target of the chemical warfare, and that changes with the point of view of the ecologist. All plants exude material, and some of that exudate suppresses growth. If the exudate is undirected happenstance that damages self, friend, and foe alike, then such allelopathic activity is so local as to be evolutionarily uninteresting. At the other extreme, some definitions would only admit allelopathy so long as there is demonstrable Darwinian selection to tolerate the chemical on the part of an exposed competitor. In between are conditions where the chemical exudate variously damages the allelopath's own offspring relative to the competing species and conditions where plants of other species are damaged, but not in a way that changes the competitive regime of populations.

There are too many different sorts of things happening in an ecological setting for us to find many general scaling formalities, like the Reynolds number of fluid mechanics. In the behavior of fluids there are regularities that pertain to viscosity, density, and rate of flow of the fluid, and the size of the obstruction. The nice thing about an obstruction is that it could be a paddle, a fish, an airplane, or anything else, but the name "obstruction" defines the entity with respect to the particular scaling activity one wishes to quantify. The problem with seeking such neat regularities in ecology is that ecological material does not possess such simply quantifiable attributes as do fluids or solid objects that obstruct fluids. Of course a particular organism has a mass, and even a drag coefficient, but that is special to its unique identity, and does not generalize to other organisms, even of the same species. A fluid is a robust condition that we find in nature often. We find lots of organisms, too, but the class "organism" is a fuzzy set to which things variously belong. Organism is a helpful concept, but it is too rich a class, too much reflective of quirks of the observer, for it to have consistent, quantifiable attributes like viscosity and density. Furthermore, as ecological concepts go, "organism" is very well behaved, so most ecological entities will be even more problematic.

These problems hark back to the softness of ecology as a discipline that we raised in an earlier section. Remember that the soft sciences study changes in human attitudes, values, and perceptions. The nature of the perception and conception themselves becomes the object of study in soft systems. Unfortunately, in ecology we seem to have no access to a hard system in nature waiting for us to find it. True, hard sciences also only deal with perceptions, namely data, and not ontological reality. However, the regularity of the material they study leads quickly to data collection on unambiguous entities. In ecology, it is the irregularities themselves that constitute the phenomena of interest, much as in economics price only becomes interesting if people's values are changing because of a complex of influences. When it comes to number and power of general scaling properties, ecology is about as depauperate as personality research in psychology or value research in home economics.

Even economics has many more scaling principles than ecology, probably because it is built upon relatively unambiguous primitives, such as a given currency. Only island biogeography stands out in ecology as having predictive scaling laws, probably because it is grounded in relatively unambiguous events like extinction of species or unequivocal values like the distance to the mainland. In soft sciences like ecology, it is the untidiness of the mess which is the object of study. We hasten to add that this untidiness of the material studied is not an invitation for ecologists to be vague or mystical. There are rigorous ways to deal with untidy messes, as we established in the prior section.

It is important to remember that the underlying conceptions we bring to an ecological observation very much color what we find. We need terms like community, ecosystem, and population, but they are all the product of an arbitrary process of defining things. Let us keep what is allowed in the structures embedded in our observational framework separate from what is possible in the undefined world of continuously scaled fluxes and processes of nature, independent of observation (figure 9.11).

We only see certain things when we look at a certain scale, and presumably that aspect of nature still exists whether we actually look at it or not. However, there are very many processes pertinent to given ecological experience, and they operate from very small to very large ranges in both time and space. A particular scale of observation interacts with

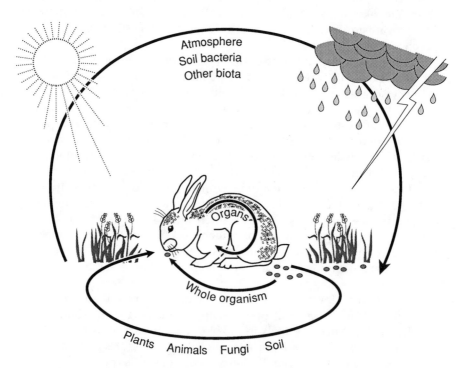

Figure 9.11. The cycling of nitrogen occurs at many scales from those inside an organism up to the scale of nitrifying and denitrifying bacteria interacting with biologically available nitrogen formed by electrical discharges in the atmosphere.

each one of those processes and material fluxes in a different way. Change the scale and one sees something different almost every time; that is one of the few things one can say about scaling in ecology in general. By itself, that is not a very helpful generality. Any scale of observation in particular is a very local consideration, only a special case. Science seeks generalities, not a cataloging of local happenings, so an agenda that hopes to find the real scales at which ecological systems work tackles the problem from the wrong end. Anyone starting out by seeking the real scale of nature will generate much detail with little general application. The better way to go about gaining insights starts with defining the universe of discourse and the type of question to be addressed. Then let nature manifest the scales that it will in that defined universe.

Scaling Inside a Criterion

The specialties inside ecology represent natural subdivisions that form because of the different types of question that each subdiscipline asks. There is plenty of room within ecology as a whole for even a majority to remain inside their respective specialties. However, even those who wish to remain focused narrowly, who are not interested in efforts to ask unified questions across ecology at large, are still touched by the relationship between criteria and scale. Scaling allows a linking between special cases within communities alone, populations alone, or ecosystems alone.

Robert Rosen has put together some general principles of scaling, along with notions of similitude, that should be helpful in comparing ecosystems with each other, or any ecological entity within its class of structures. He uses the van der Walls gas equation to advance the argument. Although gas laws would appear, on the face of it, to have little to do with ecology, Rosen is able to translate the conceptual devices for gases into biologically meaningful analogies. Gases have analogs for genomes, phenotypes, mutations, and environments once one reaches the powerful level of abstraction Rosen employs. Accordingly, we ask the reader's indulgence as we lay out Rosen's case.

The van der Waals equation is based on the equation of simple pressure, temperature, and volume equations for gases, the ideal gas law:

$$pV = rT$$

The van der Waals equation is:

$$(p + a/V_2)(V - b) = rT$$

The van der Waals equation has two new parameters: a quantity a, which pertains to the attraction between gas particles, and b, which relates to the volumes of the particles.

The behavior of a gas is not related directly to the volume and pressure that we measure. With regard to pressure, we need a correction for the difference between the pressure measured on the gas as a whole (the only sort of pressure we can measure) and the pressure felt by the individual particles. That correction is made by a in the van der Waals equation.

We also need a correction for the bulk volume due to the volume of the gas particles themselves. There is a difference between measured volume of the gas and the volume of the space between the particles. With little pressure, the volume of the particles is small in comparison to the volume of the whole gas. Therefore, with little pressure, the ideal gas law is predictive, even if it is uncorrected for volume of the particles. With increasing pressure, the particles themselves do not get smaller, although the gas as measured does. Under pressures close to that sufficient to cause liquefaction, the volume of the gas particles themselves is a significant proportion of the measured volume. The proximity of the particles governs critical changes of state; therefore, close to liquefaction the volume of the particles themselves, b, is an essential part of a predictive equation.

Rosen makes an elegant translation of the van der Waals equation to give it intuitive meaning for ecologists. He couches it in terms of biological orthodoxy. The prevailing biological paradigm views biological form, or phenotype, as the result of the interaction between the genome of the organism and the environment in which it develops. Mutation is viewed as a rare event that affects only a small part of the genome. Transfer of DNA between organisms is viewed as a special case—sex— or as a pathological departure from the reference conditions, or viruses. The genetic material in the organism is taken to be fixed for most purposes. Each gas is a special case because, depending on the gas, it takes different pressures to cause liquefaction at a given temperature. The terms a, b, and r identify a given gas and make it distinct from all others. Rosen points out that these three parameters of the van der Waals equation are the genome of the gas that determine its response to its environment.

Now consider the variables in biological systems. They are the environment, on the one hand, and the phenotype, on the other. Of the three critical gas variables, pressure and temperature refer to the ambient conditions with which the gas is equilibrating. Accordingly, they are environmental variables. Volume is something the gas manifests once the environmental variables are given. So it follows that volume is the phenotype of the gas under the environmental conditions in question. Thus there are two independent variables of control, and one dependent variable on the dimension of behavior, as well as a set of genomic parameters.

Note the cubic relationship posited in the van der Waals equation. Rosen points out that "the behavior of a gas at equilibrium depends on whether this cubic term has one or three real roots. The gas/liquid phase transition resides precisely here." In geometric terms, phase transition corresponds to a fold in the surface that relates volume to temperature and pressure. It is positioned so as to correspond to the pressure and temperature values at which liquefaction occurs. For any given gas, the van der Waals equation describes a phenotypic or behavioral surface that has a pleated fold under certain environmental pressures and temperatures (figure 9.12). The fold is what Thom has called a catastrophe cusp. As the environment moves across the region of the edge of the cusp, the system becomes unstable, and reorganizes by either liquefaction or volatilization into the other phase.

A model for the gaseous phase has nothing to do with the liquid phase. The behavior of gas molecules is almost independent, whereas in liquids the molecules are highly correlated in their behavior. When a gas becomes liquid, the constraints on the individual particles change. The rules for being a liquid appear "as if from nowhere," Rosen says. We are talking here not of a continuous change that can be mapped dy-

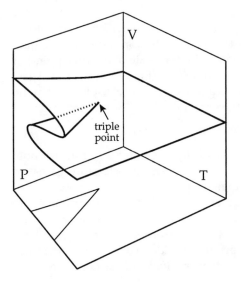

Figure 9.12. The folded surface defined by the van der Waals' equation. In the region of the fold the gas liquifies (after Rosen 1989).

namically. Rather the change is to a new structure and is instantaneous. When a liquid becomes a gas, the constraints that hold particles in the highly correlated liquid state are broken. There is nothing about the high correlation of particles in a liquid that tells anything about this new uncorrelated gaseous state, for the constraints that gave the old degree of correlation are instantaneously irrelevant. That is why we need a completely new model for liquids that has nothing to do with the model for gases.

Consider again the pressure, temperature, and volume space for a gas. There is the gaseous region and a liquid region, with transitional regions in between. Within a given region of the behavior space, behavior is continuous, such that a very small change in temperature and pressure always makes only a small difference in volume. This small change in volume could be accommodated by a small rescaling of the behavioral axis; we could rescale the units on the volume axis and remove the difference resulting from the small change in environmental variables. A similarly small change in environmental variables that goes across into another region will give a radically different state on the behavioral axis, so different that it could not be accommodated by a rescaling of the behavioral axis. This is because, with a change in region giving a change of phase, there are new constraints in place, so the underlying rules of behavior are different. A simple rescaling can accommodate a small change within a region because within the region where the gas phase occurs, the same basic rules of behavior underlie what is found. In this sense the entire liquid phase is similar. The same applies within the gas phase.

Rosen goes further with his analysis of the van der Waals equations. There is a singular point where the cubic relation specified by the equation has three coincident real roots (otherwise called the triple point). It is at the base of the pleat in the behavior space (figure 9.12). For each gas, that triple point is singular. The location of the critical triple point in the p, V, T space is determined by the genome a, b, r. The relationship is:

$$p_c = a/27b^2; \; V_c = 3b; \; T_c = 8a/27rb.$$

Translating this to a more biologically meaningful setting we note that more than one gene contributes to a given character; thus gases are polygenic. Note also that some genes play a role in the expression of more than one character; thus gases display pleitropy. So even in a system as homogeneous as gases, the sort of complexity that can make biological systems challenging manifest themselves. This can complicate scaling considerably, as well as produce decidedly unexpected misleading behavior, when an error of system specification has happened.

It is possible to express the behavior of the gas in the three-dimensional space of p, V, and T not on directly scaled axes of pressure, volume, or temperature, but in relation to the critical triple point. Thus pressure, which might be expressed in pounds per square inch, can be expressed alternatively by a dimensionless variable derived from P/P_c. This is a rescaling of the variables relative to a point which is a singular characteristic of every gas. Such a rescaling of the behavior space allows direct comparison of the behavior of all gases. Each gas has its own values for a, b, and r, and so occupies a particular place in a (a, b, r) genome space for all gases. We have already rescaled the individual behavior spaces of all gases into comparable terms. Accordingly, it is possible to plot different gases as points in the genome space, and then plot their states as cross-sections of corresponding states above the genome space (figure 9.13).

Since we can only draw in three dimensions, the three-dimensional space of the genomes is represented as only the two-dimensional plane

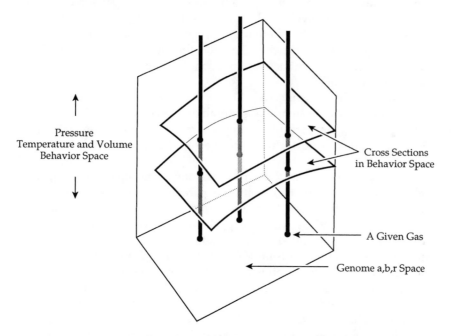

Figure 9.13. The plane at the bottom of this figure is the three-dimensional a, b, r genome space of all gases. The vertical axes of the cube is the three-dimensional space of the redefined pressure, temperature, and volume dimensionless variables. In this renormalized space, the catastrophic fold disappears, making all gases a simple rescaling of all other gases (after Rosen 1989).

on the bottom of the box, and the three-dimensional variable space of rescaled pressure, temperature, and volume has to be collapsed into a single behavioral axis, but that is purely a matter of representation. Figure 9.13 is a compromised representation of a six-dimensional space, but it will serve our purpose. Note that the cross-sections of the genomic space corresponding to each behavioral state are not folded. This is because the folds in their state spaces have been normalized away in the relativization to their respective triple points at the base of each gas's respective fold.

Being able to express the behaviors without the folds is important because it makes all parts of the entire behavior space similar, in the way that all states in the gaseous phase were similar. Remember how the discontinuity of the catastrophe cusp meant that all gaseous states were dissimilar to all liquid states, so that the change to or from a liquid phase could not be accommodated by a scale change. Having found a way to express the behaviors in terms of the genome of the gas without the manifestation of the instability around the cusp, a comparison between gases now only involves a scale change. A change of gas is only a change of scale, so long as the reference between them is singular (the triple point). By reference to the critical point in the state space of all gases, it is possible to summarize them all in a simple scaling operation inside a unified framework.

Rosen says that the message here is that there may be an enormous number of states in a state space of high dimensionality (e.g., the states of all gases) but there are interesting generalizable states of small dimensionality that offer powerful predictive summaries. The trick is to find them.

In the above case, if we were to take one of the state variables, say temperature, and try to use it as a parameter, then the new four-parameter family (a, b, r, T) would be unstable and the folds which had been normalized away would reappear. This is what happens when a system is ineptly specified so that it manifests middle number unpredictability. Catastrophic folds indicate a radical change in constraints, and it is the changes in constraints that make for middle number uncertainty. In a middle number system one cannot tell when one of the system parts will impose constraints on the whole; in the same way, going over a catastrophic cusp, the constraints in the domain one is leaving tell nothing of the new constraints that will operate on the new surface, when the system reaches that new state. In both cases the problem is the unexpected imposition of new constraints for which old constraints are not predictors.

The whole discussion here focuses on problems of what it means to be similar. To summarize the above ideas, Rosen says, "Similarity means precisely that we can completely compensate for a change of control by means of a scale change (a coordinate transformation) of behavior alone," which would not be possible if the behavioral surfaces were folded. Remember that one cannot cross folds by merely changing scale, because the fold indicates a change in the rules. Such is the change from liquid constraints on particles to gas constraints on particles.

This translates into strictly ecological terms when we notice how the central words of ecological parlance are scale-independent. As we have stated repeatedly earlier in this book, this scale-independence arises because the words are scale-relative terms. Competition is not fixed to a given scale until the actors are identified; the actors bring scale with them as baggage, whereas ecological concepts do not have a scale independent of ecological actors. When the scale changes have been made, competition between bacteria can be compared to competition between whales. Of course there are differences between bacterial and balanid competition, but there had better be some relationships that translate between the two cases, otherwise the term and the concept of competition are vacuous. It is no accident that terms like competition can be formulated into fairly straightforward algebraic statements, although there is fighting between the schools of students of competition as to what the exact formula should contain. Some of the arguments turn on what must be part of the relationship for it to become or remain competition. Those arguments can be characterized as discussions of how different can a situation be, and yet still be similar. As long as there is only a scale change, then the systems described are similar.

All this highlights the importance of having strict definitions of the major types of entities that ecologists study. We have emphasized that our definitions are not fundamentally correct, but they are explicit. Only when ecologists know what is included in a definition can they work out whether the situation being described to them is similar to the one they have in mind. This is so obvious as to be banal, except that by now it should be clear that even subtle differences in the nuance associated with a term can have radical consequences for scaling and what is meant to constrain what. Even slight changes of specification, as when temperature is introduced as a parameter instead of as a variable, can introduce catastrophic consequences, in the strictly mathematical sense. A small change that brings with it a catastrophe cusp in the state space can make the situation dissimilar to what was specified before. Differences in details of descriptions are one thing, for they can be accommodated by rescaling the system, but dissimilar situations are much more problematic. We need to pay enough attention to ecological descriptions so that we can tell the differences between ecological systems only rescaled and ecologically dissimilar situations. It should be clear by now that telling that difference is not trivial.

Fundamental Dimensions

Biologists have long used the idea that some systems are models of others, where the difference is one of scale. Microcosms are an obvious example. However, the use of models is very widespread and includes the substitution of animal models for humans in medical research.

The difference between model systems and the systems they model should be one of scale alone, if the findings in the model system are to be applicable in any general way. The central questions here revolve around the notions of similarity, and they are as fundamental as any questions in science. Consider the situation when a modeled system changes, that is to say when a model is applied to a related system, but not the one for which it was created (figure 9.14). In such a case, what is the relationship to the new system? If a new model is created to deal with the modified observed situation, to what extent is the new model a model of the old model and vice versa? Finally, to what extent is the new

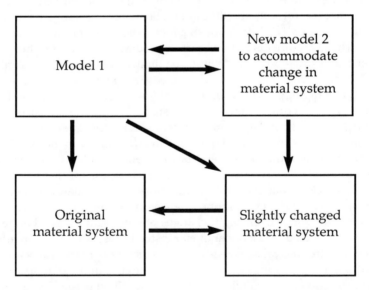

Figure 9.14. Rosen's consideration of the van der Waals equation addresses questions of new models being required for slightly new physical circumstances, and the extent to which the new model and the new situation represent models themselves of the old model and the old situation.

situation a model of the old situation? The law of corresponding states asserts that all van der Waals equations with constants fixed for the gases in question are models of each other. All gases are models of each other. Elsewhere Rosen points out that geometry is a systems enterprise, in that all isosceles triangles are models of each other. So these questions of the relationships between models have a long heritage.

Earlier in this century, well before the current interest in biology, D'Arcy Thompson was explicit about the scaling of relationships in biology. He suggested that closely related species are similar. Note that the closeness of the relationship is a matter of genomes, while the similarity is a matter of phenotypes, so the statement is not trivial. As we shall see in the ensuing discussion, Thompson's model is not always correct. He outlined a standard fish form on a grid, and then distorted the grid to transform the fish into a new form, an angelfish. He did the same for the face of an ape and that of the human (figure 9.15). He formally called these ideas the "principle of transformations." We now stretch these until they snap. Major ideas, like D'Arcy Thompson's principles, are most instructive of all only when they fail. That is one of the ironies of science.

Rosen concludes his paper by going well beyond the pragmatic or heuristic and relativistic position that we have taken in organizing this book. However, his position is still an epistemological one and should not be mistaken for a simple research for the scale of the real world. Without Rosen's brilliant moves to high levels of abstraction, we have not until now dared to emphasize the scaling of the material system as he does. The critical step Rosen takes is to note that there are other ways to classify parameters and variables than phenotypes, genotypes, and environment. The alternative classification is into fundamental and derived magnitudes.

When dealing with a set of magnitudes, or dimensions, some quantities can be derived from others by mere algebra. At the heart of the issue is a set of fundamental quantities which provides the dimensions for the derived quantities. The fundamental set should be dimensionally independent, while the derived set will have dimensions algebraically derived from the fundamental set. Rosen points out that if these dimensions are related, say through laws of a material system, then separation of dimensions into fundamental and derived has great significance. In van der Waals' gases, the genome a, b, r is fundamental in this important way, although this is by no means necessary in other systems. The fact that a, b, r form a fundamental family is responsible for all gases being models of each other, and there is only one similarity class for gases. Gas laws embody a peculiarly powerful set of systems in this regard.

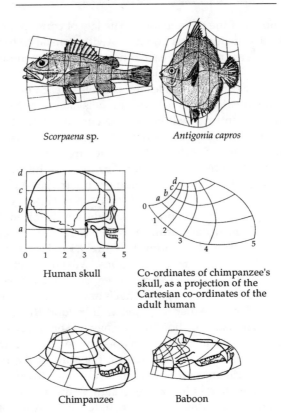

Scorpaena sp.   *Antigonia capros*

Human skull   Co-ordinates of chimpanzee's skull, as a projection of the Cartesian co-ordinates of the adult human

Chimpanzee   Baboon

Figure 9.15. D'Arcy Thompson (1942) viewed the relationship between species as a rescaling operation in his principle of transformation (after Thompson).

Remember how all gases can be put into a genome space, and how the behaviors in state space can be plotted on derived pressure, temperature, and volume axes. The derived variable space is expressed as relative to the triple point for the gas in question, and the triple point has been derived from the a, b, r genome. Thus the expression of pressure, temperature, and volume is derived, not fundamental. Remember that the cross-sections in figure 9.13 are not folded, that is to say, the space which compares the gases is behaviorally stable. When we take one of the derived variables and include it in the parameter space, the behavior space above does fold, it manifests instability. The stable space is defined by fundamental dimensions. In the case of the gases, the stable space was the a, b, r genome space that applies to all gases.

A stable space defines similar situations. In unstable parameter spaces, at the point where the system becomes unstable the system passes into a new similarity class, and manifests dissimilarity with the old behavior. Thus, two closely related species can be dissimilar. It is here that D'Arcy Thompson's principle of transformations fails. This would be because the small metric distance between the species puts them into different classes of similarity; the change, which we could call a mutation, happens to cross the fold. Thus some mutations may create a still closely related species, but one which is dissimilar. Inside a stable space, or inside a region in an unstable space, mutations generate similar species. While the above may apply to literal mutation and literal species in Darwinian evolution, the same arguments apply to the more general use of species, genome, and mutation as applied to gases, for example.

This line of argument can be applied to the vociferous debates within narrow facets of ecological subdisciplines. Sometimes such discussions can degenerate into assertions about this or that definition being correct. More enlightened altercations are less concerned with the true definition of, say, competition, and focus upon the meaning of choosing this as opposed to that definition. One might expect various dialects in competition theory to be only matters of style within a class of similar conceptions of competition. In such cases, a simple rescaling can translate one dialect of competition into another by the inclusion or exclusion of some small conceptual device, or by changing the measure of competition. The differences can be transformed away. Those speaking not just closely related dialects should form schools of thought.

Plant demographers working with seed box experiments may not be perfectly unanimous as to working definitions of competition, but they do form a school of thought existing inside a stable parameter space. Then there are rival schools of thought; one might expect these to have critically different definitions. Note that the difference in conception here may not be large, but it is critical. A small difference renders the old set of concepts unstable, so that the new definition is not just different, albeit in some small manner, but it actually belongs to a different similarity set. Note here that rescaling is not possible because a different set of constraints applies. For example, neighborhood competition is not congruent with population competition. This is because neighborhood competition is dissimilar from population competition. It is possible to look at large neighborhoods and small populations, such that the population competition under investigation is smaller-scale than the extended neighborhood competition. It would seem that the introduction of breeding systems changes the logical type of the competition and makes it dissimilar to competitions in large neighborhoods.

Nevertheless, both neighborhood and population competition are of the same type if we are only concerned with competition as a general phenomenon. It is still competition: individual with individual, population with population. There is some correspondence between individuals competing and populations competing, even though it is not possible simply to rescale population competition down and reach competition of constituent individuals. Put another way, population competition is not the simple sum of individual competition; if it were then the summation would be a simple rescaling operation.

Thus questions of scaling and criteria for comparison and scale change are not limited to the pursuit of our agenda, which is the unification of ecology as a cohesive body of thought. Ecologists who would readily define themselves as distinctly inside one of the major disciplines and who are happy to focus their entire professional effort inside that specialty still have to be aware of the issues raised in our efforts to unify ecology. Even population biologists uninterested in finding community mechanisms still have to pay attention to criteria and scale if their discussions with like-minded colleagues are to bear full fruit.

All of this is a matter of degree. Rosen points out that while stability in his sense is all or nothing, there are degrees of bifurcation, that is, degrees of stability, degrees of dissimilarity. All this is reflected in the standard biological taxonomic order where there is more similarity within a species than within a genus or a family. There are degrees of dissimilarity between different communities, or between different ecosystems, and all levels of difference are worthy of investigation. The discussion above about neighborhood and population competition being dissimilar at the level of the distinction between individuals and populations, but similar at the level where both are cases of competition, is an example.

Already Rosen's strategy of renormalization has been used in a strictly ecological model by Bruce Milne. It is only a first step, but shows great promise. Milne created a two-phase landscape derived from a remotely sensed landscape. On this landscape he randomly placed simulated animals weighing between two and six kilograms. There are well-known allometric relationships for herbivores that link body mass, home range size, population density, and metabolic rate. The animals could only use one phase of the landscape for food. There was a simple rule: if the animal had enough to eat it remained at the pixel; if not, it moved.

Milne found that all the species of animals followed the same general sequence of resource use, but the different sizes of the animals made each species' use of resources distinct within that general pattern. Across all species, at the outset a large percentage of animals moved because they were in unfavorable habitat. This movement led some animals to favorable habitat, so the percentage of moving animals declined. As might be expected, animals began to aggregate on favorable habitat so that the resource at those sites was consumed. This reversed the trend which was originally toward less movement as time passed to one where the percentage of animals moving rapidly increased. Furthermore, for all species, the reversal occurred at a sharply defined point. The differences between the animals was the timing and local population densities at which the reversal occurred. Milne used the population parameters at that critical point of reversal to rescale the animals. Thus the parameters for each species changed from absolute amounts to quantities relative to the population parameters at the critical reversal point. The animals thus rescaled became identical.

It is not surprising that a rescaling to a universal herbivore was possible in this case because they were all created from a set of allometric scaling rules. However, a rescaling of animals from field data may not be possible. That would be very instructive, for it would tell us when a slight increase in size leads to something not just bigger, but fundamentally different. It might be able to tell us that eating is not necessarily the same thing for different sized animals. In some senses there will be a correspondence, but in other ways an elephant eating vegetation is not just a big version of a mite preying on a smaller mite. Translating Rosen's instabilities in the manner of Milne will lead us to understand what ecological terms mean, and will allow us to expose new logical types. At last ecology has a tool for making qualitative decisions. We have been able to quantify rationally for a long time, but heretofore we have made most qualitative decisions by trial and error.

Our predilection has been for the differences between major ecological conceptual devices, but our strategy would still have utility inside a much more focused discourse. The message is that students at all levels need to find the critical inclusions that change logical types. All ecologists need to find the small differences that radically change the discussion. After Rosen's landmark work, the phrase "fundamentally different" now takes on a more modest meaning, but one which can be the fine forceps in the dissection of ecological problems.

Conclusion

This book has been an attempt to introduce critical points of tension that can only be resolved by a new way of thinking about ecological problems. Business as usual in ecology has been distinctly incremental in recent years. Theory has been barely relevant to field practitioners, who have therefore been forced to take up the slack as best they can. As a remedy, let us find the critical issues in ecology and let us drive our thinking onto the knife edge of the dilemmas that the critical issues pose. Enough of patching over irritating inconsistencies, enough of poking ecological material to see what happens in some general way. Ecology needs to identify the critical points of tension, and then the empiricists need to test predictions coming from explicit theory.

While this book only makes relatively few references to chaos and fractals, it is part of the same intellectual revolution. A new theory of complexity is emerging. The "Toward" in the title of this book is important. The ideas in this book are incomplete, and as a whole this volume represents some first halting steps toward a more thoughtful conception based on observables rather than a reification of tangibles. Computers have obviated retreat to lines of reason that appeal to what would happen in principle if we could only observe and take into account more detail. Bill Wimsatt refers disparagingly to "in principle arguments" that reify the system. In the modern world, complexity is hard to manage not because there are too many parts to track, but because of our limitations in conceiving the whole.

Until recently we have muddled the material and the modeled system. This book has been an attempt to develop a protocol to preempt that class of errors. We have tried to operationalize general systems theory for ecology with as little fanfare as possible. Sometimes we may have appeared esoteric, but our bottom line is practice and action, so we hope field ecologists and experimentalists have had the patience to bear with us. In the final analysis we hope to have filled a toolbox of concepts to help straighten out the unnecessary convolutions. That should free us all to tackle the necessary twists and turns of nature that were the intended objects of study all along.

We have not generated a set of explicit hypotheses that need to be tested. Our empirically talented colleagues can do that much better than us. We have not even offered a theory that might lead to hypotheses. However, we hope that both explicit theory and testable hypotheses will follow from this work. If we have made a contribution, it is a meta-theory, a way of looking at ecological material so that interesting theory about the material system can be pared away from semantic contention. Ecologists who have read this book and find it useful will continue to use standard methods of data collection and analysis. We do not intend otherwise. We just hope empty contention about ontology and "correct" definitions is put in its place. More positively, we hope that there emerges a greater flexibility of thought and boldness of inquiry, leading to sounder practice in ecology.

NOTES (BY PAGE NUMBER)

Introduction

2. The confusion between under- and overdispersion is treated in Greig-Smith (1983:11, 60). If individuals are scattered at random on the ground, then a sampling of individuals using a quadrat produces a Poisson distribution of number of quadrats against number of individuals per quadrat. A Poisson distribution has a mean equal to the variance. Another name for variance is dispersion, and so overdispersed implies more variance in the frequency distribution than the Poisson, and underdispersion implies less variance than that expected for a Poisson distribution. If individuals are evenly scattered across the ground, then all quadrats will encounter a similar number of individuals. So long as the quadrat is large enough, then no quadrats will be empty and no quadrats will hit a clump and so be especially full. Thus the variance is low, and a population of evenly spaced individuals is underdispersed. If individuals are aggregated upon the ground, then many quadrats will be empty and some quadrats will be especially full because they hit a large clump. Accordingly, very few quadrats will have the average number of individuals per quadrat within it and the dispersion away from the mean will be large. Thus a population of individuals aggregated on the ground will be overdispersed.

4. Efforts in recent years by ecologists to move upscale have included an atmospheric orientation with respect to global cycling processes and human activity. Crutzen and Graedel (1986); Dickinson (1986); McElroy (1986); Post et al. (1990).

7. J. A. McMahon et al. (1978, 1981) identified a branched biological hierarchy that moved above the organism level in one direction toward communities through populations and in the other direction to the biosphere through process-functional ecosystems. Their hierarchy is unnecessarily organism-centered, but represents a helpful early effort to characterize the distinction between certain fundamental ecological criteria. Cousins (1990) creates a new version of the conventional hierarchy where between organism and biosphere he inserts an "ecotrophic module" defined by the range of the top carnivore in the system.

8. The mechanical design of trees is discussed by T. McMahon (1975).

8. The great majority of biology textbooks start with physiology and end with large ecological systems; for example, *Biology, the Science of Life* by Wallace, King, and Saunders (1986). Less commonly, textbooks begin with large-scale ecological systems and work down to organisms and cellular processes; for example, *Introduction to Plant Science: A Humanistic and Ecological Approach* by Russell (1975).

10. A summary of chaos theory and fractals is found in Gleick (1988), Mandelbrot (1983), Milne (1988), Orbach (1986), Peitgen and Saupe (1988); Holling (1986) considers surprise; self-organization is covered by Eigen (1977) and Prigogine (1978); the basis of catastrophe theory is found in Thom (1975).

1. The Principles of Ecological Integration

14. For a discussion of the observation problem as it occurs in physics and extends to biological systems, see Pattee (1978, 1979a, b).

18. Wessman et al. (1988) link reflection to lignin to soil nitrogen.

20. The closest view to our own that we found in the literature before the development of hierarchy theory was that of Rowe (1961). On page 421 he states, "The distinction between the structurally integrated object (e.g., plant), the aerial or geographic aggregate (e.g., plant community) and the non-object, non-aggregate class (e.g., species) is important in the problem of setting up a consistent level-of-integration scheme. The thesis advanced here is that each level must be an object as defined above, and that objects at successive levels must be structurally related." On page 425, Rowe separates level of organization from what he calls "point of view." Our level of organization is a scale-defined conception. Rowe's requirement that "successive levels must be structurally related" implies a formal scaling operation like the one we recommend for distinguishing between levels. Rowe's "points of view" appear equivalent to our "criteria for observation." Nevertheless, his scheme differs from ours in that the ecosystem is a level above the organism. In our scheme, a cow's stomach is a perfectly good ecosystem, a view that is at odds with Rowe's scheme. Both Major (1958) and Egler (1942) also use the concept of "points of view."

22. For the multiple criteria analysis of North Dakota, see Wallace (1908).

29. Webster (1979) identified five criteria for hierarchical arrangement. These represent the five headings for hierarchical criteria used below.

30. Herbert Simon's comments on system inventory were made at an informal meeting of the evolution and genetics group on the campus of the University of Wisconsin-Madison.

30. Herbert Simon (1962) suggests that "if there are important systems in the world that are complex without being hierarchic, they may to a considerable extent escape our observation and our understanding. Analysis of their behavior would involve such detailed knowledge and calculation of the interactions of their elementary paths that it would be beyond our

capacities of memory or computation." We agree with Simon's general sentiments although our definition of complex invokes hierarchy a priori.

34. The observation of a dune system being constrained by the constant return of a large river was made by Joe Walker, of CSIRO, in a personal communication to Allen about an Australian dune system.

37. For an analysis of management associated with budworm, see Holling (1986) and Walters and Holling (1990).

40. Clements defines the association and the formation in Clements (1904:9–10), and Pound and Clements (1900).

40. Egler's attempt at a hierarchical approach to ecology occurred in his 1942 contribution, "Vegetation as an object of study."

40. Robert McIntosh complains explicitly about "new" ecology in McIntosh (1985:276–77).

43. Hoekstra, Allen, and Flather (1991) performed an analysis of the relationship between choice of type of organism and ecological concept.

44. The work on floating carnivorous plants as representative of an entire ecosystem was Robert Bosserman's 1979 Ph.D. dissertation.

44. Sir Arthur Tansley first published his formal presentation of the ecosystem in Tansley (1935).

45. Transeau published early work on the accumulation of energy by plants (1926). In his 1940 *Textbook of Botany* (Transeau, Sampson, and Tiffany 1940) he gives a comparison of the energy balance in an Illinois cornfield and a New York apple orchard. This comparison was made a full two years before Lindeman's (1942) classic paper on food web energetics. Tansley (1935) makes no reference to Transeau's early work on system energetics. Odum (1959) appears to be one of the first to credit Transeau. He refers to the carbon dioxide method of measuring productivity, "Ecologists have made various attempts to measure production in whole intact communities in this way, beginning with the pioneer experiments of Transeau (1926)." Colinvaux (1979) gives an account of Transeau's writing of the 1926 paper.

49. For a summary of the new framework, see Allen and Hoekstra (1990).

49. Vogel (1988) gives an extended treatment of the occasions when biological systems bump up against physical limits. Most limits, however, are strictly biological in nature and represent a subset of what is possible.

50. Rosen (1979) gives an extended account of purposeful behavior and anticipation in biological systems.

2. *The Landscape Criterion*

55. Von Humboldt and Bonpland (1807) and Grisebach (1872) are two among several expeditionary naturalists of the day. Unger (1836) performed a local biogeographical study in Europe.

55. The prime mover in the resuscitation of landscape ecology in America was Richard Forman (Forman 1979; Forman and Godron 1981; Forman, Calli, and Leck 1976).

55. Frederic Clements' doctoral dissertation was, along with Cowles (1899), the first work in America with a modern community conception. The citation to Pound and Clements (1900) is because almost all of the first edition was destroyed in a fire. Smith (1899) quoted by Sheail (1987:8) explicitly uses the word community. "Such a community made up of chief species, subordinate species, and dependent species, constitutes a Plant Association (Smith 1898)." From Sheail it would appear that the principal effort of Robert Smith was the mapping of vegetation. Cowles and Clements were less concerned with mapping and were more focused on vegetational processes. For the first decade of this century, the British ecologists were focused on mapping more than a community synthesis, as occurred in America.

55. Through the first four decades of this century there were acrimonious arguments as to the correct definition of community. The alternative view to that of Clements was stated in a running battle from 1910 until 1939 (Gleason 1910, 1917, 1929, 1939). Gleason's most eloquent statement was published in 1926. The battle between the Clementsians and Gleason reached its height in Nichols (1929) and Gleason (1929). In an autobiographical account, Gleason (1953) indicates that he felt the attacks were somewhat personal. "George Nichols pulverized my theory. Worse than that he ridiculed it." Gleason felt himself identified as "an ecological outlaw" and "a good man gone wrong." In retrospect, a reading of the original literature suggests there is very little difference of opinion as to the facts of vegetational pattern and behavior.

   The defense of Clements' definition of community was conducted as a classic Kuhnian (Kuhn 1970) paradigm defense. The scant logic of a fight between correct definitions is even openly admitted at one point in the commentary to Gleason's (1939) presentation. There Conard, speaking of the Clementsian view said, "It is therefore so useful that whether logical or not, I am for it." Much of the confusion through the discussion stems from a pervading philosophy of the naive realism that the present authors suggest is a poor starting point for scientific discussion. For example, in the same discussion in Just (1939) following Gleason's paper Emerson says, "I think that the species has objective reality as well as the individual. If you say that "Homo sapiens" is a figment of the mind, I think you will be forced by logic to say that we are all individually figments of the mind."

55. The subtitle to Wiens and Milne (1989) is "Landscape ecology from a beetle's perspective."

56. For general references on fractals refer to Mandelbrot (1983), Milne (1988), and Peitgen and Saupe (1988).

61. The fractal nature of progressive human activity is seen in the patterns of enclosure at Newnham, a village in Northamptonshire. The smallest enclosures are gardens; the next larger enclosures are late-medieval fields; the large fields came later with acts of parliamentary enclosure. By contrast, nucleated settlements in Donegal in 1834 show large patterns of enclosed land. By 1850 the area is more finely dissected by subdivision within the

larger fields. Both these and the patterns reported by Dury (1961) show how human activity appears to produce simple patterns that accommodate to prior simple patterns of earlier human endeavors. Krummel et al. (1987) show a low fractal dimension across scales of human endeavor but a jump to a higher fractal dimension at scales larger than local human activity. O'Neill et al. (1988) show that the greater the human impact in a region, the lower was the fractal dimension in examples from across the United States.

62. The changes in the fractal dimension over time in Georgia is found in Turner (1987).

63. For an extended treatment of how the stability of system parts affect predictability, see Allen and Hoekstra (1985).

65. The first explicit use of the terms small, middle, and large number systems is attributable to Weinberg (1975).

65. For the account of the unpredictability of middle-sized woodlots with respect to number of birds, see Urban and O'Neill (in press).

66. The four fundamental patterns of nature are published in Stevens (1974).

73. References to neutral landscapes, percolation theory, and critical thresholds are all found in Gefen, Aharony and Alexander (1983), O'Neill and Gardner (1990), Milne (in press, a), Milne et al. (1989), Gardner et al. (1987) and Turner et al. (1988, 1989). The figure of the 30 percent threshold is not explicitly stated but can be extracted from the graphs (R. H. Gardner, personal communication).

74. When Forman was rekindling the interest in landscape ecology, there was an extensive coining of terms for landscape patterns. A similar phase in community ecology is exemplified by Clements (1905). Forman (1981) begins this process for landscapes. Five years later, Forman and Godron (1986) have a thirteen-page glossary that consists principally of landscape terminology. Figure 4.14 in Forman and Godron diagrams no less than twenty-seven types of drainage patterns.

75. The modification of landscapes by corridors changing fractal landscape, into the Euclidean landscapes was a personal communication of B. Milne.

76. For a discussion of the size of organisms and their scale of landscape occupancy, see Milne (1990), Milne et al. (1989), and Milne (in press a, b).

77. For an account of the Alaska pipeline interfering with animal movements, see Kameron (1983) and Schideler (1986).

78. For an analysis of the distinction between the uses of plants and animals for various ecological concepts, see Hoekstra, Allen, and Flather (1991).

78. Forman and Godron (1986) report unpublished data of Baudry and Forman showing a significant negative correlation with distance from intersection of herb species surrounding corn and bean fields in New Jersey.

79. McCune's study of the Bitterroot Mountains is reported in McCune and Allen (1985a, b).

79. Prigogine has written extensively on the emergence of higher levels of order in Nicolis and Prigogine (1981), Prigogine (1978, 1982), Prigogine et al. (1969), and Prigogine and Nicolis (1971). Gleick (1987:311) discusses the emergence of pattern in snowflakes.

82. Allen and Wyleto (1983) discuss analysis of prairie vegetation with fire inside and outside the system as defined by data transformations.
84. Musick and Grover (1990) discuss the use of textural measures as indices of landscape pattern in ecological systems. These workers were responsible for the comparison of LTER sites with respect to texture. The textural measures for landscapes were developed by Haralick (1979), Haralick and Shanmugan (1974), and Haralick, Shanmugam, and Dinstein (1973).

3. The Ecosystem Criterion

89. O'Neill et al. (1986) use a definition for ecosystem that is significantly different from the one used in the present work.
92. Shugart, West, and Emanuel (1981) summarize the implications of simulation studies performed by Emanuel, Shugart, and West (1978) and Emanuel, West, and Shugart (1978), coming to the conclusion that the inclusion of American chestnut as a viable species radically altered the periodicities of stand production. Chestnut did this by its shade tolerance and sprouting capabilities interfering with other dominant species, thus shifting the dynamics of stand composition.
92. For an account of the disruption of the forest ecosystem by the lead smelter, see Jackson and Watson (1977). For a more general account and hard data on the effects of heavy metal insults on ecosystems, see O'Neill et al. (1977).
93. Hyatt and Stockner (1985) refer to a program of nutrient enhancement in lakes as a means of controlling the size of fish populations in Canada conducted through the 1960s and 1970s. Explicit positive results are reported by Hyatt and Stockner (1985), Stockner (1981), and LeBrasseur and Parsons (1979). It is of note that the death of Atlantic salmon immediately after spawning occurs at a much lower rate than in Pacific salmon. If the nutrient arguments are correct, then a plausible explanation for this difference in salmon life cycles is the nutrient load that occurs in water bodies in the eastern United States because greater rainfall produces deciduous trees that, in turn, produce a suitable nutrient pulse for maintaining enough biomass to support the growing salmon in the eastern streams. The western seaboard of North America that serves the Pacific salmon has much less rainfall, and so has streams surrounded by less terrestrial biomass. Accordingly, the dying of the salmon is the only solution to nutrient depletion in headwaters because of a one-way transport of nutrients out of the system with no significant input from a terrestrial source. See Walters and Holling (1990) for an analysis of management strategies for Pacific salmon.
96. A formal presentation of airsheds is to be found in Summers (1989).
102. Some of the earliest work using radioactive traces to identify boundaries of ecosystems in the manner of biochemists was performed at Oak Ridge National Laboratory (Reichle and Crossley 1965).
102. Hubbard Brook was a center for ecosystem experimentation at the watershed scale through the 1960s and 1970s (Likens et al. 1967, 1977, 1978; Bormann et al. 1974).

103. For an elaboration of the concept of spiraling resources in stream eco-systems, see Elwood et al. (1983).
104. Harris et al. (1977) inferred two periods of vigorous below ground growth in yellow poplar, once in the fall and then again in the spring.
104. The forest simulation that calculated residence times for carbon at 54 years and nitrogen at 1,810 years was performed by O'Neill et al. (1975).
104. The work on turnover times of carbon and nitrogen on the grasslands is found in Parton et al. (1987). The model carbon flow has a hierarchy of carbon compartments with turnover times of .5, 1.5, 2.5, 3.0, and 1,000 years because of the different reaction rates. Grazing pressure does not much change the status of the faster-moving compartments because they merely draw on the resource base in the slow-moving compartments. See also Sala et al. (1988).
105. The analysis of the literature search in BIOSIS is found in Hoekstra, Allen, and Flather (1991).
106. The review of scaling effects in plant compensation is found in Brown and Allen (1989). The argument between McNaughton and Belsky first occurred in Belsky (1986) and McNaughton (1986).
108. The work on zooplankton correlation with phytoplankton being negative for a short period but positive for longer periods is found in Carpenter and Kitchell (1987).
109. Parton et al. (1987) model organic matter and nitrogen in compartments that operate at different speeds. See also Sala et al. (1988).
111. The work in investigating BIOSIS is reported in Hoekstra, Allen, and Flather (1991).
113. Margalef's pioneering work of broad scope in aquatic systems was published as Margalef (1968).
114. John Magnuson has published a series of papers that consider the relative size of a water body and the characteristics that can be used to predict the fish within it. He compares oceans and lakes, having collated the significant data set on small temperate lakes (Magnuson 1988, in press; Magnuson et al. 1989).
115. May (1981:219) states, "This notion [complexity implies stability] has tended to become part of the folk wisdom of ecology."
117. Ideas of the economy of nature with everything connected to everything else goes back to Linnaeus. Speaking of Linnaeus' view, Larson (1971) suggests that "the task of the naturalist was to represent this economy of nature in each of the three kingdoms of nature." The system of publication for Linnaeus was peculiar by modern standards. The title of I. J. Biberg (1749) is "Academic essay on the economy of nature submitted by I. J. Biberg." While Biberg was a student of Linnaeus, his principal purpose was to get together the money to publish the above work of Linnaeus himself.
117. In the context of diversity and stability relationships, various authors have identified that increase in diversity in a system that is connected at random will decrease stability in the final analysis (Levins 1974; Margalef 1972; MacArthur 1972; May 1974).

117. A general discussion of community matrices and stability with respect to the loop structure of the subsystems is found in O'Neill et al. (1986). Other pertinent discussions occur in Tansky (1978), Levins (1974), McMurtrie (1975), Austin and Cook (1974), and DeAngelis (1975).

121. Pattee (1972) discusses the difficulty of developing adaptive behavior in large complex systems. His argument is that when the system crosses a certain level of complexity, any adaptive response has linked to it a set of maladaptive responses. Improvement in such systems is impossible.

121. The stimulus for the headings "Stability and Diversity Wars" and "The Empirical Strikes Back" comes from Martin Burd.

122. The work on the ten microcosms performed at Oak Ridge is reported in Van Voris et al. (1980).

123. Platt and Denman (1975) report extensive use of power spectra in ecological systems.

125. For a further discussion of the reduction of ecosystems not usually leading to organisms, see Allen, O'Neill, and Hoekstra (1987).

4. The Community Criterion

126. The center of the contentious debate between Gleason and the followers of Clements was the paper by Gleason (1926). The height of the argument was reached in Nichols (1929) and Gleason (1929).

127. Hugh H. Iltis reports a conversation between himself and John Curtis toward the end of Curtis' life. Iltis reports that Curtis felt the community could be captured in an analogy where there is tension between all community members as if they were connected by elastic bands. More than this, Curtis felt that the tension on the elastic bands was constantly shifting because, as one member of the community changes, its relationship to immediate others changes and the tension passes through the web of rubber bands. Thus a notion of tension and shifts in tension underlie Curtis' synthesis of his ideas on community. Allen and Hoekstra (1990) relate the notion of community to others. "While many ecologists have not thought of communities in exactly this way, their definitions of community often bear some similarity to ours (e.g., Cooper 1926; Curtis 1959; Greig-Smith 1983; Oosting 1948). Accordingly, their communities emerge, like ours, as having at least something of an intangible component. Both Cooper and Oosting talk of communities as concrete patches on the ground, but then go on to refer to an open "abstract" community that includes all stands of the community dominant. Cooper extends this abstract community over geological time with the poetic image of a braided stream. Although Greig-Smith and Curtis aim to quantify the community relationships on the ground, they are the first to say that it is extremely difficult, fraught with ambiguity and arbitrary decisions of measurement. Our definition appears to solve some of the dilemmas recognized by other authors, and it does this by categorically refusing to put communities on the landscape as simple patches."

127. Bradshaw's work on the evolution of communities on toxic mine tailings is synthesized in Bradshaw and McNeilly (1981). Begon, Harper, and Townsend (1986:71) summarize the situation found in recently formed communities on toxic substrates, "When it [the pollutant] is newly arisen or is at extremely high concentrations, there will be fewer individuals of any species present (the exceptions being naturally tolerant variants or their immediate descendants). Subsequently, however, the polluted area is likely to support a much higher density of individuals, but these will be representative of a much smaller range of species than would be present in the absence of the pollutant. Such newly evolved, species-poor communities are now an established part of man's environment."

128. For a summary that emphasizes the inconclusiveness of the experimental data as to the mechanisms of allelopathy, see Kershaw (1973:94–96).

129. See Moorhead (1969) for background on Darwin. In his *The Descent of Man*, Charles Darwin acknowledges on page 19, "The conclusion that man is the co-descendant with other species of some ancient lower and extinct form is not new in any degree. Lamarck long ago came to this conclusion that has lately been maintained by several naturalists and philosophers." The assertion that Clements died a Lamarckian is made by Worster (1977).

130. A full account of the German work on plant physiology that was the extension of Darwin's evolutionary ideas is found in Cittadino (1990).

130. Examples of early biogeographical exploration are von Humboldt and Bonpland (1807), Grisebach (1838, 1872), and Unger (1836). The work of Unger is a detailed biogeographical study of the alpine valley in which he was the medical doctor. This work is distinctly ecological, identifying plants that grow on limestone as opposed to those on slate. Unger maps 93 species with dot maps in great detail, predating any other use of dot maps by a full 20 years. The Unger work has been forgotten until it was rediscovered by Hugh Iltis. A paper is in preparation by H. H. Iltis and J. Camerini that notes the significance of this early use of dot maps.

130. Cittadino (1990) explicitly talks about imperialism as the driving force behind the late nineteenth-century German tropical biology. Examples of German tropical biologists mentioned by Cittadino are A. F. W. Schimper studying coastal vegetation in Indo-Malaysia, Georg Volkens using Egyptian desert plants to test hypotheses developed in studies of native European vegetation, and Heinrich Schenk who studied tropical climbing lianas. The Germans who most influenced the young Nebraskan ecologists were Warming (1896) and Schimper (1898). Of particular importance here is Drude (1890, 1896). Direct evidence for this influence is found in Pound (1896), entitled "The plant geography of Germany."

130. The essential tension between geographical influences and physiological influences is captured beautifully in the title of Blackman and Tansley's (1905) "Ecology in its physiological and phyto-topographical aspects," a review of Clements (1905). The essential discomfort they feel comes through in what is essentially a very positive review. The full integration of the organismal and landscape perspectives was only achieved in America. The

British found the notion of the formation a bit mystical. The British either performed physiology or plant geography, not both at the same time.

130. The Botanical Seminar at the University of Nebraska was an informal group begun by Bessey and his students in 1886. One of the principal activities of the seminar was the Botanical Survey of Nebraska. This work culminated in Pound and Clements (1900), a joint doctoral thesis. The first community work coming from that group was the thesis of MacMillan (1892, 1899).

130. Cowles (1899) uses the term "plant society," while Pound and Clements (1900) used the word "association." All these workers used the term "formation," to imply something a little more inclusive than the association or plant society. Their prime influence appears to be Drude, although the original use of formation goes back to Grisebach (1838). Smith (1899) talks about the plant association as being a community.

131. In his large work on plant succession, Clements (1916) is very aware of the importance of the processes of invasion through proximity, and ecesis (the process of the filtering of plants by environment so that not all arrivals end up as part of the association).

131. Pound and Clements (1898) are explicit in their use of quantification of plants in quadrats.

132. The principal schools of phytosociology are the Zurich-Montpellier School (Braun-Blanquet 1932) and the Uppsala School (du Reitz 1921). For a clear accounting of the Zurich-Montpellier School, see Mueller-Dombois and Ellenberg (1974). The associations produced by these subjective approaches to vegetation classification are particularly suitable for vegetation mapping.

132. Gleason (1953) attributes his interest in the prairie-forest border region to his father having homesteaded on the unbroken prairies. Frederic Clements arrived as a local boy at the University of Nebraska at the age of 16. The respective formative years of these two scientists appears to have colored their interests in plant ecology and the differences in their two perspectives.

132. Clements (1905), and in his subsequent writing, indicated that the association between plant species on the ground was more than the sum of its parts and could be considered a local "superorganism." Gleason stood alone for the first quarter of this century as a moderating force for the excesses of metaphor and analogy of the superorganism concept.

134. Cooper's braided stream analogy is the centerpiece of his paper (1926).

134. In the early 1950s several workers independently developed techniques for quantitative analysis of vegetation. Curtis and Greene (1949) tried to reproduce the sort of discrete groups of vegetation type that might be implied by Clements' conception of community. They were unable to identify discrete breaks, and from that negative result Curtis went on to develop methods of analysis that would reflect Gleason's concept of community. Curtis' quantitative methods were published as Curtis and McIntosh (1951) and Bray and Curtis (1957). Meanwhile Whittaker (1956) and Goodall (1954) devel-

oped their own methods of quantitative analysis. Goodall (1953) developed one of the first quantitative, so-called objective methods of vegetation classification. The development of all these methods gave a protocol to implement a Gleasonian approach to community studies.

135. A distinctly dynamic view of vegetation emerged at the turn of the century (Pound and Clements 1900; Cowles 1899).

135. Before the development of mathematical gradient analyses of vegetation, soil scientists were conscious of continuous changes in soil that could be ordered in sequences. In Curtis (1959) there is an explicit citation of Milne (1935).

136. Two explicitly geometric expressions of vegetation analysis occur in Gittins (1969) and Allen and Skagen (1973).

140. For contrast between a coordinate system and a vector mode of description, see Gittins (1969).

140. For general discussions of clustering techniques, see Wishart (1969) and Rohlf (1974). For an elementary introduction to the notion of clustering and its underlying geometry, see Allen and Skagen (1973). For an advanced treatment of modern techniques for clustering, see Legendre and Legendre (1987).

140. For a general discussion of normal and inverse analyses, see Greig-Smith (1983:215ff.).

141. For an elementary account of the geometry of point projection, see Allen and Skagen (1973).

142. Curtis was acutely aware that choice of vegetation units greatly influences the outcome of a study. In contrast to the subjective decisions made by European schools of phytosociology, Curtis (1959) erected criteria for the selection of sites before going to the field. Furthermore, these criteria did not determine which species associations were acceptable and which were not. For example, in discussing the southern Wisconsin upland forests on 92, he says, "A mixture of royal palms and magnolias would have been acceptable, had it occurred." His stands had to be undisturbed, 15 acres in size, and a site should be of upland nature. Given these criteria, he analyzed the first 95 stands that qualified. The randomness of the site selection procedure depends only upon the first 95 samples being selected at random relative to the species in the vegetation. It is important that the vegetation should not guide the site selection process, for then there is a circularity of argument.

143. At the height of the debate between Gleason and the Clementsians, Nichols (1929) and Gleason (1929) appeared to agree on most of the facts of variation vegetation.

144. It is probably no accident that the first attempts at ordering vegetation by Curtis and his colleagues came from an attempt at classification, since that would involve less of a departure from the then conventional Clementsian view of plant communities. Curtis and Greene (1949) appear to have been making a good faith attempt to verify the view that supposed discrete classes of vegetation. Despite their best effort, they found the vegetation to

vary continuously, more in line with Gleason's view of vegetation over Clements. From this early study, Curtis and McIntosh (1951) developed a method that explicitly displayed continuous gradients of vegetational change.

144. John Harper's reference to certain mystical qualities in community ecology was made in a personal communication to Allen.

144. In his doctoral thesis, David Roberts (1984) relied heavily upon the ideas of Sukachev and Dylis (1964).

144. Curtis' thoughts that plants are better indicators of environment than any measure ecologists could devise was personally communicated to Allen by Grant Cottam.

145. For direct environmental gradients, see Whittaker (1956).

145. For an accounting for Roberts' linking between environmental and species space, see Roberts (1984, 1989b).

145. Ellenberg's approach to the analysis of gradients is distinctly experimental. In this way he hoped to show not only patterns of changes in species composition, but he also wished to demonstrate the underlying physical causal factors. He focused particularly upon changes in soil pH, giving changes in species vigor. Along a contrived experimental gradient he would identify the environmental conditions at which species grew most vigorously. He would then use this relationship between environment and optimal performance as a weighting function in characterizing vegetation. Curtis and McIntosh (1951) obtained weights for species, but found them by ordering stands directly observed from the field with no underlying experimentation. Ellenberg used his weights in a similar fashion to Curtis and McIntosh. Ellenberg's early work on vegetation analysis occurs in Ellenberg (1948, 1950, 1952). A summary of the results of his competition work is found in Ellenberg (1963).

149. For a summary of the various European schools of phytosociology, see Mueller-Dombois and Ellenberg (1974).

150. Both Curtis (1959) and Whittaker (1956) imply a model of environmental determinism. The behavior in the vegetation space is driven by the environmental space.

150. At the time of writing, the only formal extended treatment of the use of cliques by David Roberts in the identification of community constraints is in his Ph.D. dissertation (Roberts 1984). A formal treatment is in preparation for the open literature and will be submitted to the journal *Ecology*.

154. Hutchinson (1957) is the first conception of the niche as a hypervolume space.

154. The analysis of the BIOSIS literature search is in Hoekstra, Allen, and Flather (1991).

155. DeVito's (1988) work on birds was his Ph.D. thesis. No open literature citation is available at this time.

157. The advantages of scaling down communities until all members function in an equivalent fashion is found in Root (1975). For the central work on island biogeography, see MacArthur and Wilson (1967).

5. The Organism Criterion

160. The term "holon" was coined by Koestler (1967).
161. See Schrodinger (1959) and Polanyi (1968) on the essential purposiveness of biology. For a further, more recent discussion, see Rosen (1979).
163. The theory of endosymbiosis is now generally accepted, occurring in most elementary college-level biology textbooks. For a broad statement of the theory, see Margulis (1981).
163. Atsatt (1988), without presenting new data, pulls together a diverse literature of host and parasite coevolution and genetic mingling to suggest "that the nucleus of vascular plants is a dual entity in which a fungal genome makes varying contributions to the differentiation of highly specialized cells." He asks if vascular plants are not "inside-out" lichens.
164. The perfect species is one where members all look the same and are fully capable of breeding freely within the species, but have difficulty breeding with other species. In plants, the perfect species is almost the exception rather than the rule, so plant species can be grouped into different types that are departures from the perfect conventional species. For an account of the different criteria for recognizing species, see Heslop-Harrison (1967:114–15).
165. In the case of the Rh blood group, if a fetus is Rh+ and the mother is Rh−, the mother can become allergic to the small quantities of blood passing across the placenta from the blood system of the developing child. With a first child this is usually not a serious problem, but if in subsequent pregnancies the Rh− mother carries an Rh+ fetus, the mother produces copious quantities of antibodies that damage the blood of the developing child. The mother's immune system mistakes the baby for an invading pathogen.
165. In his "living systems theory," Miller (1978) contrasts and compares in extended detail seven levels of organization from cell to global political systems.
168. In the case of the reflex arc associated with the hand, the input nerve indicating some trauma to the hand enters the spinal column. At that same level in the spinal column is a motor nerve that responds immediately with an output signal that causes the muscle on the arm to pull the hand away from the cause of the distress. These reflexes occur over a much shorter time than if the brain had to process a signal and then decide to pull the hand out of danger. The eyes may see that the foot has experienced a trauma, and the owner of the brain winces knowing that in a short amount of time, the foot is going to hurt. However, there is a critical waiting period in which further damage could be done unless the reflex arc were to pull the foot away from the damaging influence. There are, therefore, several levels of decision makers in the human body.
169. Clements (1905) indicates that the association of species in communities forms a superorganismal structure.
169. The ideas of Jakob von Uexkull concerning the self-world of animals are recognized for their systems application by Bertalanffy (1968) (Uexkull

1920, 1929; Uexkull and Kriszat 1934). Uexkull and Kriszat are translated into English as Uexkull (1957).

185. Corner (1981) develops the idea of polarity of the environment driving the polarity of organisms in a discussion of the evolution of plants from the open ocean to the intertidal zone, to the land.

189. Haldane (1963) gives a general account of the consequences of the increase of size in animals and the relative importance of gravity, weight, and surface tension.

189. For an account of the exponents associated with the thickening of trees, see McMahon (1975).

190. For a general account of the fluid dynamics of wind with respect to organisms and drag coefficient, see Vogel (1981).

191. For an extended account of the Reynolds number in a biological context, see Vogel (1981). For a simple and brief account see Stevens (1974).

193. A comprehensive summary of Schmidt-Nielsen's work on physiology and scaling is found in Schmidt-Nielsen (1984).

193. For citations to Potts' work on coral, see Potts (1983, 1984a, 1984b, 1985, 1987), Potts et al. (1985), Potts and Swart (1984) and Smith and Potts (1987). For a full translation of coral reef formation into hierarchical terms, see Miller (1991).

195. The history of the paradigm fight between *Planaria* and *Drosophila* is found in Fausto-Sterling and Mitman (in press).

196. For an extended discussion of genets and ramets, see Harper (1977). Harper discusses mortality of genets and ramets on pages 191—194. The original work to which Harper refers is Kays and Harper (1974).

6. The Population Criterion

208. In the oral presentation of the paper abstracted in Prince (1982), it was stated that the most southern population of the mopane tree in southern Africa was a vigorous small population in a favorable habitat with healthy adult individuals undergoing effective reproduction. Other favorable unoccupied sites were identified only two to three miles further south, but these were unoccupied by the species. Prince (1982) asserts that this is an example of a distribution that can be explained by epidemic models for biogeographical limits (Carter and Prince 1981). The models are not to deal with disease per se, but liken occupied sites to infected individuals. In that infection that can be described as an all-or-nothing phenomenon, sites at the limit of a species range are either fully infected/occupied or not. Salisbury (1932, 1964) asserts that many species are abundant at the limits of their range.

208. Harper (1977:708) reports data from G. R. Sagar that show changes in floristic richness in a grassland community across a pattern of ridges and furrows that repeats every 10 meters. Two species are found in the furrows, "both of which are stoloniferous species in which the genets explore and 'sense' the habitat over a considerable distance, whereas the species on the

ridges are mainly forms with no clonal growth (*Ranunculus bulbosus*) or with very localized intravaginal tillering (*Lolium perenne, Cynosurus cristatus*). A statement about the floristic richness of a pasture as a whole has no real meaning in relation to the chances of an interspecific encounter." Harper (1977:125) reports the work of Harper, Williams, and Sagar (1965) where different species in the genus *Plantago* were shown to be germinating in different sites according to fine details of soil texture. Harper, Williams, and Sagar (1965) offer an intuitive thermodynamic input-output model where the contact of the seed with the soil differs depending upon seed shape, which in turn determines the suitable soil texture for germination. In this case topography and local humidity are interrelated in biologically important ways at a scale of about 1 mm.

209. For an excellent introduction to the difference between difference and differential equations and the calculation of instantaneous growth rates, see Pielou (1974). May (1981c) also gives an excellent treatment of the development of difference and differential equations, and contrasts the simple logistic equation with one that more realistically separates mortality from recruitment. The former, simple equation is:

$$dN/dt = rN(1 - N/K)$$

where r is the instantaneous growth rate of the population. K is the carrying capacity, representing the limit of numbers that can be indefinitely supported by the environment. When N < K, then N/K < 1 and population growth occurs. When N > K, then N/K > 1 and its inverse (1 − N/K) is negative, so causing a decline in the population. The more realistic equation that separates mortality from recruitment is:

$$dN/dt = - \mu N + R[N(t - T)]$$

The μ term refers to mortality while the R refers to recruitment. A further complication is the term (t − T). This refers to a lag in the application of the equation. Thus the N associated with the (t − T) is the number in the population at a discrete time in the past. Quoting May, "Here losses due to deaths (at a constant per capita mortality rate (μ) depend simply on N(t), but recruitment into the adult population [described by some nonlinear function of population size, R(N)] at time t depends on the population, N(t − T), at time T earlier; T is the typical time taken to attain maturity."

211. Peilou (1974) develops the notion of lag in difference and differential equations at a more leisurely pace than does May (1981c). May immediately introduces notions of relative lag in difference equations and uses that to display chaotic behavior.

213. Gleick (1987) quotes the discussion of May (1981c) in the context of other developments in the study of chaotic systems.

216. For an account of the general patterns of system resilience, see Holling (1973). For a detailed accounting of the modeling of a host-parasite system with different degrees of parasite efficiency, see Holling and Ewing (1971).

220. For an extended account of the ecology of the Tsembaga tribe in New Guinea, see Rappaport (1968). For an abstracted account of the energy balance in that system, see Rappaport (1972).
221. For an accounting of Holling's view of system fragility in human managed schemes, see Holling (1986).
222. For a creative synthesis of prey-predator relationships of different types, see Colinvaux (1979).
224. Steven Bartell could identify whether it was the plankton or the fish that were controlling the system at any one time by looking at the sensitivity of the plankton to manipulation of their part of the model structure. If a change in the model for phytoplankton produced little change in phytoplankton performance in the model, then something else must be controlling them other than photosynthesis or nutrient uptake. That would occur when the fish controlled the system and the grazing was the dominant factor. If, however, a change in the phytoplankton section of the model produced a change in phytoplankton performance, then the system would be controlled from the bottom, with phytoplankton behaving as autonomous, unconstrained entities. Until recently, it was generally held that the control of production in lake systems was primarily by abiotic factors constraining the algae in the system. However, experiments by Carpenter et al. (1987) have shown that the food web above the algae can be the controlling factor on algal production. For example, big fish eat small fish, which therefore cannot eat zooplankton, which in turn increase in numbers, and so crop down the phytoplankton. General considerations here are discussed by Carpenter and Kitchell (1988). The notion of switching control from top-down to bottom-up and back again is now generally accepted. For a discussion of these issues, see Carpenter (1988).
224. The extended discussion of middle number systems occurs in Weinberg (1975). For a general discussion of Holling's work on budworm and other epidemics, see Holling (1986).
230. The lizard work of Jonathan Roughgarden appears as Roughgarden and Pacala (1989), Roughgarden, Gaines, and Pacala (1987), and McLaughlin and Roughgarden (1989).
232. For a full account of positive feedback systems and in particular the mathematics of mutualism, see DeAngelis et al. (1986).
233. For a clear accounting of de Wit replacement series, see Harper (1977:255–58, 302–4). The experimental design was introduced by de Wit (1960) from W. H. van Dobben. There have recently been attacks on the design, but Cousens (1990) indicates that these refer to misuses. The method still applies to the sort of qualitative use suggested here.
236. Harper (1977:291–95) summarizes the work of Harper and McNaughton (1962) that shows four species of poppy (*Papaver*), where all were constantly present as weeds in fields, but at four very different densities. "In 16 of the 20 cases, the survivorship of a species was poorer when it preponderated in a mixture. This effect may be summarized by saying that density-

dependent mortality was largely species specific—each species reacted primarily to and suffered most from its own density in a mixture. The four cases where this was not true involved *P. apulum* (3 cases) and *P. dubium* in the presence of *P. lecoqii*. *P. apulum* was the odd man out in not being a member of the communities in which the others are found" (Harper 1977). Harper's interpretation of the experimental results was that the common species interferes with itself more than it interferes with the rare species just before the rare species would be driven to extinction. Therefore rare species survive.

7. The Biome and Biosphere Criteria

239. Roberts (1984, 1987b, 1989b) identifies the relationship between vegetation and physical environment as a relation instead of a function.

240. Holling's suggestion that animals and fire are critical components that are fast enough to move across landscapes so as to allow large areas to become coherent was a personal communication to Allen.

241. Neilson's research that leads him to think of vegetation physiognomy as a stable wave interference pattern between climate periodicity and tolerances of critical life stages in the dominant forms is found in Neilson (1986, 1987a, b) and Neilson and Wulstein (1983, 1986).

241. Curtis (1959) reports early observers of Wisconsin oak openings as saying that fire was required to keep them open. A. W. Schorger published several papers in the transactions of the Wisconsin Academy of Sciences Arts and Letters where he used newspaper reports from the nineteenth century to document the observations of local people.

242. Neilson (1986) identifies in some detail the effect of the Rockies on local climates as global climate changes. Neilson (1987b) reports on the paleohistory of the pinyon-juniper woodlands of the Great Basin.

242. Tobey (1981) suggests that the eventual decline of Clements' view of the plant community began when, to the surprise of adherents to the theory, the stable climax vegetation of the prairies began to break up in the droughts of the 1930s.

242. Neilson and Wulstein (1986) report growth chamber data that identify the drought limits in which oak trees can establish.

243. Huntley and Webb (1989) suggest that the different migration times of species after an ice age make present-day communities entities of unique composition. The effect of this is to put a severe time constraint on species-specific accommodation within temperate forests.

244. O'Neill et al. (1986) talk of a dual hierarchy where constraint flips between biotic and physical factors.

244. Schrodinger (1959) and Polanyi (1968) suggest that life can only be properly understood as a goal-directed system. See Lovelock (1979) for an account of the Gaia hypothesis.

245. Shea (1989) gives a thorough review of the literature of the damage to the ozone layer.

247. The insight that all predictions are that nothing will happen comes from Kenneth Boulding in a personal communication to Allen.

251. For an analysis of surprise in ecological management, see Walters and Holling (1990).

8. Management of Ecological Systems

256. Other advocates of a rapprochement of management and basic research are Walters and Holling (1990).

256. The notion of the noosphere is developed in Teilhard de Chardin (1959).

258. For an account of the German school of evolutionary biology at the end of the last century, see Cittadino (1990).

258. The image of the braided stream occurs in Cooper (1926).

258. Pound (1896) was clearly influenced by the German physiological plant ecologists. Ecology became a self-conscious discipline in America at the turn of the century (Cittadino 1990). During the second decade of this century in Britain the British Ecological Society, and in America the Ecological Society of America were established. Meanwhile the continental Europeans stagnated in the development of highly contrived classification systems (Braun-Blanquet 1932; duReitz 1921). The flowering of modern quantitative methods of vegetation analysis occurred in both Britain and particularly the United States (Curtis 1959; Whittaker 1956; Greig-Smith 1983 is the third edition: the 1964 second and the 1957 first edition give a better reflection of the paradigm shift underway at that time). Presently, in England and the United States there is a systematic replacement of vegetation scientists with plant population biologists who use plant demography in the style of Harper (1977). However, the quantitative methods of vegetation science developed in America and Britain have now been transferred back to Europe where much of the vegetation science in the world is presently being performed. A new quantitative journal, the *Journal of Vegetation Science,* is anchored in Europe. Allen (1989) identifies a distinctly French-Canadian connection back from the New World to the Old in this regard (Legendre and Legendre 1987).

265. A new, self-conscious science of restoration ecology has recently emerged, and is embodied in Jordan, Gilpin, and Aber (1987).

266. A fine example of research on a shifting mosaic habitat is that of Mark Davis. He has collected data on a perennial herb, *Penstemon grandiflorus,* growing in an oak woodland. The herb grows in openings and has a love-hate relationship with gophers. The gophers cause disturbance, which seems to be required for establishment of the herb on their mounds. However, intense activity of the gophers buries and kills young plants. The gophers move into areas opened by fire and appear to keep them open for an extended time. When the gophers move away, then the trees begin to close the gap. The herb can persist for some years even in semi-shaded habitats in which it could not possibly establish. Davis is linking all of this to the landscape patch mosaic dynamics. Davis (1989, 1990), Davis et al. (1991, in press).

268. The workshop held in December of 1988 on (ecology for a changing earth) is reported in Brown and Roughgarden (1990). The working subgroup on human affairs consisted of Timothy Allen, James Brown, Joel Cohen, Jared Diamond, Edward J. Rykiel, Jr., Robert Howarth, and Stefan Somer.

269. Around the Great Lakes are several sites that have been impacted by human activity, designated as "areas of concern." A series of "Remedial Action Plans" are in preparation for those regions. Preliminary indications are that the Clinton River suffers significantly from a non-point source diffuse input that places a limit on the extent to which oxygen depletion in the system can be remedied. The case of the Clinton River and other areas of concern are cited in Great Lakes Water Quality Board (1987).

269. From simulation models that projected presettlement figures for a now urbanized landscape, Watson et al. (1981) identified a significant springtime pulse coming from suburban and urban landscapes. They cite some preliminary data from the University of Wisconsin Arboretum across the lake from the urbanized area that suggests their analysis conservatively underestimated the impact of urbanization.

270. The figure of 1,810 years mean residence time for a nitrogen atom in a forested ecosystem comes from a simulation of O'Neill et al. (1975).

271. For an account of peasant life in Western Europe and the impact of early agriculture upon Atlantic soils, see Evans (1956).

272. For an account of the incorporation of disturbances in the development of complex systems, see Allen and Starr (1982), O'Neill et al. (1986), and Prigogine and Nicolis (1971).

277. For an accounting of the estimation of recovery times from disturbance for avifaunas of Pacific Islands in terms of island biogeography theory, see Diamond (1972) and Diamond and May (1981).

278. For an accounting of Holling's gentler style of ecosystem management, see Holling (1986).

278. The gentler style of tropical slash-and-burn agriculture compared to Western technological agriculture can be seen in Rappaport (1968, 1972).

278. The plan for the Forest Service management of wolves in the boundary waters area of Minnesota was presented by James H. Rogers, staff officer in the Superior National Forest, to a Forest Service workshop on land management strategies held in Madison, Wisconsin in November 1989. Mech and Nowak (1981) claim a return of the gray wolf to Wisconsin. Mech (1973, 1977) gives a general accounting of wolves in the Superior Forest area. Theil (1985) reports telemetry data on the Minnesota-Wisconsin border suggesting that high densities of roads appear to work against the establishment and persistence of wolf populations.

279. For a general accounting of the use of FORPLAN, see Hoekstra, Dyer, and LeMaster (1987).

280. For a definition of the use of the term "ecosystem approach" by the International Joint Commission through the 1980s, see Great Lakes Science Advisory Board (1980).

281. Lynton Caldwell's comment that we do not manage human-influenced system but rather manage people who act on the system, was made to Allen in a personal communication.
281. For an extended account of the importance of esthetics in management, see Ackoff (1981).

9. *A Unified Approach to Basic Research*

285. Scenario 1, where animals move through a landscape in an orderly sequence is described in Bell (1971).
285. Scenario 2, Brown and Allen (1989) refer to the confusion associated with scaling of herbivory depending upon longevity of leaves and replacement rates.
285. Scenario 3, for the complexities of dune and slack vegetation, see Ranwell (1959, 1960). Morton (1974a, b) used pattern analysis to tease apart the differently scaled processes controlling the vegetation.
285. Scenario 4, for the presentation of triangular ordinations of stress, competition, and disturbance, see Grime (1974, 1979).
286. Scenario 5, when DeAngelis et al. (1986) discuss mutualism at length, they are at pains to identify whether they are talking about the positive feedback that builds the mutualism or the higher level at which the entire system is a competitive unit.
287. The reference to island biogeography in the design of natural reserves is Diamond and May (1981).
290. The critical reference for field theory is Walker et al. (1989).
292. Harper (1977:168) discusses the work of Ross and Harper (1972) where the influence of nearest neighbor is identified as being more significant than the influence of any other plant.
293. For the work on landscape position and biotic and physical/chemical variability, see Kratz et al. (1991).
295. The first development of fuzzy set theory was published in Zadeh (1965) with a more recent extended introduction in Kaufman (1975). The explicit use of fuzzy sets in community ecology has been in the work of David Roberts (1984, 1986, 1989a, b).
295. Bray and Curtis (1957) ordinations have been recast as a fuzzy set operation by Roberts (1986).
296. At Hubbard Brook, Melillo and Bormann (1976) show that pin cherry was particularly important in sealing the gaps of nutrient loss in a clearcut ecosystem.
297. For an account of fish working as ecosystem components as opposed to community members, see Hyatt and Stockner (1985), Stockner (1981), LeBrasseur and Parsons (1979).
299. For a full listing of the Forest Service publications associated with the Fraser Experimental Forest from 1940 to 1977, see Alexander (1978).
299. The reference to Huchins is to his presidential addresses at the University of Chicago, quoted in Rosen (1979).

300. For a discussion of successively higher levels of cycling, see Eigen (1977).
300. For references to Prigogine's work on emerging structure in nonequilibrial systems, see Prigogine (1978, 1982), Prigogine et al. (1969), Prigogine and Nicolis (1971), and Nicolis and Prigogine (1977).
306. For the work behind Wedin's synthesis, see Wedin (1990), Wedin and Tilman (1990), Tilman and Wedin (1991a, b).
308. The development of Checkland's scheme for soft-systems analysis is Checkland (1981).
311. The hierarchical view of DNA functioning at a low level merely as a replicator and at a high level as a controller of cell function occurs in Gould (1981).
315. The reference to Becky Brown's work is Brown and Allen (1989).
318. Pattee (1978, 1979a, b) emphasizes the difference between restrictions on what is possible as opposed to a subset that is what is allowed. Pattee makes the important distinction between arbitrary rules for obtaining structures as opposed to inexorable laws that pertain to the dynamics of the system.
319. Rosen's discussion of van der Waal's equation occurs in Rosen (1989).
321. For an accounting of catastrophe theory and folded surfaces, see Thom (1975).
327. The reference to a geometric basis to systems models occurs in Robert Rosen's presidential address to the Society of General Systems Research (Rosen 1981).
327. For an explication of D'Arcy Thompson's theory of transformations, see Rosen (1962).
330. For an account of the distinction between neighborhood and population competition, see Harper (1977) and Ross and Harper (1972).
330. Although Bruce Milne has performed renormalization on about a dozen different systems, the only published accounts of a renormalization in ecology are Milne et al. (1990); Milne et al. (in press).
332. The discussion in this book is part of the same revolution as described in Gleick (1987).
332. The ridicule of "in principle" arguments occurs in Wimsatt (1982).

BIBLIOGRAPHY

———◆◆◆———

Ackoff, Russell L. 1981. *Creating the Corporate Future: Plan or Be Planned For*. New York: Wiley.

Alexander, Robert R., compiler. 1978. *Gen Tech Rep. Rm-40-A*. Fort Collins, Colo.: USDA Forest Service.

Allen, T. F. H. 1987. Hierarchical complexity in ecology: a noneuclidean conception of the data space. *Vegetatio* 69:17–25.

Allen T. F. H. 1989. Numerical ecology: the French connection. *Ecology* 70:284–85.

Allen, T. F. H., and T. W. Hoekstra. 1985. The instability of primitives and unpredictable complexity. In B. Benathy, ed., *Systems Inquiring*. Proceedings of the 25th annual meeting of S.G.S.R. May 27–31, Los Angeles. Seaside, Cal.: Intersystem.

Allen, T. F. H., and T. W. Hoekstra. 1990. The confusion between scale-defined levels and conventional levels of organization in ecology. *J. Vegetation Science* 1:5–12.

Allen, T. F. H., R. V. O'Neill, and T. W. Hoekstra. 1987. Interlevel relations in ecological research and management: some working principles from hierarchy theory. *J. Applied Systems Analysis* 14:63–79.

Allen, T. F. H., and S. Skagen. 1973. Multivariate geometry as an approach to algal community analysis. *Br. phycol. J.* 8:267–87.

Allen, T. F. H., and Thomas B. Starr. 1982. *Hierarchy: Perspectives for Ecological Complexity*. Chicago: University of Chicago Press.

Allen, T. F. H., and E. P. Wyleto. 1983. A hierarchical model for the complexity of plant communities. *J. Theor. Biol.* 101:529–40.

Atsatt, Peter R. 1988. Are vascular plants "inside-out" lichens? *Ecology* 69:17–23.

Austin M. P., and B. G. Cook. 1974. Ecosystem stability: a result from an abstract simulation. *J. Theor. Biol.* 45:435–58.

Begon, Michael, John L. Harper, and Colin R. Townsend. 1986. *Ecology: Individuals, Populations and Communities*. Sunderland, Mass.: Sinauer.

Bell, R. H. V. 1971. A grazing system in the Serengeti. *Sci. Am.* 225:86–94.

Belsky, A. J. 1986. Does herbivory benefit plants? A review of the evidence. *Am Nat.* 127:870–92.

Bertalanffy, Ludwig von. 1968. *General Systems Theory*. New York: Braziller.

Biberg, I. J. 1749. *Specimen academicum de Oecomia Naturae.* Uppsala; English translation: "The oeconomie of nature," in Benjamin Stillingfleet, ed., *Miscellaneous tracts relating to natural history, husbandry and physick.* 4th ed., printed for J. Dodsley, London (1791).

Blackman, F. F., and A. G. Tansley. 1905. Ecology in its physiological and phyto-topographical aspects. *New Phytol.* 4:199—203, 232–53.

Bosserman, Robert W. 1979. The hierarchical integrity of *Utricularia* - periphyton microecosystems. Ph.D. dissertation, University of Georgia. Reproduced as *Okefenokee Ecosystem Investigations.* Technical report no. 4, University of Georgia, Athens.

Bormann, F. H., T. G. Siccama, R. S. Pierce, and J. S. Eaton. 1974. The export of nutrients and recovery of stable conditions following deforestation at Hubbard Brook. *Ecol. Monog.* 44:255–77.

Bradshaw A. D., and T. McNeilly. 1981. *Evolution and Pollution.* London: Arnold.

Braun-Blanquet, J. 1932. *Plant Sociology: The Study of Plant Communities.* Translated by G. D. Fuller and H. S. Conard. New York: McGraw-Hill.

Bray, J. R., and J. T. Curtis. 1957. An ordination of the upland forest communities of Southern Wisconsin. *Ecol. Monogr.* 27:325–49.

Brown, B. J., and T. F. H. Allen. 1989. The importance of scale in evaluating herbivory impacts. *Oikos* 54:189–94.

Brown, James H., and Jonathan Roughgarden. 1990. Ecology for a changing earth. Final Report of a workshop held in Santa Fe, December, 1988. *Bull. Ecol. Soc. of Am.* 71:173–88.

Carpenter S. R., ed. 1988. *Complex Interactions in Lake Communities.* New York: Springer-Verlag.

Carpenter, Stephen (1989) Temporal variance in lake communities: blue-green algae and the trophic cascade. *Landscape Ecol.* 3:175–84.

Carpenter, Stephen, and James F. Kitchell. 1987. The temporal scale of variance in limnetic primary production. *Am. Nat.* 129: 417–33.

Carpenter S. R., and J. F. Kitchell. 1988. Strong manipulations and complex interactions: consumer control of lake productivity. *Bioscience* 38:764–69.

Carpenter, S. R., J. F. Kitchell, J. R. Hodgson, P. A. Cochran, J. J. Elser, D. M. Lodge, D. Kretchmer, X. He, and C. N. von Ende. 1987. Regulation of lake primary productivity by food web structure. *Ecology* 1863–76.

Carter R. N., and S. D. Prince. 1981. Epidemic models used to explain biogeographical limits. *Nature* 293:664–65.

Center for Biotic Systems. 1974. Environmental analysis of the Kickapoo river impoundment. A report to the US Army Corps of Engineers. *IES Report: 28.* Madison: University of Wisconsin.

Chabot, Brian, and Harold A. Mooney. 1985. *Physiological Ecology of North American Plant Communities.* New York: Chapman Hall.

Chardin, Pierre Tielhard de. 1959. *The Phenomenon of Man.* New York: Harper and Row.

Checkland, Peter. 1981. *Systems Thinking, Systems Practice.* New York: Wiley.

Cittadino, Eugene. 1980. Ecology and the professionalization of botany in America, 1890–1905. *Stud. Hist. Biol.* 4:171–98.

Cittadino, Eugene. 1990. *Nature as the Laboratory: Darwinian Plant Ecology in the German Empire, 1880–1900.* Cambridge: Cambridge University Press.

Clements, F. E. 1904. Developments and structure of vegetation. *Rep. Bot. Survey of Nebraska*, 7.

Clements, F. E. 1905. *Research Methods in Ecology.* Lincoln, Neb.: University Publishing Co.

Clements, F. E. 1916. Plant succession. *Carnegie Inst. Wash. Publ.* 242:1–512.

Colinvaux, P. 1979. *Why Big Fierce Animals are Rare.* Princeton: Princeton University Press.

Cooper, W. S. 1926. The fundamentals of vegetational change. *Ecology* 7:391–414.

Corner, E. J. H. 1981. *The Life of Plants.* Chicago: University of Chicago Press.

Cousens, Roger. 1990. Letters to the editor: On replacement series methodology. *Brit. Ecol. Soc. Bull.* 21:242–44.

Cousins S. H. 1988. Fundamental components in ecology and evolution. In W. F. Wolff, C. J. Seoder, and D. R. Drepper, eds., *Ecodynamics.* Berlin: Springer.

Cousins, S. H. 1990. Countable ecosystems deriving from a food web entity. *Oikos* 57:270–75.

Cowles, H. C. 1899. The ecological relations of the vegetation of the sand dunes of Lake Michigan. *Bot. Gaz.* 27:95–117, 167–202, 281–308, 361–91.

Crutzen, P. J., and T. E. Graedel. 1986. The role of atmospheric chemistry in environment-development interactions. In William C. Clark and R. E. Munn, eds., *Sustainable Development of the Biosphere.* Cambridge: Cambridge University Press.

Curtis, J. T. 1959. *The Vegetation of Wisconsin.* Madison: University of Wisconsin Press.

Curtis, J. T., and H. C. Greene. 1949. A study of relic Wisconsin prairies by the Species presence method. *Ecology* 30:83–92.

Curtis, J. T., and R. P. McIntosh. 1951. An upland forest continuum of the prairie-forest border region of Wisconsin. *Ecology* 32:476–96.

Darwin, Charles. 1874. *The Descent of Man and Selection in Relation to Sex.* 2d ed. New York: Hurst.

Davis, Mark. 1989. Patterns of distribution in an oak woodland and the demography of a perennial herb. *Bull. Ecol. Soc. Am.* 70:93.

Davis, Mark. 1990. Patterns of extinction and landscape dynamics in a oak savanna habitat. *Bull. Ecol. Soc. Am.* 71:133.

Davis, Mark A., Jeff Villinski, Kirsten Banks, Jodi Buckman-Fifield, Jon Dicus, and Susan Hofman. 1991. Combined effects of fire, mound-building, root loss and plant size on growth and reproduction in *Penstemon grandiflorus. Am. Midland Nat.* 125:150–61.

Davis, Mark A., Jeff Villinski, S. McAndrew, H. Scholtz, and E. Young. In press. Survivorship of *Penstemon grandiflorus* in an oak woodland: combined effects of fire, pocket gophers and plant characteristics. *Oecologia.*

DeAngelis, D. L. 1975. Stability and connectance in food web models. *Ecology* 56:238–43.

DeAngelis D. L., W. M. Post, and C. C. Travis. 1986. *Positive Feedback in Natural Systems*. New York: Springer-Verlag.

DeVito, Emile D. 1988. Landscape Scale and Avian Habitat in the New Jersey Pine Barrens: Foraging for context-dependent resources within a forest mosaic. Ph.D. dissertation, University of Wisconsin, Madison.

Diamond, J. M. 1972. Biogeographical kinetics: estimation of relaxation times for avifaunas of southwest Pacific islands. *Proc. Nat. Acad. Sci. USA* 69:3199–203.

Diamond, Jared M., and Robert M. May. 1981. Island biogeography and the design of natural reserves. In R. M. May, ed., *Theoretical Ecology: Principles and Applications*. 2d ed. Sunderland, Mass.: Sinauer.

Dickinson, R. E. 1986. Impact of human activities on climate—a framework. In William C. Clark and R. E. Munn, eds., *Sustainable Development of the Biosphere*. Cambridge: Cambridge University Press.

Drude, O. 1890. *Handbuch der Pflanzengeographie*. Stuttgart: Engelhorn.

Drude, O. 1896. *Deutschlands Pflanzengeographie*. Stuttgart: Engelhorn.

DuReitz, G. E. 1921. Zur methodologischen Grundlage der modernen Pflanzensoziologie. *Akad. Afh., Uppsala*.

Dury, G. H. 1961. *The British Isles*. New York: Norton.

Egler Frank E. (1942) Vegetation as an object of study. *Philosophy of Science* 9:245–60.

Eigen, M. 1977. The hypercycle: Principle of natural self-organization. *Naturwissenschaften* 64:541–65.

Eldredge, N. 1985. *Unfinished Synthesis*. Oxford: Oxford University Press.

Ellenberg, H. 1948. Unkrautgesellschaften als Mass fur den Sauregrad, die Verdichtung und andere Eigenschaften dei Ackerbodens. *Ber. u. Landtechnik* 4:130–46.

Ellenberg, H. 1950. *Landwirtschaftliche Pflanzensoziologie. I. Unkrautgemeinschaften als Zeiger fur Klima und Boden*. Stuttgart: Eugen Ulmer.

Ellenberg, H. 1952. *Landwirtschaftliche Pflanzensoziologie. II. Wiesen und Weiden und ihre standortliche Brewertung*. Stuttgart: Eugen Ulmer.

Ellenberg, H. 1963. *Vegetation Mitteleuropas mit den Alpen*. 2d ed. Stuttgart: Eugen Ulmer.

Elwood, J. W., J. D. Newbold, R. V. O'Neill, and W. Van Winkle. 1983. Resource spiraling: an operational paradigm for analyzing lotic ecosystems. In Thomas D. Fontaine and Steven M. Bartell, eds., *Dynamics of Lotic Systems*. Ann Arbor, Mich.: Ann Arbor Press.

Emanuel W. R., H. H. Shugart, and D. C. West. 1978. Spectral analysis and forest dynamics: long term effects of environmental perturbations. In H. H. Shugart, ed., *Time Series and Ecological Processes*. Philadelphia: Society of Industrial and Applied Mathematics.

Emanuel, W. R., D. C. West, H. H. Shugart. 1978. Spectral analysis of forest model time series. *Ecol. Modeling* 4:313–26.

Evans, Estyn. 1956. The ecology of peasant life in Western Europe. In W. L. Thomas, Jr., ed., *Man's Role in Changing the Face of the Earth*. Chicago: University of Chicago Press.

Fausto-Sterling, Anne, and Gregg Mitman. In press. Whatever happened to *Planaria?* C. M. Child and the physiology of inheritance. In A. Clarke and J. Fugimura, eds., *The Right Tools for the Job: Instruments, Materials, Techniques and Work Organization in the Twentieth Century Life Sciences.* Princeton: Princeton University Press.

Forman, R. T. T., ed. 1979. *Pine Barrens: Ecosystem and Landscape.* New York: Academic Press.

Forman, R. T. T. 1981. Interactions among landscape elements: a core of landscape ecology. In S. P. Tjallingii and A. A. van Veer, eds., *Proc. Int. Congr. Neth. Ecol.* held in Veldhoven, April 6–11, 1981. Wageningen: Pudoc.

Forman, R. T. T., A. E. Galli, and C. F. Leck. 1976. Forest size and avian diversity in New Jersey woodlots with some land use implications. *Oecologia* 26:1–8.

Forman, R. T. T., and M. Godron. 1981. Patches and structural components for a landscape ecology. *Bioscience* 31:733–40.

Forman, R. T. T., and M. Godron. 1986. *Landscape Ecology.* New York: Wiley.

Gardner, Robert H., Virginia H. Dale, and R. V. O'Neill. 1990. Error propagation and uncertainty in process modeling. In R. K. Dixon, R. S. Meldahl, G. A. Ruark, and W. G. Warren, eds., *Process Modeling in Forest Growth Responses to Environmental Stress.* Portland, Ore.: Timber Press.

Gardner R. H., B. T. Milne, M. G. Turner, and R. V. O'Neill. 1987. Neutral models for the analysis of broad-scale landscape pattern. *Landscape Ecology* 1:19–28.

Gefen, Y., A. Aharony, and S. Alexander. 1983. Anomalous diffusion on percolating clusters. *Phys. Rev. Lett.* 50:77–80.

Gittins, R. 1969. The application of ordination techniques. In I. H. Rorison, with assistance from A. Bradshaw, eds., *Symposium on Ecological Aspects of Mineral Nutrition in Plants. British Ecological Society Symposium* 9. London: Blackwell.

Gleason, H. A. 1910. The vegetation of the inland sand deposits of Illinois. *Bull. Ill. St. Lab. Nat. Hist.* 9:21–174.

Gleason, H. A. 1917. The structure and development of the association. *Bull. Torrey Bot. Club* 43:463–81.

Gleason, H. A. 1926. The individualistic concept of the plant association. *Contrib. NY Bot. Gard.,* no. 279.

Gleason, H. A. 1929. Plant associations and their classification: a reply to Dr. Nichols. *Proceedings of the International Congress of Plant Sciences;* at Ithaca, August 1926, vol. I, 624–41. Banta, Menasha, Wisconsin.

Gleason, H. A. 1939. The individualistic concept of the plant association (with discussion). *Am. Midland Nat.* 21:92–110.

Gleason, H. A. 1953. Autobiographical letter. *Bull. Ecol. Soc. Am.* 34:40–42.

Gleick, James. 1988. *Chaos: The Making of a New Science.* New York: Penguin.

Goodall, D. W. 1953. Objective methods for the classification of vegetation. I. The use of positive interspecific correlation. *Austral. J. Bot.* 1:39–63.

Goodall, D. W. 1954. Objective methods for the classification of vegetation. IV. An essay in the use of factor analysis. *Austral. J. Bot.* 2:304–24.

Gould, Stephen Jay. 1981. The ultimate parasite: what happens to bodies if genes act for themselves. *Nat. Hist.*, November.

Great Lakes Science Advisory Board. 1980. The ecosystem approach: scope and implications of an ecosystem approach to transboundary problems in the Great Lakes Basin. *Special Report to the International Joint Commission.* Presented July 1978, second printing December 1980. Windsor, Ont.: International Commission.

Great Lakes Water Quality Board. 1987. *1987 Report on Great Lakes water quality: Appendix A, Progress in developing Remedial Action Plans for Areas of Concern in the Great Lakes Basin.* Presented at Toledo Ohio, November 1987. Report to the International Joint Commission, Windsor, Ontario.

Greig-Smith. 1983. *Quantitative Plant Ecology.* 3d ed. London: Blackwell.

Grime, J. P. 1974. Vegetation classification by reference to strategies. *Nature* 250:26–31.

Grime, J. P. 1979. *Plant Strategies and Vegetation Processes.* Chichester: Wiley.

Grisebach, August. 1838. Ueber den Einfluss des climas auf die Begranzung der natur-lichen Flora. *Linnaea* 52:159–200.

Grisebach, August. 1872. *Die Vegetation der Erde nach ihrer klimatischen Anordnung: Ein Abriss der vergeleichenden Geographie der Pflazen.* 2 vols. Leipzig: Engelmann.

Haldane, J. B. S. 1963. On being the right size. In R. M. Hutchins, M. J. Adler, and C. Fadiman, eds., *Gateway to the Great Books.* #8, Natural Science. Chicago: Encyclopedia Britannica.

Haralick, R. M. 1979. Statistical and structural approaches to texture. *Proceedings of the IEEE* 67:786–804.

Haralick, R. M., and K. S. Shanmugam. 1974. Combined spectral and spatial processing of ERTS imagery data. *Remote Sensing of Environment* 3:3–13.

Haralick, R. M., K. S. Shanmugam, and I. Dinstein. 1973. Textural features for image classification. *IEEE. Transactions on systems, man and cybernetics SMC* 3:610–21.

Harper, John L. 1977. *Population Biology of Plants.* Chicago: Academic Press.

Harper, J. L., and I. H. McNaughton. 1962. The comparative biology of closely related species living in the same area: VII. Interference between individuals in pure and mixed populations of *Papaver* species. *New Phytologist* 61:175–88.

Harper, J. T., J. T. Williams, and G. R. Sagar. 1965. The behavior of seeds in soil. Part 1. The heterogeneity of soil surfaces and its role in determining the establishment of plants from seed. *J. Ecol.* 53:273–86.

Harris, W. Frank, R. S. Kennerson, and Nelson T. Edwards. 1977. Comparison of below ground biomass of natural deciduous forest and Loblolly Pine plantations. *Pedobiologia* 17:369–81.

Heslop-Harrison J. 1967. *New Concepts in Flowering Plant Taxonomy.* Cambridge, Mass.: Harvard University Press.

Hoekstra, T. W., T. F. H. Allen, and C. Flather. 1991. Implicit scaling in the choice of organism and ecological concept: on when to make studies of mice and men. *Bioscience* 41:148–54.

Hoekstra, Thomas W., A. A. Dyer, and Dennis C. LeMaster, eds. 1987. FOR-PLAN: an evaluation of a forest planning tool. Proceedings of a Symposium held November 4–6, 1986, Denver. *USDA Forest Service General Technical Report RM-140.* Fort Collins, Colo.: USDA Forest Service.

Holling, C. S. 1973. Resilience and stability in ecological systems. *Annual Review of Ecology and Systematics* 4:1–23.

Holling, C. S. 1986. The resilience of terrestrial ecosystems: local surprise and global change. In William C. Clark and R. E. Munn, eds., *Sustainable Development of the Biosphere.* Cambridge: Cambridge University Press.

Holling C. S., and S. Ewing. 1971. Blind man's bluff:exploring the response space generated by realistic ecological simulation models. In G. P. Patil, E. C. Pielou, and W. E. Waters, eds., *Statistical ecology.* Vol 2. *Proceedings of the International Symposium on Statistical Ecology.* University Park: Penn State University Press.

Humboldt, Alexander von, and Aime Bonpland. 1807. *Essai sur la geographie des plantes.* Paris: F. Schoell; reprint New York: Arno Press, 1977.

Huntley, B., and T. Webb III. 1989. Migration: species' response to climate variation caused by changes in the earth's orbit. *J. Biogeogr.* 16:5–19.

Hutchinson, G. E. 1957. Concluding remarks. *Cold Spring Harbor Symp. Quant. Biol.* 22:414–27.

Hyatt, Kim D., and John D. Stockner. 1985. Responses of sockeye salmon *Onchrhynchus nerka* to fertilization of British Columbian coastal lakes. *Can. J. Fish. Aqu. Sci.* 42:320–31.

Jackson, D., L. DeAngelis, B. T. Milne, M. G. Turner, B. Zygmunt, S. W. Christensen, V. H. Dale, and R. L. Graham. 1987. Indices of landscape pattern. *Landscape Ecology* 1:153–62.

Jackson, D. R., and A. P. Watson. 1977. Disruption of nutrient pools and transport of heavy metals in a forested watershed near a lead smelter. *J. Env. Qual.* 6:331–38.

Jordan, William R., Michael Gilpin, and John D. Aber, eds. 1987. *Restoration Ecology, A Synthetic Approach to Ecological Research.* Cambridge: Cambridge University Press.

Just, Theodore, ed. 1939. Plant and animal communities. *Am. Midl. Nat.* 21:1–255.

Kameron, R. D. 1983. Issue: Caribou and petroleum development in Arctic Alaska. *Artic* 36:227–31.

Kaufman, A. 1975. *Introduction to the Theory of Fuzzy Subsets. Vol 1: Fundamental Theoretical Elements.* New York: Academic Press.

Kays, S., and J. L. Harper. 1974. The regulation of plant tiller density in a grass sward. *J. Ecol.* 62:97–105.

Kershaw, K. A. 1973. *Quantitative and Dynamic Plant Ecology.* 2d ed. New York: Elsevier.

Koestler, Arthur. 1967. *The Ghost in the Machine.* New York: Macmillan.

Kolasa J., and D. Strayer. 1988. Patterns in the abundance of species: a comparison to two hierarchical models. *Oikos* 53:235–41.

Kratz, T. K., B. J. Benson, E. Blood, G. L. Cunningham, and R. A. Dahlgren.

1991. The influence of landscape position on temporal variability in four North American ecosystems. *Am. Nat.* 138:355–78.

Krummel, J. R., R. H. Gardner, G. Sugihara, and R. V. O'Neill. 1987. Landscape patterns in a disturbed environment. *Oikos* 48:321–24.

Kuhn, T. S. 1970 *The Structure of Scientific Revolutions.* Chicago: University of Chicago Press.

Larson, James L. 1971. *Reason and Experience.* Berkeley: University of California Press.

LeBrasseur, R. J., and T. R. Parsons. 1979. Addition of nutrients to a lake leads to greatly increased catch of salmon. *Environ. Conserv.* 6:187–90.

Legendre, Pierre, and Louis Legendre, eds. 1987. Developments in numerical ecology. *NATO ASI Series G: Ecological Sciences* Vol. 14. New York: Springer-Verlag.

Levin, S. (1986) The icosahedron as the three-dimensional finite element in biomechanical support. In J. Dillon, ed., *Proceedings International Conference on Mental Images, Values and Reality.* Salinas, Cal.: Intersystems Publication.

Levins, R. 1974. Discussion paper: The qualitative analysis of partially specified systems. *Ann. NY Acad. Sci.* 123:38.

Likens, G. E., F. H. Bormann, N. M. Johnson, and R. S. Pierce. 1967. The calcium, magnesium and potassium budgets for a small forested ecosystem. *Ecology* 48:772–85.

Likens, G. E., F. H. Bormann, R. S. Pierce, and N. M. Johnson. 1977. *Biogeochemistry of a Forested Ecosystem.* New York: Springer-Verlag.

Likens, G. E., F. H. Bormann, R. S. Pierce, and W. A. Reiners. 1978. Recovery of a deforested ecosystem. *Science* 199:492–96.

Lindeman, R. L. 1942. The trophic-dynamic aspect of ecology. *Ecology* 23:399–418.

Lovelock, J. E. (1979) *Gaia: A New Look at Life on Earth.* Oxford: Oxford University Press.

MacArthur, R. 1972. Strong, or weak, interactions? In E. S. Deevey, ed., *Growth by Intussusception: Ecological Essays in Honor of G. Evelyn Hutchinson. Trans. Conn. Acad. Arts Sci.* 44:177–88.

MacArthur R. H., and E. O. Wilson. 1967. The theory of island biogeography. *Mono. Pop. Biol.* 1:1–215.

McClark, William C., and R. E. Munn, eds. 1986. *Sustainable Development of the Biosphere.* Cambridge: Cambridge University Press.

McCune, B., and T. F. H. Allen. 1985a. Will similar forests develop on similar sites? *Can. J. Bot.* 63:367–76.

McCune, B., and T. F. H. Allen. 1985b. Forest dynamics in the Bitterroot Canyons, Montana. *Can. J. Bot.* 63:377–83.

McElroy, M. B. 1986. Change in the natural environment of the earth: the historical record. In William C. Clark and R. E. Munn, eds., *Sustainable Development of the Biosphere.* Cambridge: Cambridge University Press.

McIntosh, R. P. 1975. H. A. Gleason, "individualistic ecologist," 1882–1975: His contributions to ecological theory. *Bull. Torrey Bot. Club* 102:253–73.

McIntosh, R. P. 1985. *The Background of Ecology, Concept and Theory.* Cambridge: Cambridge University Press.

McLaughlin, John, and Jonathan Roughgarden. 1989. Avian predation on *Anolis* lizards in the Northeastern Caribbean: an interisland contrast. *Ecology* 70:617–28.

McMahon, J. A., D. L. Phillips, J. V. Robinson, and D. J. Schimpf. 1978. Levels of organization in biology: an organism-centered approach. *Bioscience* 28:700–4.

McMahon, J. A., D. J. Schimpf, D. C. Anderson, K. G. Smith, and R. L. Bayn, Jr. 1981. An organism-centered approach to some community and ecosystem concepts. *J. Theor. Biol.* 88:287–307.

McMahon, Thomas A. 1975. The mechanical design of trees. *Sci. Am.* 233:93–102.

MacMillan, Conway. 1892. *The Metaspermae of the Minnesota Valley: A List of the Higher Seed-Producing Plants Indigenous to the Drainage Basin of the Minnesota River.* Minneapolis: Harrison and Smith.

MacMillan, Conway. 1899. Minnesota plant life. *Report of the Survey, Botany Series III.* Saint Paul: University of Minnesota.

McMurtrie, R. F. 1975. Determinant of stability of large randomly connected systems. *J. Theor. Biol.* 50:1–11.

McNaughton, S. J. 1986. On plants and herbivores. *Am. Nat.* 128:765–70.

Magnuson, John J. 1988. Two worlds for fish recruitment: lakes and oceans. *Am. Fish. Soc. Sym.* 5:1–6.

Magnuson, John J. 1991. Fish and fisheries ecology. *Ecol. Applications* 1:13–26.

Magnuson, John J. Cynthia A. Paszkowski, Frank J. Rahel, and William Tonn. 1989. Fish ecology in severe environments of small isolated lakes in Northern Wisconsin. In R. R. Sharitz and J. W. Gibbons, eds., *Freshwater Wetlands and Wildlife, Conf-8603101, DOE Symposium Series No. 61.* Oak Ridge: USDOE Office of Scientific and Technical Information.

Major J. 1958. Plant ecology as a branch of botany. *Ecology* 39:352–63.

Mandelbrot, B. 1983. *The Fractal Geometry of Nature.* New York: Freeman.

Margalef, R. 1968. *Perspectives in Ecological Theory.* Chicago: University of Chicago Press.

Margalef, R. 1972. Homage to Evelyn Hutchinson, or why there is an upper limit to diversity. In E. S. Deevey, ed., *Growth by intussusception: Ecological essays in honor of G. Evelyn Hutchinson. Trans. Conn. Acad. Arts Sci.* 44:213–35.

Margulis L. 1981. *Symbiosis in Cell Evolution.* San Francisco: Freeman.

May, R. M. 1974. Stability and complexity in model ecosystems. *Mono. Pop. Biol.* 6:1–265, 2d ed.

May, R. M. 1981a. *Theoretical Ecology: Principles and Applications.* 2d ed. Sunderland, Mass.: Sinauer.

May, R. M. 1981b. Patterns in multi-species communities. 197-227 In R. M. May, ed., *Theoretical Ecology: Principles and Applications.* 2d ed. Sunderland, Mass.: Sinauer.

May, R. M. 1981c. Models for single populations. In R. M. May, ed., *Theoretical Ecology: Principles and Applications.* 2d ed. Sunderland, Mass.: Sinauer.

Mech, L. D. 1973. Wolf numbers in Superior National Forest of Minnesota. *US For. Serv. Res. Pap. NC-97.*

Mech, L. D. 1977. A recovery plan for the eastern timber wolf. *Natl. Parks. Conserv. Mag.*, January.

Mech L. D., and R. M. Nowak. 1981. Return of the gray wolf to Wisconsin. *Amer. Midl. Nat.* 105:408–9.

Melillo, J., and F. H. Bormann. 1976. Reorganization of a terrestrial ecosystem following perturbation. In Orie L. Loucks, ed., *The Study of Species Transients, Their Characteristics and Significance for Natural Resource Systems.* Indianapolis: TIE.

Miller, James G. 1978. *Living Systems.* New York: McGraw-Hill.

Miller, William, III. 1991. Hierarchical concept of reef development. *N. Jb. Palaont. Abh.* 182:21–35.

Milne, B. T. 1988. Measuring the fractal geometry of landscapes. *Applied Mathematics and Computation* 27:67–79.

Milne, B. T. 1990. Lessons from applying fractal models to landscape patterns. In M. G. Turner and R. H. Gardner, eds., *Quantitative Methods in Landscape Ecology.* New York: Springer-Verlag.

Milne, B. T. In press, a. Spatial aggregation and neutral models in fractal landscapes. *Am Nat.*

Milne, B. T. In press, b. The utility of fractal geometry in landscape design. *Landscape and Urban Planning.*

Milne, B. T., Kevin M. Johnston, and Richard T. T. Forman. 1989. Scale-dependent proximity of wildlife habitat in a spatially neutral Bayesian model. *Landscape Ecology* 2:101–10.

Milne, Bruce, Monica G. Turner, John A. Wiens, and Alan R. Johnson. 1990. Interactions between fractal geometry of landscapes and allometric herbivory. *Bull. Ecol. Soc. Am.* 71:257.

Milne, Bruce, Monica G. Turner, John A. Wiens, and Alan R. Johnson. In press. Interactions between fractal geometry of landscapes and allometric herbivory. *Theoretical Population Biology.*

Milne, G. 1935. Some suggested units for classification and mapping, particularly for East African soils. *Soil Res.* 4:1–27.

Moorhead, Alan. 1969. *Darwin and the Beagle.* New York: Harper and Row.

Morton, A. J. 1974a. Ecological studies of a fixed dune grassland at Newborough Warren, Anglesey. The structure of the grassland. *J. Ecol* 62:253–60.

Morton, A. J. 1974b. Ecological studies of a fixed dune grassland at Newborough Warren, Anglesey. II. Causal factors of the grassland structure. *J. Ecol* 62:261–78.

Mueller-Dombois, D., and Heinz Ellenberg. 1974. *Aims and Methods of Vegetation Ecology.* New York: Wiley.

Musick, Brad, and Herbert D. Grover. 1990. Image textural measures as indices of landscape pattern. In M. G. Turner and R. H. Gardner, eds., *Quantitative Methods in Landscape Ecology.* New York: Springer-Verlag.

Neilson, R. P. 1986. High-resolution climatic analysis and Southwest biogeography. *Science* 232:27–34.

Neilson, R. P. 1987a. Biotic regionalization and climatic controls in Western North America. *Vegetatio* 70:135–47.

Neilson, R. P. 1987b. On the interface between current ecological studies and the paleobotany of Pinyon-Juniper woodlands. In *Proceedings, Pinyon-Juniper Conference. USDA Forest Service, General Technical Report INT-215.*

Neilson, R. P., and L. H. Wullstein. 1983. Biogeography of two American oaks in relation to atmospheric dynamics. *J. Biogeogr.* 10:275–97.

Neilson, R. P., and L. H. Wullstein. 1986. Microhabitat affinities of Gambel Oak seedlings. *Great Basin Naturalist* 46:294–98.

Nichols G. E. 1929. Plant associations and their classification. *Proceedings of the International Congress of Plant Sciences;* at Ithaca, August 1926, vol. I, 624–41.

Nicolis, G., and I. Prigogine. 1977. *Self-Organization in Non- Equilibrium Systems: From Dissipative Structures to Order through Fluctuations.* New York: Wiley.

Odum, E. P., in collaboration with H. T. Odum. 1959. *Fundamentals of Ecology.* Philadelphia: Saunders.

O'Neill, R. V., B. S. Ausmus, D. R. Jackson, R. van Hook, P. van Voris, C. Washburne, and A. P. Watson. 1977. Monitoring terrestrial ecosystems by analysis of nutrient export. *Water, Air and Soil Pollution* 8:271–77.

O'Neill, R. V., D. L. DeAngelis, J. B. Waide, and T. F. H. Allen. 1986. A hierarchical concept of ecosystems. *Monographs in Population Biology* 23:1–272.

O'Neill, R. V., and R. H. Gardner. 1990. Pattern, process, and predictability: the use of neutral models for landscape analysis. In M. G. Turner and R. H. Gardner, eds., *Quantitative Methods in Landscape Ecology. The Analysis and Interpretation of Landscape Heterogeneity.* Ecological Studies Series. New York: Springer-Verlag.

O'Neill R. V., W. F. Harris, B. S. Ausmus, and D. E. Reichle. 1975. A theoretical basis for ecosystem analysis with particular reference to element recycling. In F. G. Howell, J. B. Gentry, and M. A. Smith, eds., *Mineral Cycling in South Eastern Ecosystems. DOE Symposium Series, Conf. 740513.* Oak Ridge: Technical Information Center, ORNL.

O' Neill, R. V., A. R. Johnson, and A. W. King. 1989. A hierarchical framework for the analysis of scale. *Landscape Ecology* 3:193–205.

O'Neill, R. V., J. R. Krummel, R. H. Gardner, G. Sugihara, B. Jackson, B. T. Milne, M. G. Turner, B. Zymunt, S. W. Christensen, V. H. Dale, and R. L. Graham. 1988. Indices of landscape pattern. *Landscape Ecology* 1:153–62.

O'Neill, R. V., B. T. Milne, M. G. Turner, and R. H. Gardner. 1988. Resource utilization scales and landscape pattern. *Landscape Ecology* 2:63–69.

Oosting, H. J. 1948. *The Study of Plant Communities.* San Francisco: Freeman.

Orbach, R. 1986. Dynamics of fractal networks. *Science* 231:814–19.

Parton, W. J., D. S. Schimel, C. V. Cole, and D. S. Ojima. 1987. Analysis of factors controlling soil organic matter levels in the Great Plains grasslands. *Soil Sci. Soc. Am. J.* 51:1173–79.

Pattee, H. H. 1972. The evolution of self-simplifying systems. In E. Lazlo, ed., *The Revelance of General Systems Theory.* New York: Braziller.

Pattee, H. H. 1978. The complementarity principle in biological and social structures. *J. Soc. Biol. Structures* 1:191–200.

Pattee, H. H. 1979a. Complementarity vs. reduction as an explanation of biological complexity. *Am. J. Physiol.* 236(5):12241–46.

Pattee, H. H. 1979b. The complementarity principle and the origin of macromolecular information. *Biosystems* 11:217–26.

Peitgen, Heinz-Otto, and Dietmar Saupe, eds. 1988. *The Science of Fractal Images.* New York: Springer-Verlag.

Pichat, E. 1969. Algorithms for finding the maximal elements of a finite universal algebra. *Processing* 68:214–18.

Pielou, E. C. 1974. *Population and Community Ecology, Principles and Methods.* New York: Gordon and Breach.

Platt, T., and K. L. Denman. 1975. Spectral analysis in ecology. *Ann. Rev. Ecol. Syst.* 6:189–210.

Polanyi, M. 1968. Life's irreducible structure. *Science* 160:1308–12.

Post, W. M., Tsung-Hung Peng, William Emanuel, Anthony W. King, Virginia H. Dale, and Donald DeAngelis. 1990. The global carbon cycle. *Amer. Sci.* 78:310–26.

Potts, D. C. 1983. Evolutionary disequilibrium among Indo-Pacific Corals. *Bull. Mar. Sci.* 33:619–32.

Potts, D. C. 1984a. Generation times and the Quaternary evolution of reef-building corals. *Paleobiology* 10:48–58.

Potts, D. C. 1984b. Natural selection in experimental populations of reef-building corals (Scleractinia). *Evolution* 38:1059–78.

Potts, D. C. 1985. Sea-level fluctuations and speciation in Scleractinia. *Proceedings Fifth International Coral Reef Congress, Tahiti* 4:127–32.

Potts, D. C. 1987. Population biology and evolution of clonal organisms. *Quarterly Rev. Biol.* 62:212–13.

Potts, D. C., T. J. Done, P. J. Isdale, and D. A. Fisk. 1985. Dominance of a coral community by the genus *Porites* (Scleractinia). *Mar. Ecol. Prog. Ser.* 23:79–84.

Potts, D. C., and P. K. Smith. 1984. Water temperature as an indicator of environmental variability on a coral reef. *Limnol. Ocenaogr.* 29:504–16.

Pound, Roscoe. 1896. The plant-geography of Germany. *Am. Nat.* 30:465–68.

Pound, Roscoe, and Frederick E. Clements. 1898. II. A method of determining the abundance of secondary species. *Minn. Bot Studies,* 2d series, Part 1:19–24.

Pound, Roscoe, and Frederick E. Clements. 1900. *The Phytogeography of Nebraska.* 2d ed. Lincoln, Neb.: The Seminar.

Prigogine, I. 1978. Time, structure, and fluctuations. *Science* 201:777–85

Prigogine, I. 1982. Order out of chaos. In W. J. Mitsch, R. K. Ragade, R. W. Bosserman, and J. A. Dillon, Jr., eds., *Energetics and Systems.* Ann Arbor, Mich.: Ann Arbor Science Publishers.

Prigogine, I., R. Lefever, A. Goldbeter, and M. Herschkowitz-Kaufman. 1969. Symmetry breaking instabilities in biological systems. *Nature* 223: 913–16.

Prigogine, I., and G. Nicolis. 1971. Biological order, structure, and instabilities. *Quart. Rev. Biophys.* 4:107–48.

Prince, S. D. 1982. The southern boundary of the mopane tree *Colophospermum mopane* in Botswana. *Brit. Ecol. Soc. Bull.* 13:187.

Ranwell, D. 1959. Newborough Warren Anglesey. I. The dune system and dune slack habitat. *J. Ecol.* 47:571–601.

Ranwell, D. 1960. Newborough Warren Anglesey. II. Plant associes and succession cycles of sand dune and dune slack vegetation. *J. Ecol.* 48:117–41.

Rappaport, Roy A. 1968. *Pigs for the Ancestors: Ritual in the Ecology of a New Guinea People.* New Haven: Yale University Press.

Rappaport, Roy A. 1972. The flow of energy in an agricultural society. In *Biology and Culture in Modern Perspective.* San Francisco: Freeman.

Reichle D. E., and D. A. Crossley. 1965. Radiocesium dispersion in a cryptozoan food web. *Health Physics* 11:1375–84.

Roberts, David W. 1984. Forest vegetation and site relations: theory, methods, and application to the forests of Montana. Ph.D dissertation, University of Wisconsin, Madison.

Roberts, David W. 1986. Ordination on the basis of fuzzy set theory. *Vegetatio* 66:123–31.

Roberts, David W. 1987a. An anticommutative difference operator for fuzzy sets and relations. *Fuzzy Sets and Systems* 21:35–42.

Roberts, David W. 1987b. A dynamical systems perspective in vegetation theory. *Vegetatio* 69:27–33.

Roberts, David W. 1989a. Analysis of forest succession with fuzzy graph theory. *Ecological Modelling* 45:261–74.

Roberts, David W. 1989b. Fuzzy systems vegetation theory. *Vegetatio.*

Rohlf, F. J. 1974. Methods of comparing classifications. *Ann. Rev. Ecol. Syst.* 5:101–13.

Root, Richard B. 1975. Some consequences of ecosystem texture. In S. A. Levin, ed., *Ecosystem Analysis and Prediction.* Proceedings of a conference on ecosystems held at Alta, Utah, July 1–5, 1974. Philadelphia: Society for Industrial and Applied Mathematics.

Rosen, Robert. 1962. The derivation of D'Arcy Thompson's theory of transformations from the theory of optimal design. *Bull. Math. Biophys.* 40:549–79.

Rosen, Robert. 1979. Anticipatory systems in retrospect and prospect. *General Systems* 24:11–23.

Rosen, Robert. 1981. The challenges of systems theory. *General Systems Bulletin* XI:2–4.

Rosen, Robert. 1989. Similtude, similarity, and scaling. *Landscape Ecology* 3:207–16.

Ross, M. A., and J. L. Harper. 1972. Occupation of biological space during seedling establishment. *J. Ecol.* 60:77–88.

Roughgarden, Jonathan, Stephen D. Gaines, and Stephen Pacala. 1987. Supply side ecology: the role of physical transport processes. In P. Giller and J. Gee, eds., *Organization of Communities: Past and Present.* London: Blackwell.

Roughgarden, Jonathan, and Stephen Pacala. 1989. Taxon cycling among *Anolis* lizards populations: review of the evidence. In D. Otte and J. Endler, eds., *Speciation and Its Consequences*. Sunderland, Mass.: Sinauer.

Rowe, J. S. 1961. The level-of-integration concept in ecology. *Ecology* 42:420–27.

Russell, Norman H. 1975. *Introduction to Plant Science*. New York: West.

Sala, O. E., W. J. Parton, L. A. Joyce, and W. K. Laurenroth. 1988. Primary production of the central grassland region of the United States. *Ecology* 69:40–45.

Salisbury, E. J. 1932. The East Anglian flora. *Trans. Norfolk Norwich Nat. Soc.* 13:191–263.

Salisbury E. J. 1964. *Weeds and Aliens*. 2d ed. London: Collins.

Schimper, A. F. W. 1898. *Pflanzengeographie auf physiologischer Grundlage*. Jena: G. Fischer. (*Plant Geography upon a Physiological Basis*. Translated by W. R. Fisher, edited and revised by P. Groom and I. B. Balfour. Oxford: Clarendon, 1903.)

Schmidt-Nielsen. 1984. *Scaling: Why is Animal Size so Important?* Cambridge: Cambridge University Press.

Schrodinger, Erwin. 1959. *Mind and Matter*. Cambridge: Cambridge University Press.

Shea, Cynthia Pollock. 1989. Protecting the ozone layer. In Linda Starke, ed., *State of the World 1989, A Worldwatch Institute Rreport on Progress toward a Sustainable Society*. New York: Norton.

Sheail, John. 1987. *Seventy-Five Years in Ecology: The British Ecological Society*. London: Blackwell.

Shideler, R. T. 1986. Impacts of human development and land use on Caribou: a literature review. Volume 2. Impacts of oil and gas development on the central arctic herd. *Alaska Department of Fish and Game, Technical Report 86-3*.

Shugart, Herman H., Darrell C. West, and W. R. Emanuel. 1981. Patterns and dynamics of forests: an application of simulation models. In D. C. West, H. H. Shugart, and D. B. Botkin, eds., *Forest Succession: Concepts and Application*. New York: Springer-Verlag.

Simon, H. A. 1962. The architecture of complexity. *Proc. Amer. Phil. Soc.* 106:467–82.

Smith, B. L., and D. C. Potts. 1987. Clonal and solitary anemones *Anthopleura* of western North America: population genetics and systematics. *Marine Biology* 94:537–46.

Smith, R. 1898. Plant associations if the Tay Basin. *Trans. Proc. Perthsh. Soc. Nat. Sci.* 2:200–17.

Smith, R. 1899. On the study of plant associations. *Nat. Sci.* 14:109–20.

Stevens, Peter S. 1974. *Patterns in Nature*. Boston: Little, Brown.

Stockner, J. G. 1981. Whole lake fertilization for the enhancement of Sockeye Salmon *Onchrhynchus nerka* in British Columbia, Canada. *Verh. Int. Ver. Limnol.* 21:293–99.

Sukachev, V., and N. Dylis. 1964. *Fundamentals of Forest Biogeocoenology*. Translated by J. M. McLennan. London: Oliver and Boyd.

Summers, Peter W. 1989. The atmospheric region of influence for Kejimujik,

Nova Scotia. In *Proceedings of the International Air Quality Board of the International Joint Commission First Regional Workshop on Integrated Monitoring*, St. Andrews, New Brunswick, May 31–June 2, 1988. Washington, D.C.: IJC.

Swanson, F. J., T. K. Kratz, N. Caine, and R. G. Woodmansee. 1988. Landform effects on ecosystem patterns and processes. *Bioscience* 38:92–98.

Tansky, M. 1978. Stability of multispecies predator-prey systems. *Memoirs Coll. Sci. Univ. Kyoto, Ser. B.* 7(2):87–94.

Tansley, A. G. 1904. The problems of ecology. *New Pytol.* 3:191–200.

Tansley, A. G. 1935. The use and abuse of vegetational concepts and terms. *Ecology* 16:284–307.

Theil. 1985. Relationship between road densities and wolf habitat suitability in Wisconsin. *Amer. Midl. Nat.* 113:404–7.

Thom, R. 1975. *An Outline of a General Theory of Models*. Reading, Mass.: Benjamin.

Thompson, D'Arcy Wentworth. 1942. *On Growth and Form*. Vols. I, II. Cambridge: Cambridge University Press.

Tilman, David, and David Wedin. 1991a. Plant traits and resource reduction for five grasses growing on a nitrogen gradient. *Ecology* 72:685–700.

Tilman, David, and David Wedin. 1991b. Dynamics of nitrogen competition between successional grasses. *Ecology* 72:1038–49.

Tobey, R. 1981. *Saving the Prairies: The Life Cycle of the Founding School of American Plant Ecology, 1895–1955*. Berkeley: University of California Press.

Transeau, E. N. 1926. The accumulation of energy by plants. *Ohio J. Sci.* 26:1–10.

Transeau, E. N., H. C. Sampson, and L. H. Tiffany. 1940. *Textbook of Botany*. New York: Harper and Sons.

Turner, M. G. 1987. Spatial simulation of landscape changes in Georgia. *Landscape Ecology* 1:29–36.

Turner, Monica G., Robert H. Gardner, Virginia H. Dale, and R. V. O'Neill. 1988. Landscape pattern and the spread of disturbance. In M. Ruzicka, T. Hrnciarova, and L. Miklos, eds., *Proc. VIIIth Intl. Symp. on Problems of Landscape Ecological Research, Vol 1*. Institute of Experimental Biology and Ecology, CBES SAS, Bratislava, CSSR.

Turner, Monica G., Robert H. Gardner, Virginia H. Dale, and R. V. O'Neill. 1989. Predicting the spread of disturbance across heterogeneous landscapes. *Oikos* 55:121–29.

Uexkull, J. von. 1920. *Umwelt und Imnenwelt der Tiere*. 2d ed. Berlin: Springer.

Uexkull, J. von. 1929. *Theoretische Biologie*. 2d ed. Berlin: Springer.

Uexkull, J. von. 1957. A stroll through the worlds of animals and men: a picture book of invisible worlds. In Clair H. Schiller, ed. and trans., *Instinctive Behavior*. New York: International Universities Press.

Uexkull, J. von, and G. Kriszat. 1934. *Streifuge durch die Umwelten von Tieren und Menschen*. Berlin: Springer.

Unger, F. 1836. *Ueber den Einfluss des Bodens auf die Vertheilung der Gewashse, nachgewiesen in der Vegetation des nordostlichen Tirols*. Wein: Rohrman und Schweigerd.

Urban, D. L., and R. V. O'Neill. In press. Mechanisms of avian demography: sensitivity, uncertainty, and scaling implications. *Ecology.*

Van Voris, P., R. V. O'Neill, W. R Emanuel, and H. H. Shugart. 1980. Functional complexity and functional stability. *Ecol.* 61:1352–60.

Vogel, Steven. 1981. *Life in Moving Fluids.* Boston: Grant.

Vogel, Steven. 1988. *Life's Devices.* Princeton: Princeton University Press.

Walker, J., P. J. H. Sharpe, L. K. Penridge, and H. Wu. 1989. Ecological field theory: the concept and filed tests. *Vegetatio* 83:81–95.

Wallace, Craig. 1908. North Dakota life: plant, animal and human. *Bull. Am. Geo. Soc.* 40:321–33, 401–15.

Wallace, Robert A., Jack L. King, and Gerald P. Sanders. 1986. *Biology, the Science of Life.* GLenview, Ill.: Scott, Foresman.

Waller Donald M., and David Steingraber. 1985. Branching and modular growth: theoretical models and empirical patterns. In Jeremy B. C. Jackson, Leo W. Buss, and Robert E. Cook,. *Population Biology and Evolution of Clonal Organisms.* New Haven: Yale University Press.

Walters, Carl J., and C. S. Holling. 1990. Large-scale management experiments and learning by doing. *Ecology* 71:2060–68.

Warming, Eugenius. 1896. *Lehrbuch der okologischen Pflanzengeographie: Eine Einfuhrung in die Kenntnis der Pflanzenvereine.* Translated by E. Knoblauch. Berlin: Borntraeger. (*Oecology of Plants: An Introduction to the Study of Plant-Communities.* Translated and edited by P. Groom and I. B. Balfour. Oxford: Clarendon, 1909.)

Watson, V. J., O. L. Loucks, and W. Wojner. 1981. The impact of urbanization on seasonal hydrologic and nutrient budgets of a small north American watershed. *Hydrobiologia* 77:87–96.

Webster, J. R. 1979. Hierarchical organization of ecosystems. In E. Halfon, ed., *Theoretical Systems Ecology.* New York: Academic Press.

Wedin, D. 1990. Nitrogen cycling and competition among grass species. Ph.D. dissertation, University of Minnesota, Minneapolis.

Wedin, D., and D. Tilman. 1990. Species effects on nitrogen cycling: a test with perennial grasses. *Oecologia* 84:433–41.

Weinberg, G. M. 1975. *An Introduction to General Systems Thinking.* New York: Wiley.

Wessman, C. A., J. D. Aber, D. L. Peterson, and J. M. Melillo. 1988. Remote sensing canopy chemistry and nitrogen cycling in temperate forest ecosystems. *Nature* 335:154–56.

Whittaker, R. H. 1956. Vegetation of the Great Smoky Mountains. *Ecol. Monogr.* 26:1–80.

Wiegleb, Gerhard. 1989. Explanation and prediction in vegetation science. *Vegetatio* 83:17–34.

Wiens J. A., and B. T. Milne. 1989. Scaling of "landscapes" in landscape ecology, or, landscape ecology from a beetle's perspective. *Landscape Ecology* 3:87–96.

Wimsatt, William C. 1982. Reductionistic research strategies and the biases in the units of selection controversy. In E. Saarinen, ed., *Conceptual Issues in Ecology.* Dordrecht, Netherlands: Reidel.

Wishart, D. 1969. An algorithm for hierarchical classification. *Biometrics* 25:165–70.

Wit, C. T. de 1960. On competition. *Versl. Landbouwk. Onderz.* 66:1–82.

Worster, D. 1977. *Nature's Economy: The Roots of Ecology.* San Francisco: Sierra Club.

Zadeh, L. A. 1965. Fuzzy sets. *Information and Control* 8:338.

SUBJECT INDEX

Agroecosystem, 77, 307, 335, 337; corn in, 45, 83, 271, 279, 335; and crops, 274; and landscapes, 62–63, 65, 73, 346–347; pests in, 274; tropical 220, 278, 307, 348, 351

Airsheds, 94–96, 338

Allelopathy, 128, 316, 341

Algae, 111, 185–186, 249–250

Allometry, 186–191, 193, 330–331

Amphibia, 158, 299

Animals: allometry of, 193, 330–331; in biomes, 48, 53, 239, 240, 349; in communities, 128–129, 157–158; and development, 188–189, 195–196, 346; in ecosystems, 45, 53, 105–115, and form, 186–189; and grazing, 104, 106–109, 129, 172, 240–241, 267–268, 285, 306–307, 315, 330–331, 339, 350, 352; groups of, 129, 207–210; on landscapes, 55, 65, 72, 76–78, 350; movement of, 109, 155–158, 330–331, 337, 352; and populations, 43, 330–331; and realms, 273; sessile, 161–162, 185, 193–195, 346; see also Amphibia; Birds; Budworm; Fish; Insects; Invertebrates; Mammals; Mites; Salmon; Serengeti Plain; Soil arthropods; Spiders; Wildlife; Wolves; Worms

Antelope, 207–209

Atmosphere, 123–125, 244–255, 333; and ozone layer, 245–249, 349

Biome, 47–48, 51–52, 167, 238–245

BIOSIS literature search, 43, 78, 105–115, 154–156, 195, 335, 337, 339, 344

Biosphere, 48–49, 51–52, 244–251, 255–256; and biome, 244–245; human effects on, 245–250, 271–274, 351

Birds, 37–39, 43, 65–66, 155–158, 172, 210, 226–228, 287, 344

Bitterroot Mountains, 79–83, 194–195, 337

Bond strength, 28–30, 75, 162–163, 173

Boundaries, 25–29, 74–75, 95–104, 160, 169, 187–188; of ecosystems, 94–104, 338; of foodwebs, 268–269

Bryce Canyon, 153–154

Budworm, 37–39, 86, 226–228, 240, 278, 348

Calcium, 123, 249–250

Canada, 271, 278–280, 338, 350, 351

Carbon, 104, 111, 250–251, 301, 303, 339

Carbon dioxide, 123–125, 250–251, 335

Chaos, 213–214, 223–224, 332, 334, 347

Climate, 48–49; in biome relationship, 238–239, 306, 349; change, 93–194, 250–251, 271, 274, 313; topography related to, 208–209, 238, 346–347

Clinton River, 351

Clique analysis, 150–154, 344

Clone, 15–16, 26–27, 147, 163

Community, 33, 43–44, 49, 55, 77, 126–158, 167, 265, 283, 344; accommodation between species in, 127, 132, 158, 340, 341, 349; animal versus plant, 128–129, 156–158; biogeographical origin of, 130–131; Clements/Gleason controversy over, 55, 126, 133, 142–144, 149–150, 336, 337, 340, 342–344; Clementsian view of, 40, 45, 55, 129–135, 142–144, 169, 242, 258, 307, 335, 336, 341–342, 343–344, 349; cycles in, 266, 285, 298, 301–303,

NAME INDEX

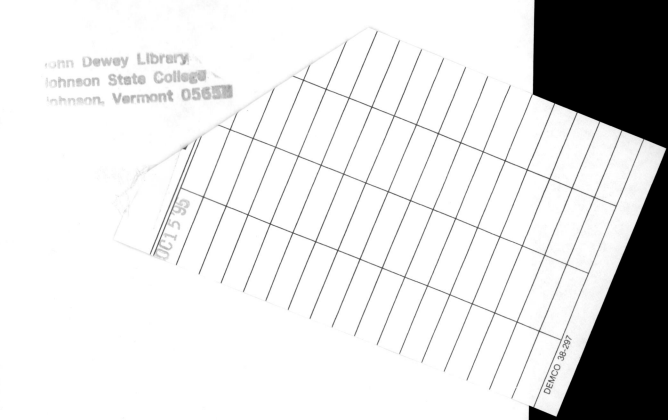

DEMCO 38-297